ARCHAEOLINGUA

A Publication Series jointly edited by the
Archaeological Institute of the Hungarian Academy of Sciences
and the Linguistic Institute of the University of Innsbruck

Editors:

SÁNDOR BÖKÖNYI and WOLFGANG MEID

Volume 1

CULTURAL AND LANDSCAPE CHANGES IN SOUTH-EAST HUNGARY

I

Reports on the Gyomaendrőd Project

Edited by
SÁNDOR BÖKÖNYI

BUDAPEST 1992

Assistant Editor:

ERZSÉBET JEREM

The Cover Photograph, done by György Csáki, represents the southern area of Endrőd, with the burial mound 'Papp-halom' in the foreground

ISBN 963 7391 60 6
HU-ISSN 1215-9239

1992

ARCHAEOLINGUA ALAPÍTVÁNY

Archaeological Institute of the Hungarian Academy of Sciences
H-1250 Budapest, Úri utca 49

Word Processing and Desktop Editing by András Kardos and Ildikó Perjés
on DTK Computers supplied by KIN-PEX Electronic Ltd

Printed by AKAPRINT, Budapest. 9220502.

INTRODUCTION

ARCHAEOLINGUA is a new publication series jointly edited by the Archaeological Institute of the Hungarian Academy of Sciences and the Linguistic Institute of the University of Innsbruck. It will publish monographs and collective volumes dedicated to what – in a very wide sense – may be called "Kulturwissenschaft", dealing with all aspects of human culture in prehistoric and early historic settings. Archaeology and Linguistics (again both understood in their widest application) will be the main supporting pillars of this interdisciplinary approach, which is also indicated by the title of the series. To prevent any misunderstanding as to the semantic content of this symbolic title the editors wish to stress that it should be understood as the abbreviation of an additive compound (the linguists would call it a "dvandva"): archaeology + language.

Archaeology is concerned with the recovery and interpretation of factual remnants of earlier cultures. For prehistoric periods, but also for historic periods in the absence of written testimony, this is often enough our only source of knowledge. Linguistics, together with philology and other associated techniques, is able to exploit language itself as a source of information, for instance where vocabulary or names reach back into prehistoric periods and thus may furnish semantic information about concepts of the past; for historical periods information which is contained in the form of written texts may be put to use. In the case of badly recorded, undeciphered or less well known languages this involves special interpretative techniques which often depend on the findings of the archaeologists. Linguistic documents such as inscriptions on material objects recovered by the archaeologist are in this sense also objects of archaeology. On the other hand, archaeological facts may need ethnic or linguistic interpretation. For these reasons archaeology and linguistics are interrelated and interdependent in more than one sense, and in many cases their findings will be mutually illuminating and will thus fit into a wider horizon of understanding cultural networks. It goes without saying that in order to obtain wide-ranging results interdisciplinary ways of approach are necessary: the archaeologist as well as the linguist will need the help of specialists in the fields of history, art, comparative mythology, ethnology, anthropology, genetics, palaeozoology, palaeobotanics, palaeoclimatology – to name only a few.

Besides such studies which – in a general or methodological way – deal with the interrelated subject of "archaeology and language" ARCHAEOLINGUA will publish studies in the prehistory and early history of central and (south)east central Europe, with special emphasis on the Danube-Alpine-Adriatic region which is interrelated in many ways, and within this wider area more specifically on Hungary in her present-day or historical confines. In accordance with the aim of arriving at a comprehensive cultural phenomenology and morphology interdisciplinary methods of research will be considered instruments vital to the task.

The series which will be supplemented by a "Series minor" for the publication of lectures and small monographs intends to become an international forum of interdisciplinary research within the outlines mentioned and will be open to all qualified researchers – individuals and teams – in the field. It will also be in the position to publish the *Acta* of international meetings dedicated to subjects within the above-mentioned range.

ARCHAEOLINGUA accepts contributions in German, English or French. If possible they should be processed by means of one of the current systems of electronic word-processing and submitted both on diskette and as a print-out. Enquiries and all further correspondence should be directed to ARCHAEOLINGUA at the address of the Archaeological Institute of the Hungarian Academy of Sciences, care of Erzsébet Jerem, Úri utca 49, H-1250 Budapest.

Budapest, January 1992 — The Editors

FOREWORD

Archaeological sites vary widely in type and furnish different kinds of information. Although burial places, in general, yield more attractive finds, they teach us comparatively little about the ordinary life situation and environmental conditions. Settlements on the other hand reflect everyday life in all its apects, e.g., settlement patterns, house types, fortifications, ways of life, religious beliefs and practices, agriculture and animal husbandry (hunting, fishing, gathering), pottery and tool making, metallurgy, diet etc. Even more important, settlements closely interact with their environment. Their inhabitants exploited several ecological niches. Certain cultures preferred well-defined environmental types, while others changed them during the course of their development (e.g., they moved from shore-lines to forested steppes) because they wanted, or were forced, to change their way of life. All in all, settlement and environment represented a unity even as they do now.

It is exactly for this reason that the Archaeological Institute of the Hungarian Academy of Sciences decided to carry out a large-scale archaeological research project on settlements in the early eighties which was started in the area of Gyomaendrőd, Békés County, in 1984.

The aim of this project is the reconstruction of changes in the settlement structure of different prehistoric and early historical periods, the determination of coeval settlement types and, finally, the study of the interactions between environment, settlements and man in this so-called "Microregion". The chronological range of these studies extends from the early Neolithic period till the end of the Turkish occupation (late 17th century).

In the selection of the Microregion, the main consideration was that it should contain a large number of archaeological sites representing, if possible, all archaeological periods. Since we knew that the systematic field surveys of the Archaeological Topography of Hungary had resulted in the discovery of an enormous number of sites, the Gyomaendrőd area seemed most suitable for our purposes. These intensive field surveys produced no less than 226 sites in this area of ca 42 km^2, that is, about 5.4 sites per km^2 which is far more than in any other surveyed region in the country (Jankovich – Makkay – Szőke 1989). The 226 sites discovered in this microregion represent eleven archaeological periods. The most commonly represented periods include the early Middle Ages (10-13th centuries) and the Neolithic. The late Middle Ages, the Turkish occupation period and the Germanic period (4-5th century A.D.) are underrepresented.

The key to the high site density is probably the rich soil that was abundantly watered by the constantly shifting branches of the Körös river. In fact the region divides into two parts: the slopes of the Maros river covered with loess and dissected by pleistocene riverbeds on the one hand, and a part of the Körös basin covered with meadow clay on the other. In fact this geographical division also played an essential role in the selection of this area. We were curious to study the effects of different environments (lying so close to each other) on settlement pattern and evolution.

This project, aimed at the reconstruction of environmental history as well as human settlement evolution in such a comparatively large region, is rather unique. There are only a few similar studies to which it can be compared, such as the German "Küstengebiet" Project (Kossack et al. 1984; Jankuhn et al. 1984) and the German Project "Siedlungsarchäologische Untersuchungen im Alpenvorland" (Becker et al. 1985). They are different in so far as, for instance, the Küstengebiet Project only covered a time span from the 5th century B.C. to the 11th century A.D.

In the Settlement Archaeology Project we have basically considered excavation as the main research method, putting special emphasis on full excavation, at least of smaller settlements from different archaeological periods. The excavation of 226 settlements would be, however, a task much too great for even a generation of archaeologists. We have excavated, therefore, only the more promising sites or, for example, settlements of the same period in different environments.

Beside excavations we also used extensive field surveys, systematic surface sherd (and other classes of find) collections, aerial photography, geophysical prospecting and core drilling. We applied these methods not only at sites but also in uninhabited areas in order to reconstruct former environments with particular concern given to hydrology.

In the Settlement Archaeology Project, the main part of the field work was done by members of the Archaeological Institute. Nevertheless, specialists from Eötvös Lóránd University, the Eötvös Lóránd Geophysical Institute, the State Geological Survey, the Museum of Natural History and the Hungarian Museum of Agriculture also participated in the field work and the evaluation of the scientific investigations. Finally we must mention the participation of Italian archaeologists led by Dr. Bruno Genito of IsMEO and geophysicists under the leadership of Dr. Mauro Cucarzi of the Fondazione Lerici in the framework of bilateral agreements between the Archaeological Institute on the one side and the two Italian institutions on the other.

As the above enumeration shows, both the excavations and their scientific evaluation have been carried out by means of a carefully planned interdisciplinary approach. Specialists from the following disciplines have been working on the project in close cooperation with archaeologists: remote sensing, geophysical prospecting, palaeohydrology, geomorphology, pedology, petrography, physical anthropology, archaeozoology, malacology, archaeobotany.

The present volume publishes a small selection of articles: in fact, the scientific evaluation of the first excavations in the Microregion. The articles cover a wide variety of topics. Among the reports of strictly archaeological nature there are preliminary and final papers on settlement excavations starting with the early Neolithic Körös culture and ending with the Árpád period and also on excavations of cemeteries of the late Migration period and of the early Middle Ages. Besides these – and emphasizing the multidisciplinary nature of both the field work and the scientific evaluation – one can find articles combining geophysical prospecting with archaeological approaches and/or the application of palaeoenvironmental techniques with geoarchaeology. There is a report of a complete settlement excavation in which archaeological, archaeozoological and -ichthyological components are integrated.

In the meantime, another interdisciplinary volume is being prepared, with special emphasis on internal evolution of a Sarmatian site.

We seriously hope that these volumes will be followed by yet others, and that as the final result of the project we shall be able to work out the outlines of the environmental and human-economic history of the area selected as the scene of the first large-scale research project in the history of Hungarian archaeology.

Budapest, January 1992 Sándor Bökönyi

Contents

Abbreviations

AAustr	= Archaeologia Austriaca, Wien
ActaAnt	= Acta Antiqua Hungariae, Budapest
ActaArchHung	= Acta Archaeologica Academiae Scientiarum Hungaricae, Budapest
ActaSzeged	= Acta Antiqua et Archaeologica, Szeged
AJA	= American Journal of Archaeology, New York
AKorr	= Archäologisches Korrespondenzblatt
ArchÉrt	= Archaeologiai Értesítő, Budapest
ARozh	= Archeologické Rozhledy, Praha
AT	= Antik Tanulmányok, Budapest
BSA	= British School of Archaeology at Athens, Annual, London
ComArchHung	= Communicationes Archaeologicae Hungaricae, Budapest
FA	= Folia Archaeologica, Budapest
MittArchInst	= Mitteilungen des Archäologischen Instituts der Ungarischen Akademie der Wissenschaften, Budapest
MTud	= Magyar Tudomány, Budapest
NyK	= Nyelvtudományi Közlemények, Budapest
OxfJA	= Oxford Journal of Archaeology, Oxford
PPS	= Proceedings of the Prehistoric Society, Cambridge
PZ	= Praehistorische Zeitschrift, Berlin
RFüz	= Régészeti Füzetek, Budapest
RivScPr	= Rivista di Scienze Preistoriche, Firenze
SA	= Sovetskaja Arheologija, Moskva
SCIVA	= Studii şi Cercetări de Istorie Veche şi Archeologie, Bucureşti
SlA	= Slovenská Archeológia, Nitra
WorldArch	= World Archaeology, London
ArchHung	= Archaeologia Hungarica, Budapest
BAR	= British Archaeological Reports, Oxford
DissPann	= Dissertationes Pannonicae, Budapest
MIA	= Materialy i Issledovanija po Archeologii SSSR, Moskva
MRT	= Magyarország Régészeti Topográfiája (The Archaeological Topography of Hungary), Budapest
StudArch	= Studia Archaeologica, Budapest
VAH	= Varia Archaeologica Hungarica, Budapest

The integrated geoarchaeological approach within the territory: the case of the Microregion

First results and further proposals

MAURO CUCARZI

Introduction

Although scientific methods related to the study of soil have long been used in the analysis and exploration of material culture, it has rarely been possible to plan and carry out investigations integrated with Geoscience in an area sufficiently large to contain more than one context and, at the same time, small enough to enable homogeneous data collection. The aim of these interventions is to study a territory as a cultural event: a space in which the signs of history are stratified.

Non-destructive study methods, applied mainly to archaeological work and described as Geoarchaeological Exploration (Cucarzi 1986), include: remote sensing, geophysics, geomorphology, analysis of stratified samples, palaeoecology, palaeobotany, etc. Operating on different working scales, these methods give an information for different levels in proportion to their definition capacity.

In order to interpret the data, and due to inevitable monetary limitations, it is necessary to analyse a variety of interventions on different scales. This is not a simple task, as it is not always possible to identify optimal parameters in advance.

Apart from these difficulties, which can partly be resolved by choosing the most suitable working scale during the ongoing investigations, geoarchaeological exploration also represents a multi-scale method.

As will be shown later, this type of approach can yield valid information on the environment, land use and settlement pattern in a region, providing a working basis for individual sites and making it possible to carry out the research in greater detail.

The Microregion

The opportunity offered to the international scientific community by the Institute of Archaeology of the Hungarian Academy of Sciences, in the early 1980-ies, i.e. to explore the Microregion, was also extremely interesting, from a geoarchaeological point of view.

The Microregion territory is, in fact, an ideal zone for this kind of research. Consisting of almost flat areas, with minimal modern environmental disturbances, the settlements and/or necropoli investigated are not very deep but all the same have a complex stratigraphy composed of different cultural facies (Jankovich et al. 1989). This provides optimal conditions for a research program integrated with geoarchaeological exploration.

Although it was clear from the very beginning that only an integrated system could provide the data required for understanding real site distribution, settlement patterns and land use, at first it was decided only to test geophysical prospection methods.

Geophysical prospection, in fact, plays a leading role in any system of geoarchaeological exploration, because – conditions permitting – it provides data on settlement or necropolis characteristics and/or on a sufficiently small scale to develop an information segment in great detail over a large area but in a short time. In this given case, however, circumstances were not particulary favourable: previous excavations in the region have so far yielded evidence on settlements which, although from different cultural phases, have characteristics similar to those of the Neolithic period, the remains of this period consisting of hut floors, pits and ditches.

Though various cases of often successful geophysical prospecting on this type of settlement have been described in literature, it must be kept in mind that it is not always possible in such evanescent situations, to attain a measurable geophysical marker.

Due to the above mentioned factors, test prospectings since 1984 have been made on sites from different periods, within the Microregion primarily in the Gyomaendrőd area, with the intention of evaluating the potentials of geophysical methods in locating sites and necropoli on the Great Hungarian Plain.

Only after identifying the most suitable procedures for this purpose, was more complex intervention initiated employing various other techniques in an area which still has to be considered experimental as it is not yet included in the integrated program launched by the Archaeological Institute of the Hungarian Academy of Sciences.

Hitherto, only a few parts the project had been put into operation: the first photointerpretation, a geophysical prospection, a geomorphological survey as well as the study of exposed sections and core samplings.

First testing campaigns (1984-1987)

Örménykút

A brief geophysical exploration campaign was conducted in 1984. In the vicinity of the village of Örménykút where Dr. Csanád Bálint was carrying out an excavation on an Avar site, as to assess the possibility of locating buried archaeological structures using the most suitable survey methods. Previous excavation campaigns had revealed the foundations of huts, fire places and ditches down to a maximum depth of about two metres, but no elevated structures. In view of the large extent of the area and the short time available together with the fact that it was merely a preliminary survey, it was decided to investigate four sample areas.

The aim was therefore to identify geophysical markers which would enable correlations to be made between anomalies and archaeological structures.

It is a well known fact that geophysical prospection is based on the possibility of measuring the contrast between the value of a given physical characteristic of the soil and the value of the same physical characteristic of a buried archaeological feature.

Initially, some rather unsatisfactory attempts were made to apply a survey system based on resistivity measurements in an attempt to trace evidence of settlement, for example, through the higher conductivity of the thin layers of charcoal in fire places situated inside huts (Seguin and Sampson 1980) or the higher resistivity of filled pits. We then proceeded to concentrate on measurements of the intensity of the earth's magnetic field in order to identify structures through their greater magnetic susceptibility caused by heating of the soil, the presence of dwellings (Tite and Mullins 1971; Le Borgne 1955, 1960; Fassbinder and alii 1990) and the abundance of potsherds in the area.

Following the suggestions of Dr. Bálint, four sample areas of 400 square metres were selected to carry out measurements of intensity of the earth's magnetic field using an ELSEC differential proton magnetometer.

As it might reasonably have been expected, the variations in field measurements were immediately found to be very slight. However, the almost total lack of causes of disturbance in the magnetic field, apart of course from diurnal variation already corrected by the application of the differential system (Weymouth and Lessard 1986), permitted the scanning of extremely small anomalies approaching the limits of the sensitivity of the instrument (1nT).

In fact, the measurements ranged between 47610 nT and 47625 nT except for one area where the range extended from 47610 nT to 47650 nT.

Despite the obviously small extent of the areas explored, we were nevertheless able to formulate some interpretative hypotheses on the origins of some anomalies, as for example, the 7.5 nT amplitude anomaly caused by a fire place (*Fig. 1*). Some trial trenches made by Dr. Bálint confirmed this interpretation, showing that fire places could be very closely pinpointed and settlement areas roughly defined through its position, although without the identification of particular structures within the areas.

Fig. 1. Örménykút: Computerised map of magnetic field intensity values: The "A_1" and "A_2" indicate anomalies produced by an Avar fire place. In this map, and in all similar maps, the associations between tones of grey and magnetic field intensity values expressed in nT are indicated in the legend. The crosses are 20 metres apart.

Gyoma 133 (1987)

During the spring of 1987, IsMEO Italian Archaeological Mission, led by Dr. Bruno Genito, conducted geomagnetic tests at the Gyoma 133 site. The method chosen was based on the experience acquired at Örménykút, and took into consideration the almost total absence of stone or brick wall structures in the building techniques of the area, from the Neolithic to the Magyar period, although environmental conditions were in contradiction to this type of approach. The site is, in fact, situated within an industrial brickyard complex, no longer active but with very many surface disturbances which might be the sort of magnetic characteristics which could overshadow possible variations caused by buried archaeological structures.

An area of 1000 square metres, near Bruno Genito's excavation area, was explored with the aid of two proton G 856 Geometrics magnetometers. On the basis of data collected at Örménykút as well as other data collected by Miklós Pattantyús at Endrőd 170 [1], and after filtering the measurement (Genito et al. in press), a map of geomagnetic isoanomalies was drawn (*Fig. 2*), in which at least 3 out of 8 ascertained anomalies seem to have been caused by buried archaeological structures.

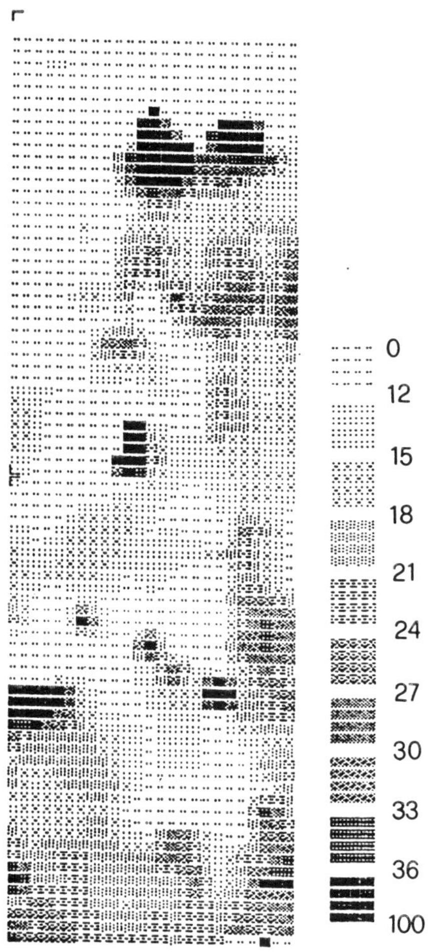

Fig. 2. Gyoma 133: Map of magnetic field intensity values of the explored area.

[1] Measurements taken together with M. Pattantyús-Á.

Endrőd 19 (1987)

During the same campaign, geomagnetic measurements were collected over an area of about 2,400 square metres inside the site of Endrőd 19 and indicated with a "b" in *Fig. 3* which had been recorded as Late Bronze Age and La Tène age (11th-4th centuries B.C.) on the basis of surface reconnaissance made by the Archaeological Institute of the Hungarian Academy of Sciences (Jankovich, Makkay and Szőke 1989, 136-137). At first, the magnetic readings did not display clear anomalies that could be related to archaeological structures, but after mathematical treatment of the original data, the computerised magnetic map (*Fig. 4*) of the area showed two main categories of anomalies. One rather large anomalies type indicated with a "B" in *Fig. 4*, is most likely to be related to differing alluvial deposits. The second sort includes many small anomalies (marked with an "A") with shapes very close to those encountered in different sites within the zone and thought to reflect archaeological features. In any case, the height of these anomalies is very low (2-6 nT).

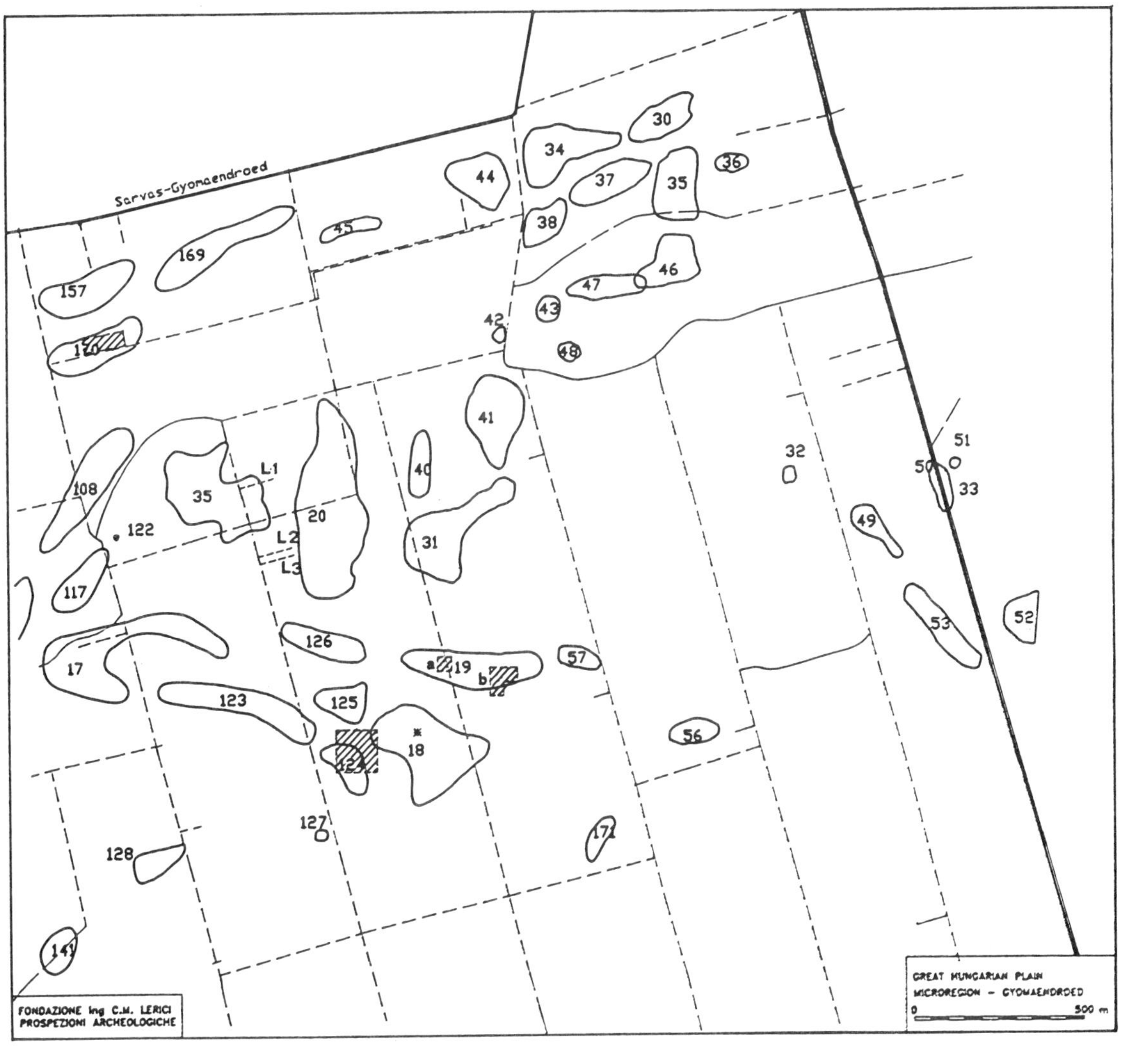

Fig. 3. Computerised map of the Gyomaendrőd area covered by the geoarchaeological investigation described in this article. The numbers are those of the sites as catalogued by the archaeologists of the Archaeological Institute of the Hungarian Academy of Sciences on the basis of surface analyses. Hatched areas are those investigated using geophysical prospection and core drillings. L1, L2, L3 are magnetic profiles.

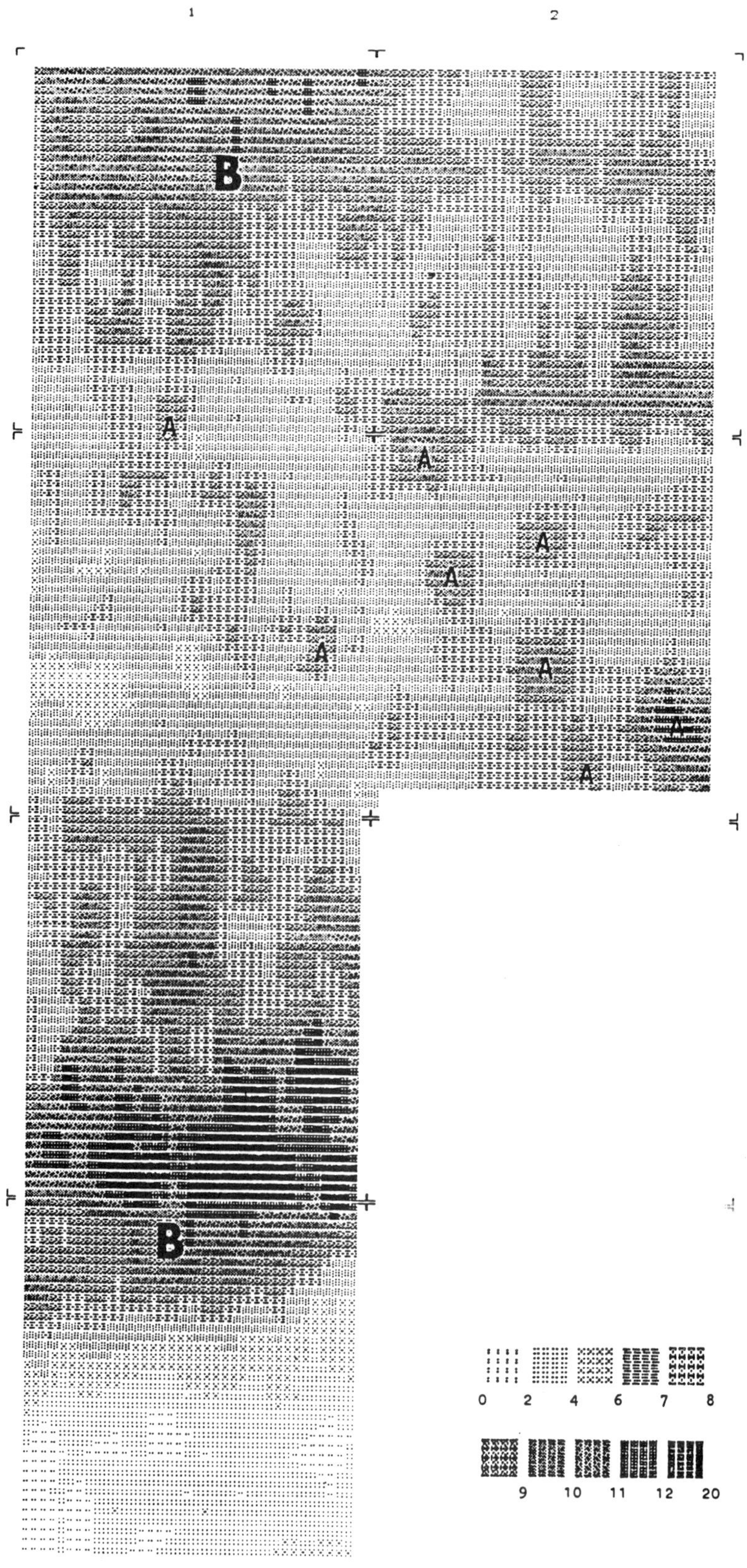

Fig. 4. Endrőd 19b: Computerised map of magnetic field intensity values, "B" indicates the magnetic Class 2 anomalies while with "A" marks those of Class 1.

First Results

On the basis of this series of tests, the first conclusions were drawn about the applicability of geomagnetic measurements in the location of settlements from various cultural phases on the Great Hungarian Plain.

The method was particularly efficacious in areas indicated by the surface survey as mainly Sarmatian (Endrőd 170, Jankovich and Pattantyús 1990). In the case of the Avar site of Örménykút, the technique of geomagnetic measurements in essence demonstrated that this technique is useful for identifying fire places. Further prospecting is required in areas with predominantly Bronze Age occupation. Nothing is known, for instance, about Copper Age and Neolithic sites.

Analyses made of wave characteristics of anomalies registered in various situations, suggest that it is possible to establish criteria to make it possible to distinguish some classes of anomalies that may be related to the settlement structures of different cultures, on the basis of some parameters of the waves produced (height, wavelength, etc).

During the spring of 1988, research based on these first conclusions was also continued together with other methods including photointerpretation and core sample analyses, while geophysical prospecting was used over much greater areas (1,5 hectars). The integrated approach mentioned in the introduction was, thus, underway.

The chosen area was that shown on the map in *Fig. 3*, with particular attention to sites 124, 18 and 19a.

Furthermore, in cooperation with Miklós Pattantyús working for the Eötvös Loránd Geophysical Institute, investigations with geomagnetic measurements were carried out on a hitherto unexplored part of site 170.

The program got underway with the acquisition and software analysis through image processing of existing aerial photographs from the area. Thereafter, some "suspect" areas were examined using prospecting methods, i.e. magnetic field intensity in the square areas near sites 124 and 19a (*Fig. 3*), and analysis of stratigraphic samples from continuous core drillings in an area of site 124.

1988 Campaign

Photointerpretation and image analyses

The use of aerial photography in general, and its use for archaeological survey in particular, has a long history and has been extensively described in the Manual of Remote Sensing (Colwell 1983) and by Ebert (1984), and in numerous other, less recent but fundamental studies (Agache 1970, Chevallier 1964, Dassie 1976, Strandberg 1967). It is, therefore, not necessary to go into great detail here. It would, however, be useful to draw attention to a few points, the first being the principles on which Remote Sensing is based: what can be seen and why in an aerial photograph?

The term "Remote Sensing" (Fisher 1975), refers to the use of electromagnetic radiation sensors in registering images of a landscape, images which yield clear information useful for many purposes.

In this specific case, the images were obtained by registering the visible region of radiation, 0.4–0.8 μm although in other cases, bands near the infra-red 0.8-3 μm, infra-red 3-14 μm, or microwaves 5-500 μm wave length are used.

The object is then to see how the electromagnetic energy, depending on the particular wave-length band used, interacts with the earth's surface and, in this case (imaging the visible radiation), how the energy of this band interacts.

Natural radiation is partly reflected, partly absorbed and partly transmitted. As there is no loss of energy in this process it may be described in the equation:

$\phi\lambda = \rho\lambda + \alpha\lambda + \tau\lambda$

where ϕ is the radiant flux

ρ is the reflected flux

α is the absorved flux

τ is the transmitted flux

The results will differ depending on the quantities involved in the equation.

Vegetation, soil, water and urban areas are the environmental components which contribute most to the interaction between electromagnetic radiation and the earth's surface.

Vegetation: The spectral capacity to reflect radiation of the wave-length used (visible light) gives predominance in the equation (1) to high absorption as well as low transmission and reflection.

The type of vegetation, the water content of the leaves, the type of soil in which it grows, the degree of ripening, the position of the sun and of the sensor, all affect the parameters of the equation.

The soil: In the case of soil, the greater part of the incident flux of the radiation of the visible light is reflected or absorbed, but very little is transmitted.

The soil's reflectivity depends substantially on five parameters, linked to its characteristics. They are, in order of importance: humidity, organic content, texture, structure and iron-oxide content (Stoner and Baumgartner 1981).

Water: The incident flux of radiation is not reflected, but absorbed or transmitted.
In the visible region, radiation on water is transmitted and very little is absorbed, whilst in the infra-red band, absorption increases and the transmission decreases.

The archaeological environment

When exploring an area for archaeological purposes, the aim is to identify traces of structures which are normally buried. The immediate question is then at what depth the structures can be discovered on an aerial photograph, using the band of visible light.

First, one must keep in mind that the penetration capacity of electromagnetic radiation in the visible region is almost negligible on the ground in the soil. The possibility of finding the trace of a buried structure, therefore, is due to the fact that the structure produces a change in the vegetation, the type of soil, major or minor humidity in the immediate surrounding areas, and not to radiation hitting the structure directly.

The selective criteria must be set, therefore, keeping the soil's reflectivity in mind. A dark trace (sign of major absorption) for instance, my be caused by a ditch, as it contains large quantities of organic material.

Under some conditions, a light trace might represent a road, which if it is paved, can contain less humidity in the layer between the present surface and that of the paving. The same result can be caused by buried walls.

These are only a few, simple examples.

If the area is cultivated, a buried structure affects plant growth and will, therefore, be identified on the basis of different cone shades.

Photointerpretation

Photointerpretation generelly means the interpretation of photograms taken from systems (platforms) flying at different heights: from balloons to satellites.

Under similar conditions, as the resolution capacity of the image is inversely proportional to the heigh at which the photograph is taken, more or less independently of the type of instrument used, it is clear that, depending on the altitude, the images will be suitable for different purposes.

Therefore, the study of aerial photographs consists of identifying all traces associated with buried structures, the evidence for which depends on the type and depth of the structure, and on later anthropic interventions under the same conditions (geological, environmental, climatic etc.) which may have changed, completely or partially, the original nature of the traces.

Areas of ancient river-beds, cones and canals can be identified, both to evaluate the presence of anthropic structures in the vicinity of these ancient transport ways and to establish the extension of the areas in which floods may have further buried anthropic remains, or in case if erosional phenomena destroyed them.

Treatment of images

In this case, a continuous image has to be obtained, with a telecamera or a scanner transforming it into a discrete image and dividing the original tones of brightness into 256 shades of grey.

Once the picture is made available, it can be refined using a series of functions, in order to "clean" it, make geometric corrections, amplify contours, develop a histogramme and transform the black and white image into one with false colours, etc. These techniques are based on the application of particular filters that employ different, potent and tested mathematical algorithms.

In this case, all necessary steps were taken to obtain a suitable image for our purpose; particular emphasis was laid on representation with false colours. This type of treatment is very efficacious in the evaluation of existing differences in the area. It is well known that the human eye can distinguish 20,000 colour nuances, but only 200 shades of grey (Ray and Fisher 1957).

Fig. 5. Photointerpretation of area 20-18-19-124-125 with many palaeoalveal traces. (Aerial photograph made by Gy. Csáki and E. Jerem)

The investigated area

Photointerpretation

The images at our disposal were photos taken from a light aircraft normally used for crop surveys etc., with a traditional device. As a result the images were not very neat, strongly inclined on the zenith and, in any case, only partially covered the area of interest. (Part of site 124/125 and the Papp-halom area, which would have been most interesting to the investigation, were not included.)

The image enhancement system consisted of an IBM PC, a specific software for image processing, and a telecamera using 256 shades of grey.

As far as possible, the images were geometrically corrected with a bilinear mathematical algorithm requiring at least 6 known points. This correction made it possible to place many palaeoalveal traces (*Figs. 5, 6, 7*) onto the topographical map.

In the image covering site Endrőd 20, there are visible traces of a fairly regular geometrical figure, well defined at some points and having a dark colour that could be interpreted as a trace of a buried structure. The chromatism could reflect a series of ditches, canals filled with soil containing large quantities of organic material or a foundation trench of building which today are totally destroyed.

Further southwards from these last traces, another faint trace can be seen. It was recognized only after the images had been treated and appeared as a shape reminescent of Neolithic ditches, Graben-Anlage. Having identified these traces on the topographic map and later on the ground, 3 magnetic profiles were made, L1, L2 and L3, each 150 metres long, in a direction perpendicular to that of the traces (*Fig. 3*).

Geophysical prospecting

Prospecting using geomagnetic methods was made in three zones:

1. Zone 1: Profiles L1, L2, L3
2. Zone 2: Including parts of sites Endrőd 124, 18
3. Zone 3: Small part of 19a

Unfortunately, as far as cases 2 and 3 are concerned, the zone was not included in the aerial photographs, but was still systematically explored because of its importance given by the various settlements with different cultural profiles known to be there.

Zone 1 – Endrőd 20

On the basis of the traces found through photointerpretation, three magnetic profiles were made in a direction perpendicular to the road and the traces.

Measurements were taken every metre, a sensor placed 0.60 metres above surface level. Profiles L1 and L2 are shown here; L3 reflects the same situation as in L2.

In *Fig. 8*, which compares the two traces, an anomaly is clearly visible in both cases with a height of approximately 10 nT, and a wavelength of approximately 60 metres, lying between 20 and 80 metres from the starting point.

Furthermore, in profile L1, there are anomalies having wavelengths between 4 and 9 metres with heights varying between 5 and 20 nT, of which the most important is the one indicated with the letter "A". This corresponds to the dark trace in, intercepting with profile L1.

Considering the chromatic characteristics of the trace on the photogramme and those of magnetic anomaly A, it is reasonable to believe that it has been produced by a buried anthropic structure, probably the foundation of a brick structure or a foundation trench later filled with soil containing a large quantity of organic material.

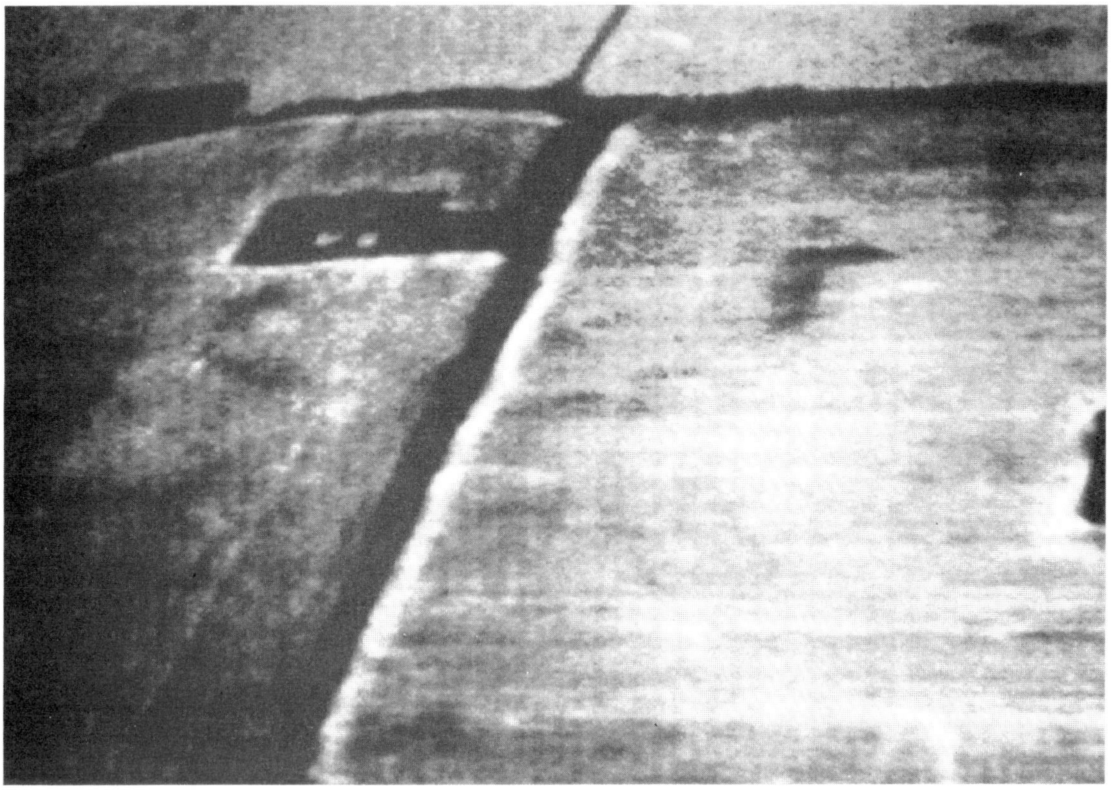

Fig. 6. Photointerpretation of area 20 with two "suspected" traces. (Aerial photograph made by Gy. Csáki and E. Jerem)

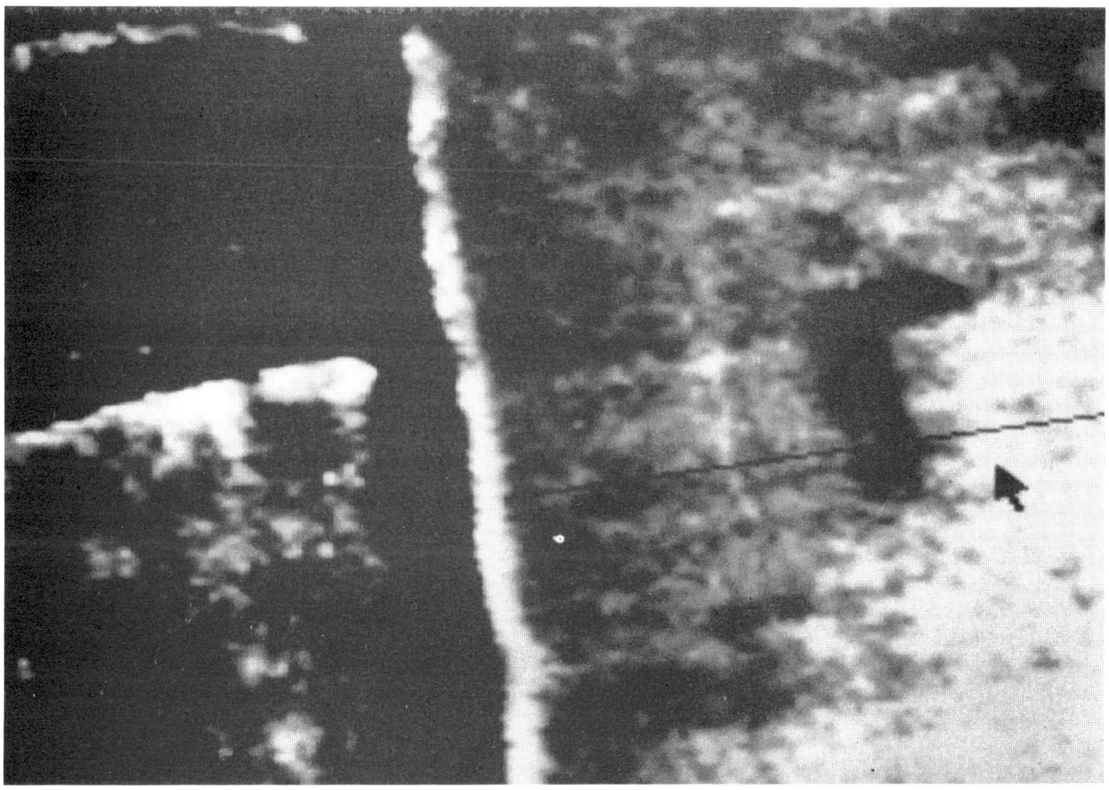

Fig. 7. Photointerpretation of area 20 zoom: geomagnetic profile L across anomaly A. (Aerial photograph made by Gy. Csáki and E. Jerem)

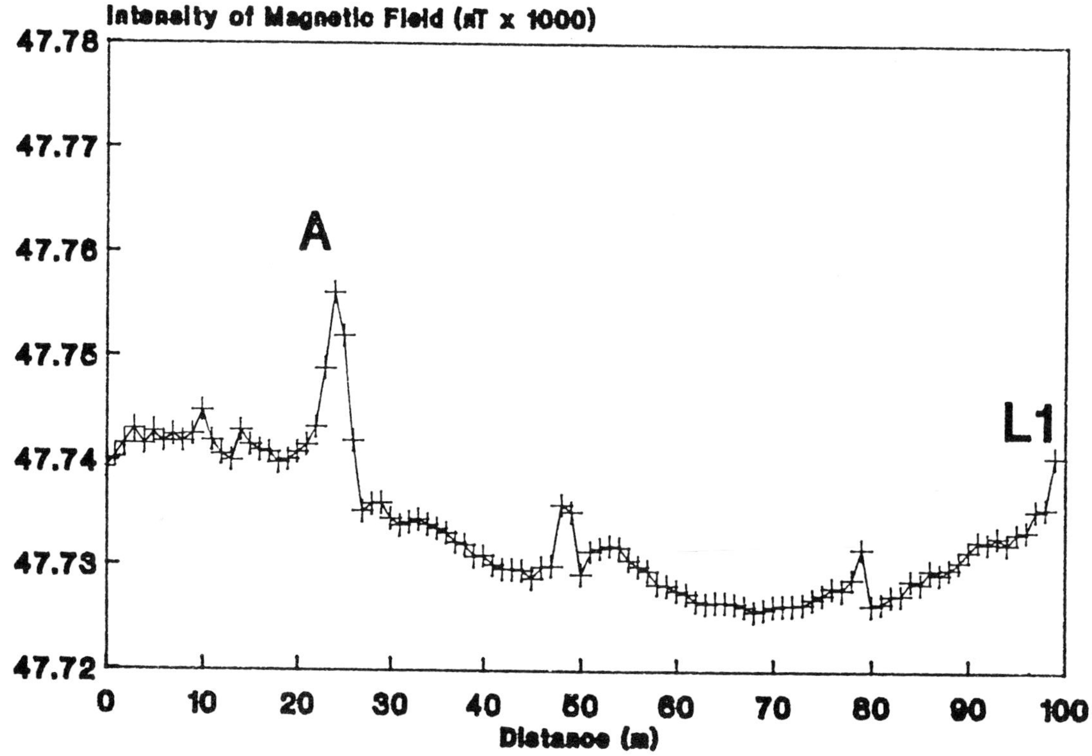

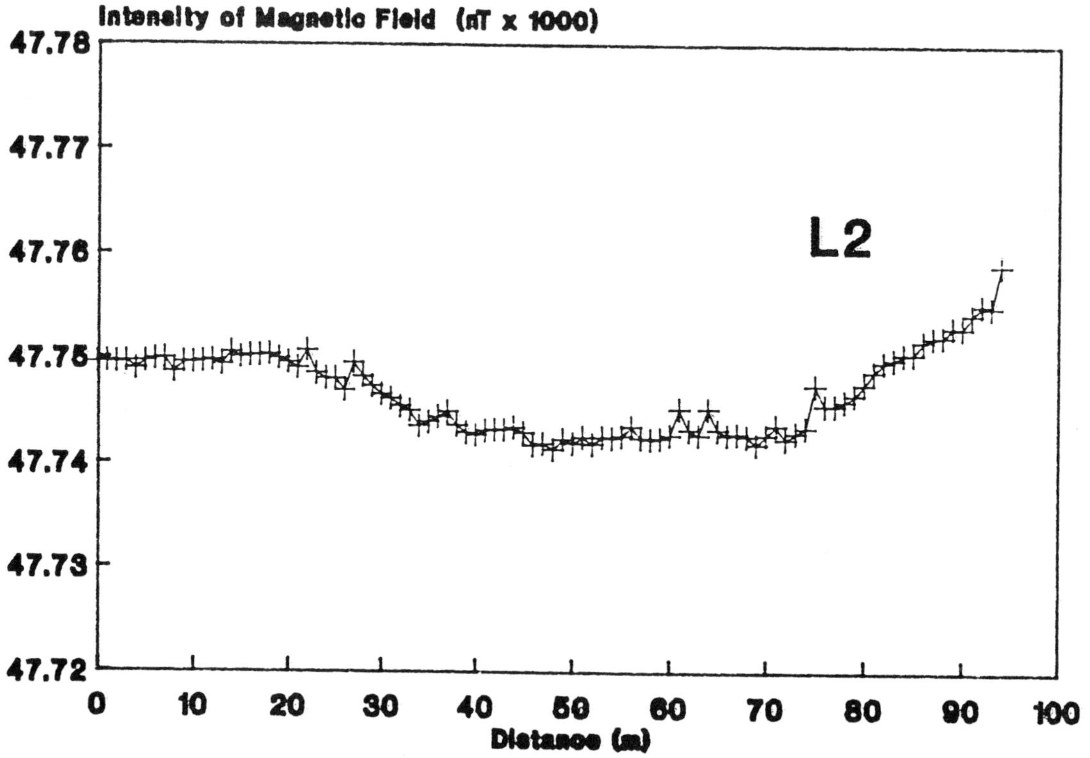

Fig. 8. Magnetic profiles L1, L2.

Trace Q, on the contrary, had no equivalent in magnetic prospection and may, therefore, be caused by a disturbance produced when taking the picture, or when treating it, rather than to buried structures. Had it really been a Neolithic Graben-Anlage, as it might have seemed at first, there would have had to be an equivalent anomaly in the magnetic prospection. It may be, that three profiles are not sufficient to form a valid piece of evidence.

The anomaly in the 60 metres wavelength may be produced by a palaeoalveal stretch along with the presence of a minor amount of organic material. Due to time restrictions, it was unfortunately not possible to check this hypothesis by examining stratigraphic samples. This latter will be done, however, during the course of future campaigns.

Zone 2 – Endrőd 124

The square area covers part of site 124, touches Kurgan 18 and extends towards site 125 (*Fig. 9*). The purpose of the prospecting was to indentify the boundaries of site 124 and, at the same time, to verify whether there were zones between the sites with no structures or whether they continued uninterrupted. First a north-south topographic grid base was installed, linking it to the 1:10.000 scale map supplied by the Archaeological Institute of the Hungarian Academy of Sciences. This was done in order to provide a basis if, in the future, explorations were to be extended. The map was then digitalised, thereby creating a computerised topographic support system.

Fig. 9 shows the area of the systematic geophysical prospecting, carried out using two proton G 856 AX magnetometers[2]. The fixed station is situated at the top of Papp-halom, marked with an asterisk on *Fig. 9*.

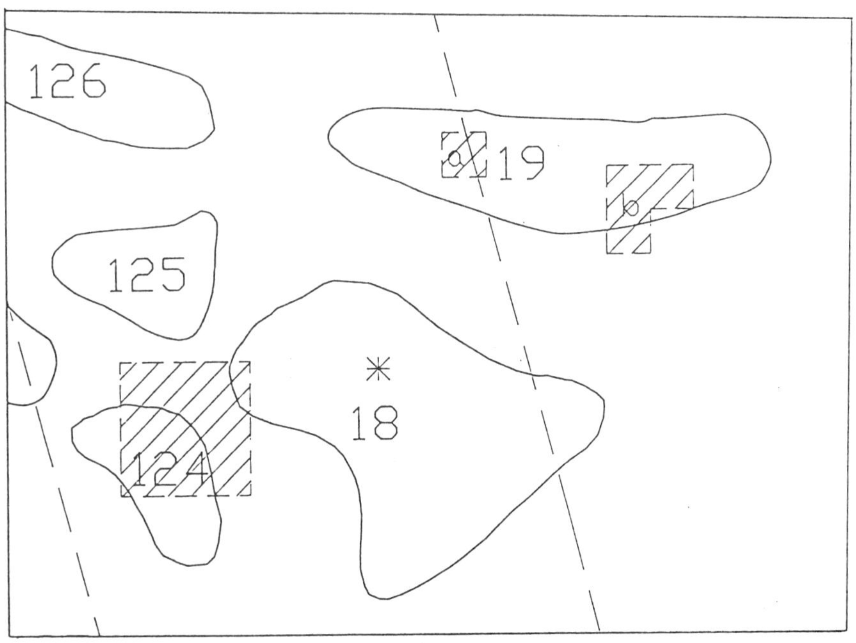

Fig. 9. Detail of the Gyomaendrőd area map, with the sites of Endrőd 124 and Endrőd 19 where the geoarchaeological investigations were carried out (hatched areas).

[2] Working with two magnetometers, one can completely eliminate daily variations in the earth's magnetic field, which sometimes can be greater than those produced by buried anthropic structures (Weymouth and Lessard 1986).

All magnetic readings were registered on a laptop computer GRID CASE II, directly, in the field, and later transferred to an IBM AT PC, in the study-laboratory at Gyomaendrőd, put at our disposal by the Hungarian Academy of Sciences.

Having eliminated the daily variation from the field, the percentage distribution of the measurements and the association between shades of grey and intervals of field values, based on the histogramme (*Fig. 10*), were programmed. This first representation revealed a vast high intensity anomaly in the north-western sector with characteristics non-referrable to archaeological events (Cucarzi 1986). The data was treated in order to eliminate regional disturbances effects with a least-squares method of interpolation and afterward with a band pass filter to reduce the noises of high frequency disturbances. The map in *Fig. 11* was created with the residuals, associating shades of grey as indicated in the legend. This map displays a number of magnetic field intensity anomalies, which can be divided into two categories, depending on their wave characteristics:

- Class 1: anomalies with an intensity between 4 and 40 nT, and a dominant wave length between 3 and 10 metres (indicated by A in *Fig. 11*).
- Class 2: anomalies with an intensity between 10 and 20 nT, and a dominant wavelength of over 10 metres (indicated by B in *Fig. 11*).

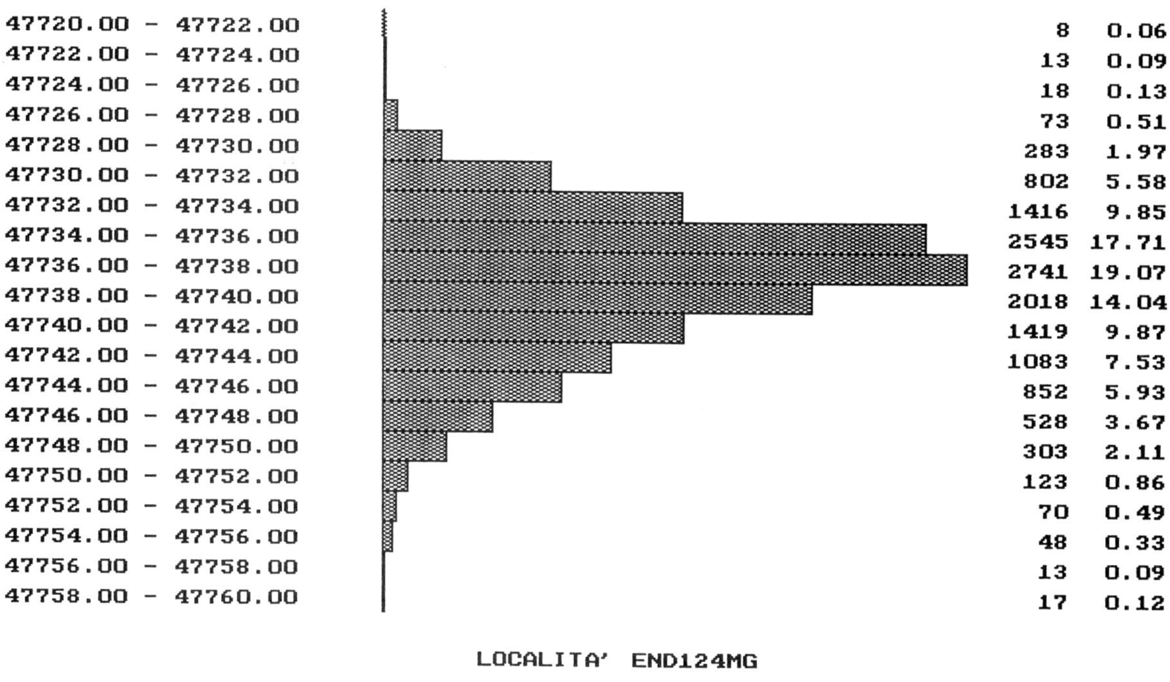

Fig. 10. Endrőd 124: Histogramm of the measurements of the intensity of the earth's magnetic field.

Within the explored area it is possible to recognize some anomalies certainly caused by non archaeological phenomena, such as that indicated by "X" on the map of *Fig. 11* and produced by trial trenches excavated previously by Dr. B. Genito.

For the interpretation of other anomalies present in the area, we have tried to develop some very simple magnetic models and compare them with some previously encountered magnetic anomaly. This part will be briefly described in the section on modelling.

Zone 3 – Endrőd 19

On a limited area (a) of site 19 (*Fig. 9*), close to the trial trench excavated by Dr. B. Genito, 2,000 geomagnetic measurements were taken using the same procedures adopted for Endrőd 124. The detailed results are reported in Genito et al, and therefore only a short summary is provided here. The geomagnetic map in *Fig. 12* exibits different anomalies that could be related to pits; the anomaly P that covers a larger area, indicates the presence of layers with higher magnetic characteristics but most likely not of anthropogenic origin.

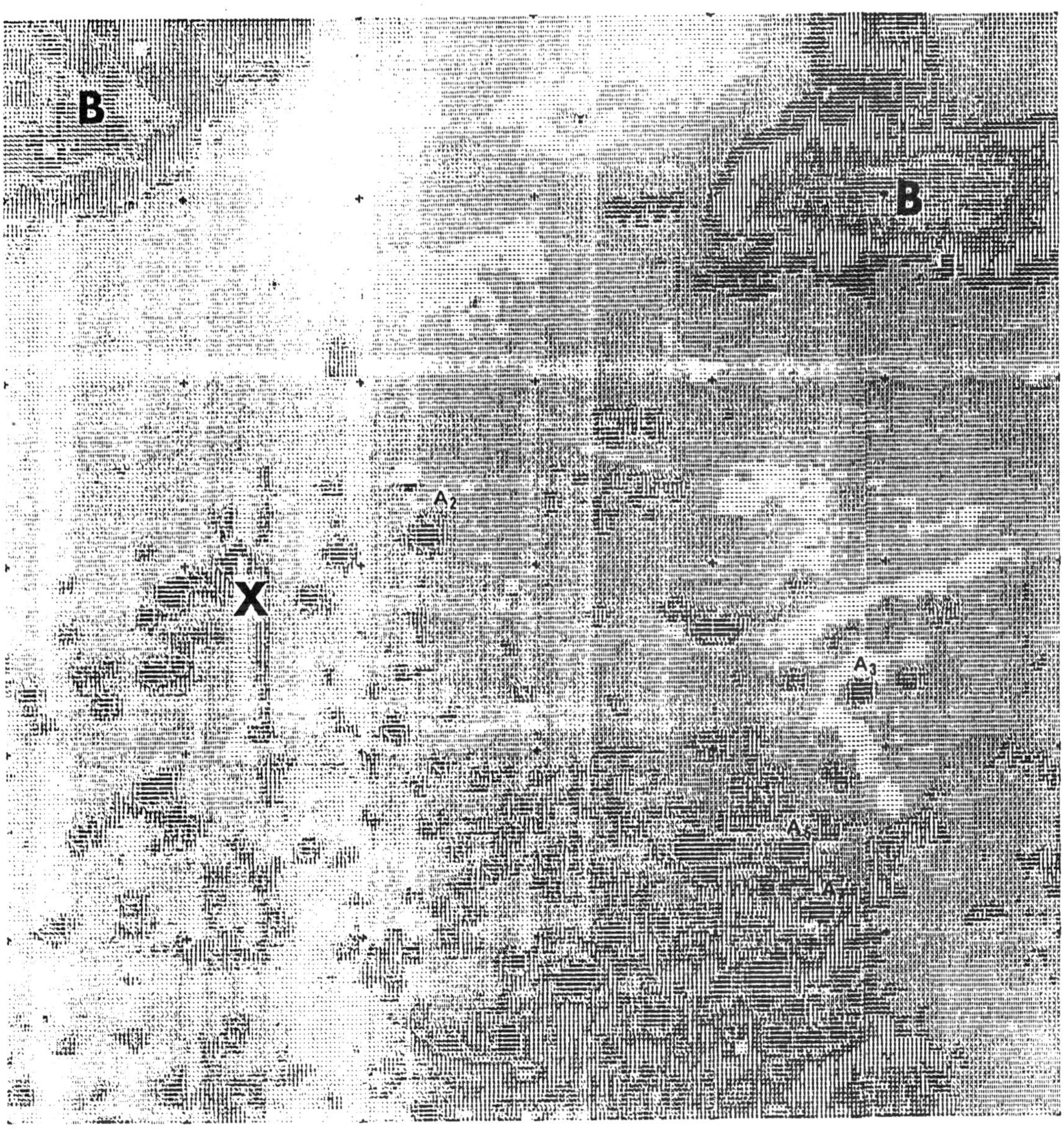

Fig. 11. Endrőd 124: Computerised map where "A" marks the magnetic anomalies of class 1 while with "B" denotes those of Class 2. A2, A3, A4, A5, A7 are the magnetic anomalies compared with the selected magnetic model. "X" indicates the anomaly caused by the trench exvacated by Dr. B. Genito.

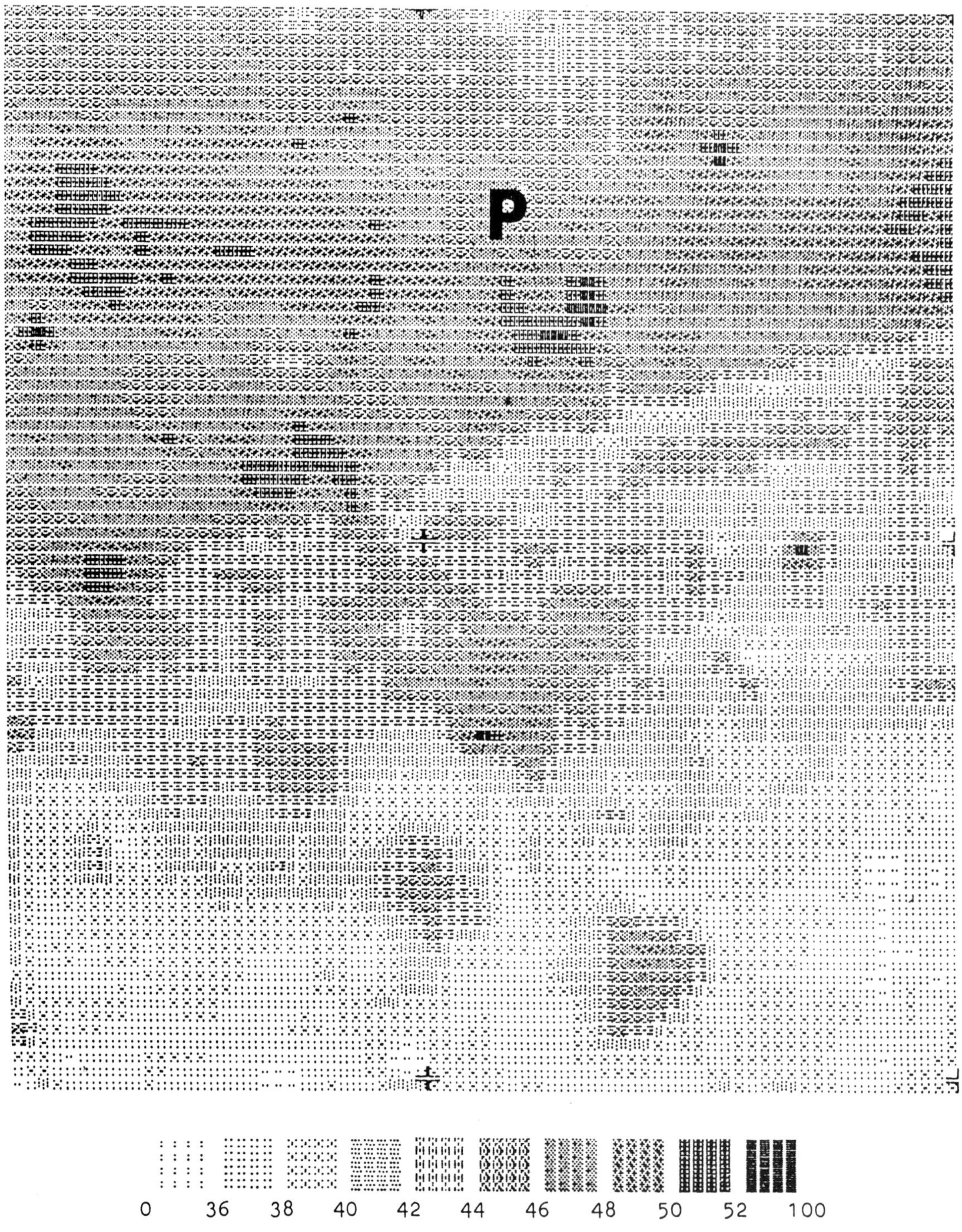

Fig. 12. Endrőd 19a: Map of magnetic field intensity values. "P" indicates a Class 2 anomaly.

Modelling

In order to better explain the archaeological meaning of the great number of magnetic anomalies recorded during the prospecting campaigns, some simple magnetic models with a certain number of observed anomalies have been attempted. The proposed models included the magnetic dipole magnetized by induction alone (and assuming the magnetization is uniform) and the horizontal tabular body.[3]

At first, numerous soil samples were taken at different depths from the trenches excavated by Dr. Bruno Genito as well as from samples of terracotta and from what is believed to be a fire place or an oven, brought to light by trench near Endrőd 124.

Apart from sedimentological analysis, these samples were used for measuring magnetic susceptibility using a Burlington M.S. 2C. The measurements showed that the magnetic susceptibility contrast between samples from alluvial unity and chernozem is approximately 10 E-6 emu, whilst the contrast between the alluvial unit and fire place fragments is around 150-250 E-6 emu.

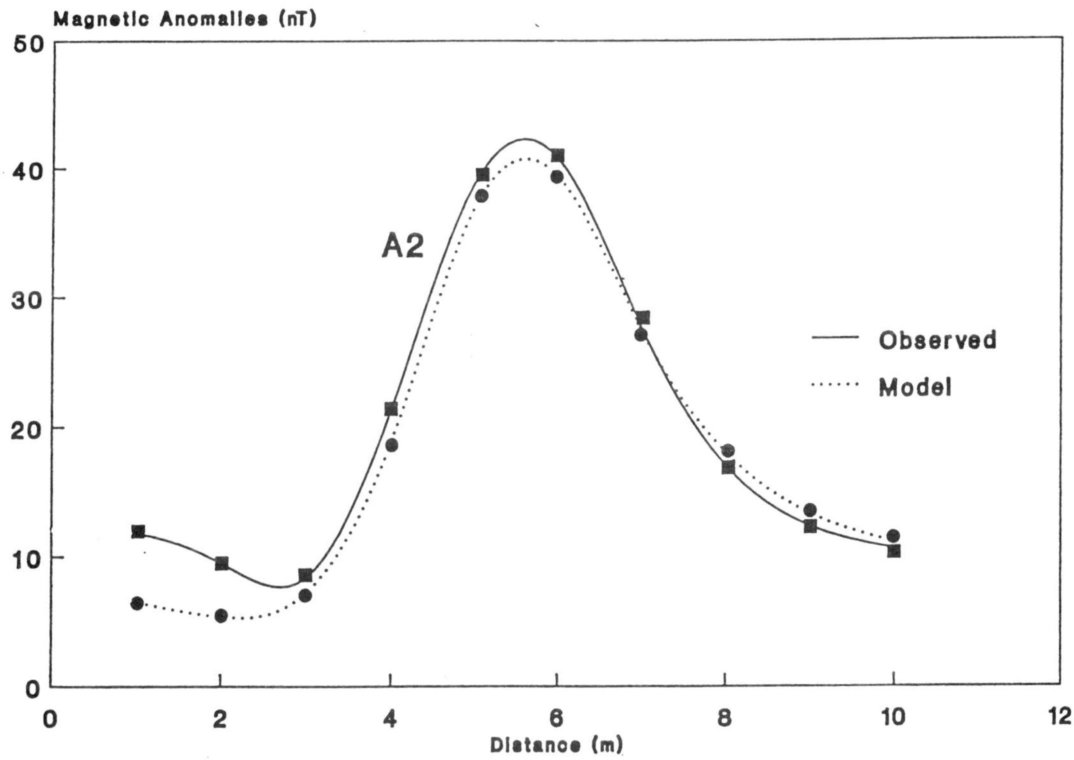

Fig. 13. Endrőd 124: Magnetic anomaly A2 and the proposed magnetic model.

[3] With this approach the inversion problem is obviously not confronted. Since the measurements are taken every metre and corrected according to diurnal variations (recorded every 60 seconds) an approximation comes about, which is not acceptable in the face of the inversion problem, although from the point of view of sensitivity use of the instrument (0.1 nT) would have made it possible.
At this stage, the question is to see whether the proposed model (1D) can help in rapidly recognizing within class 2 anomalies, some sub-classes that can somehow be associated with certain structures.
Subsequently, modelling and the inversion problem will have to be solved employing more appropriate models such as the vertical prism and assemblage figures, in order to have at least a 2D solution.

In order to set the value intervals for parameters in the equations used in these simple models, and obtain a better correlation with the observed data (Cucarzi 1986), we have developed a simple computer program that, after a chosen number of iterations, minimizes the errors between model and observed data (keeping in mind the principle of equivalence).

Attention should be paid to some anomalies of class 1x present in site 124 (*Fig. 11*). Dominant wavelengths are primarily represented by values between 4 and 6 metres, whilst the intensity varies between 5 and 45 nT. Observing the shape of the anomaly, it may be assumed that the structure which produced it may be represented by a dipole magnetic model with a length from 1.5 to 4.5 metres. The depth would be variable.

Among the numerous anomalies of this class present in the area, about ten were chosen to be studied and here A2, A3, A5, A7 will be analyzed in order to compare the observed data with theoretical onces produced by the dipole model with a wavelength varying from 1 to 5 metres and with the centre laid at 1-3 metres in depth from the sensor.

Fig. 13 shows the observed data from anomaly A2 and the theoretical curve derived from a dipole of 3.80 metres with the centre being 2.60 metres in depth and having a magnetic susceptibility of 215 E-6 emu.

Fig. 14 and *15* represent anomalies A3 and A5 and the theoretical curves of magnetic dipole models of 2.80 and 4.102 metres in length and a magnetic susceptibility of 160 and 210 E-6 emu.

These three anomalies have an intensity varying from 25 to 40 nT which is characteristic of more than 60% of all anomalies in this class present in the explored area.

Anomaly A7, *Fig. 16*, belongs to class 1 as well but represents another part of the anomalies (30-40%) with an intensity rather lower than the previous ones, varying between 5 and 15 nT.

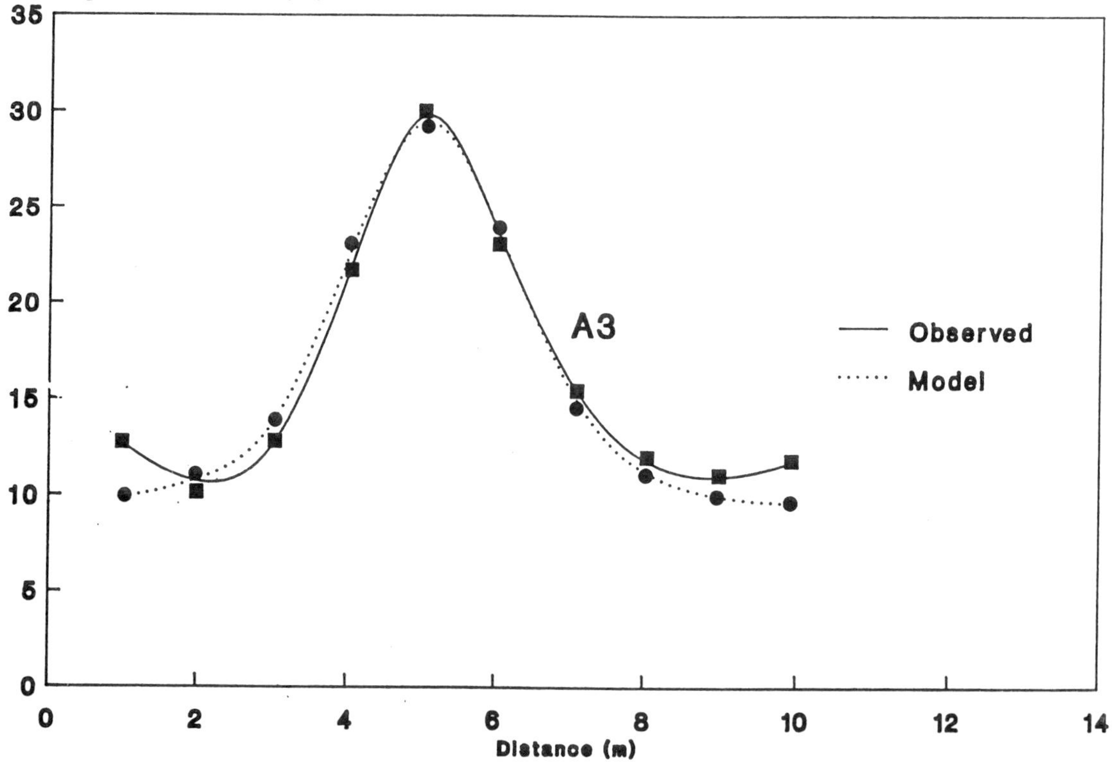

Fig. 14. Endrőd 124: Magnetic anomaly A3 and the proposed model.

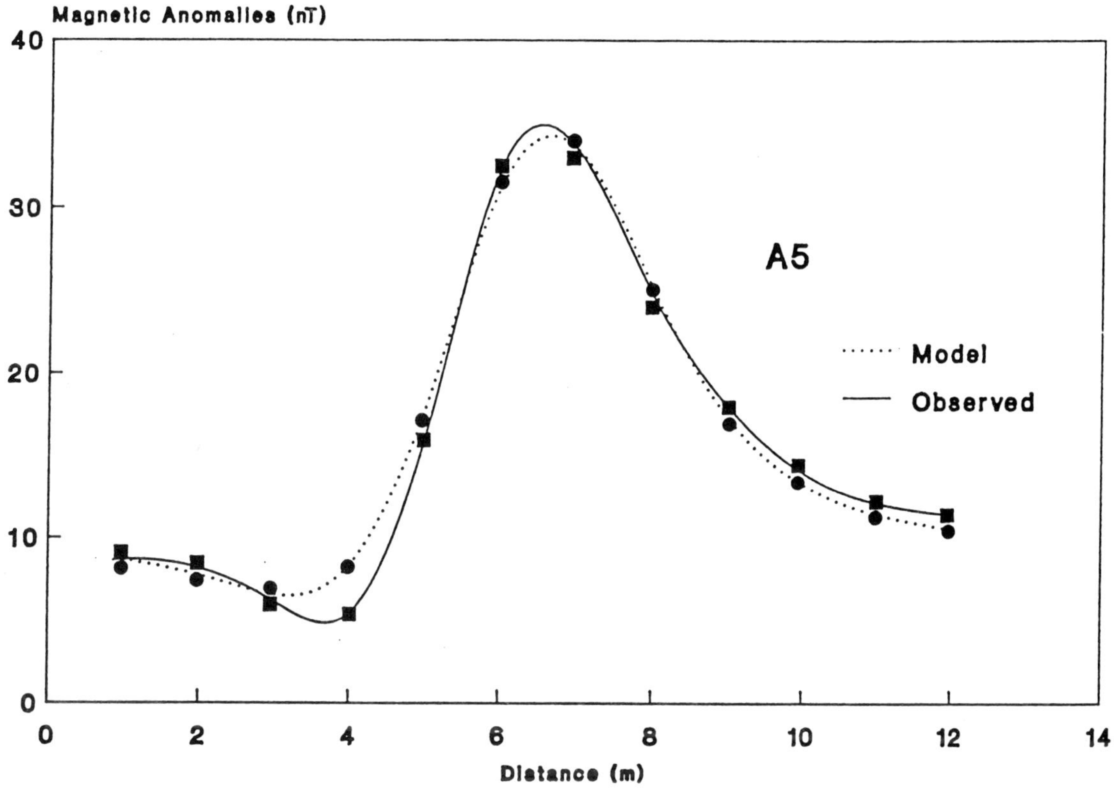

Fig. 15. Endrőd 124: Magnetic anomaly A5 and the proposed model.

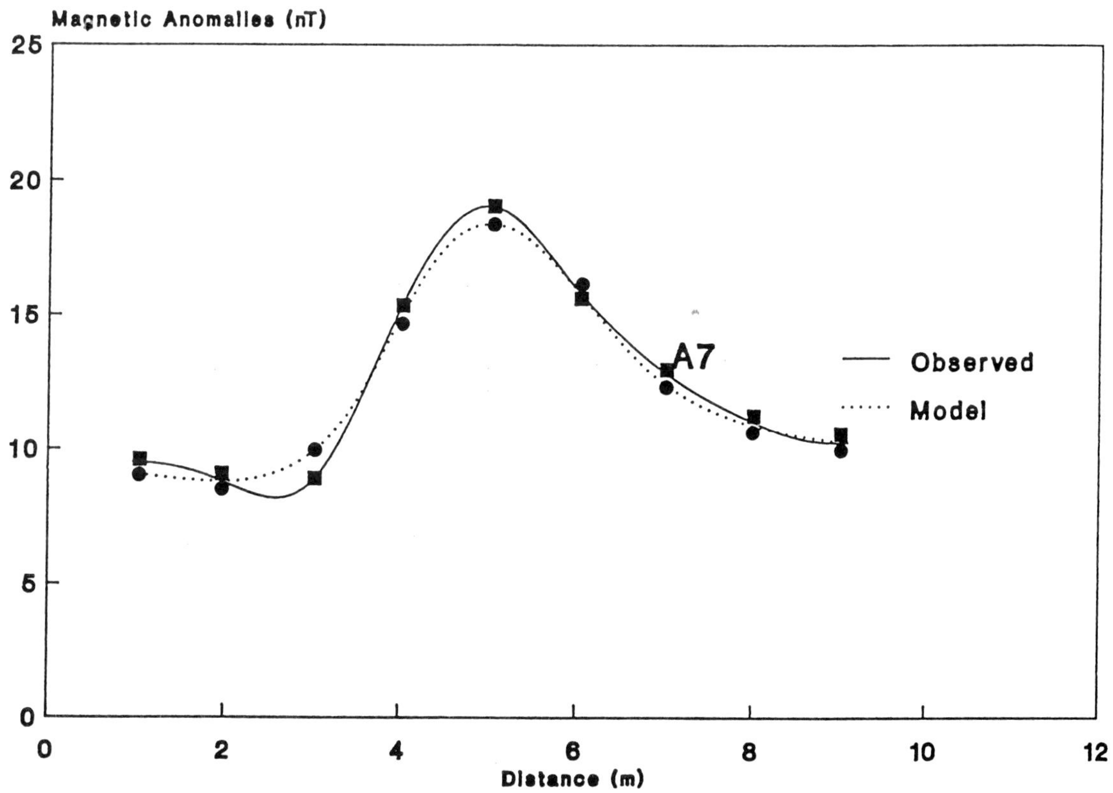

Fig. 16. Endrőd 124: Magnetic anomaly A7 and the proposed model.

Even in this case, the proposed model fit the A7 anomaly with a susceptibility of 100 E-6 emu, a wave length of 2.49 and a depth of 2.50 m. To check the reliability of the model compared to the real geometrical parameters, drillings were carried out (see paragraph on core drillings) on some anomalies. These indeed show that the buried structures are similar to pits with diameters varying between only a few centimetres and one metre below present ground level.

The log in *Fig. 17* represents the stratigraphic position of anomaly A2. Furthemore, it is interesting to see that the anomalies in this group are similar to those seen by Miklós Pattantyús at Endrőd site 170 (see Jankovich and Pattantyús this volume and Jankovich and Pattantyús 1990). M. Pattantyús-Á. checked one of these anomalies by drilling using an augering technique. He found anthropic strata with fragments of approximately three metres (see Jankovich and Pattantyús 1990).

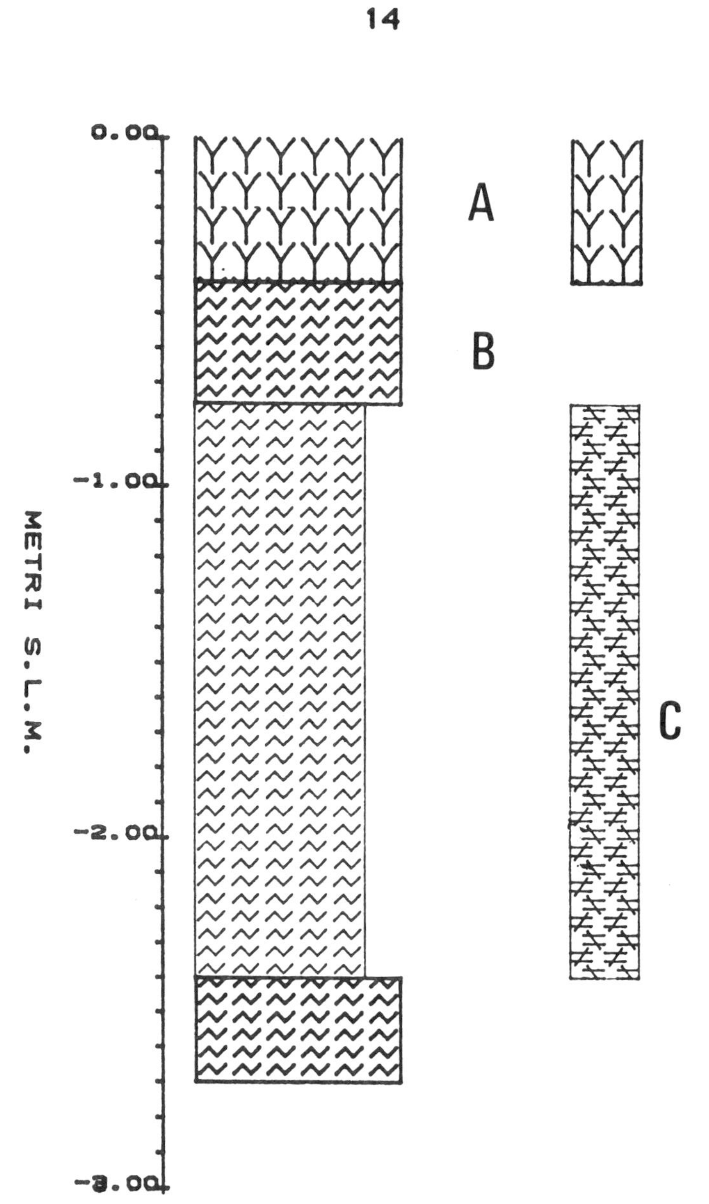

Fig. 17. Endrőd 124: Log of core-drilling 14 done in A2. "A" represents the agricultural layer, "B" silt, and "C" the anthropic layer with pottery fragments and charcoal.

In zone "a" in the Endrőd 19 site, there is a number of class 1 anomalies, of which 1 and 2 were chosen to compare with the dipping dipole model.

The best fit was, in the first case, with a dipole of 3 metres length, a depth of centre of 2.5 metres and with a susceptibility contrast of 90 E-6 emu. In the second case the best fit was with a dipole of 4.60 metres length. A depth of centre of 4.40 metres and a susceptibility contrast of 180 E-6 emu (*Fig. 18* and *19*).

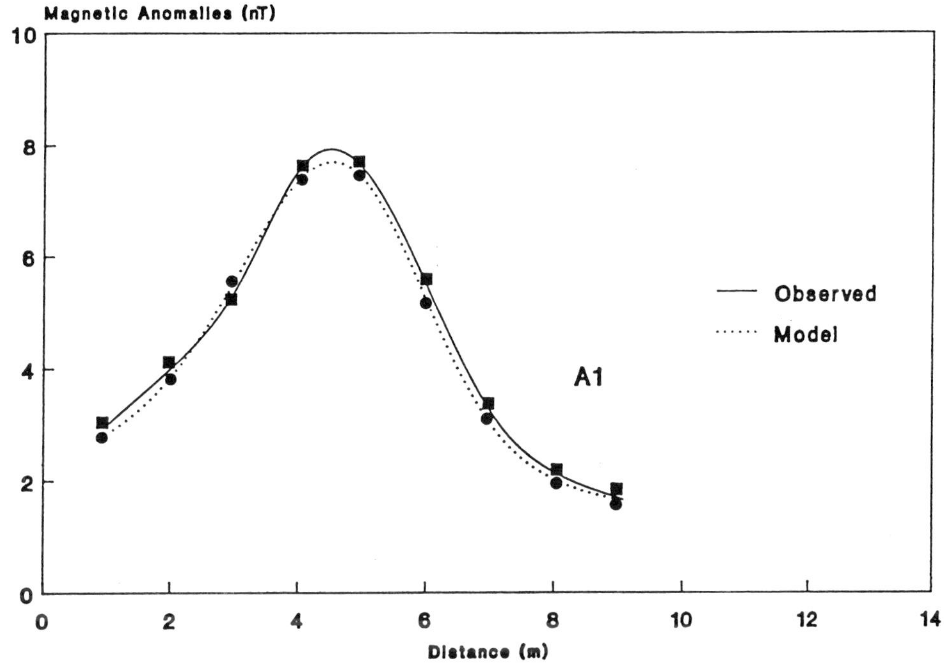

Fig. 18. Endrőd 19a: Magnetic anomaly A1 and the proposed model.

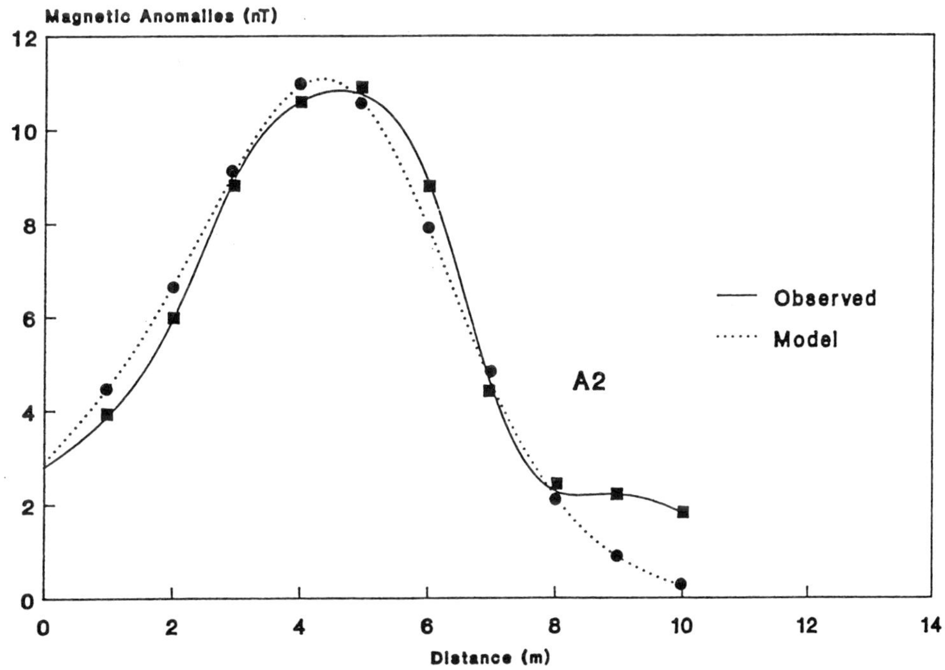

Fig. 19. Endrőd 19a: Magnetic anomaly A2 and the proposed model.

In sector "b" of the same site, there are still numerous class 1 anomalies, of which 1 and 2 were chosen. In this case, the best fit was with the selected model having a wave length of 2.40 metres a depth of centre of 3.20 metres and a susceptibility contrast 280 E-6 emu (*Fig. 20* and *21*).

There are also some class 2 anomalies (dominant wavelength more than 10 metres) in Endrőd 124 and Endrőd 19.

In Endrőd 124 two anomalies may be grouped within this class: T1 and T2 have an intensity of 10-20 nT. In both cases, the closest corresponding model is a thin, horizontal plate with a magnetic susceptibility contrast of 20-30 E-6 emu.

This might reflect a major organic component either recently carried there by man for agricultural purposes, or produced by water flows or a chernozem layer of significant thickness. These suppositions may be checked with a series of appropriate analyses of core drilling samples.

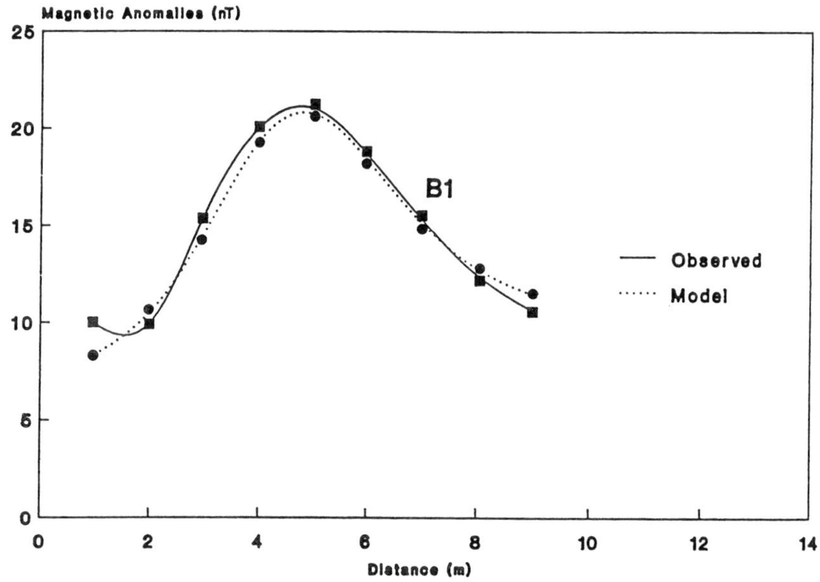

Fig. 20. Endrőd 19b: Magnetic anomaly A1 and the proposed model.

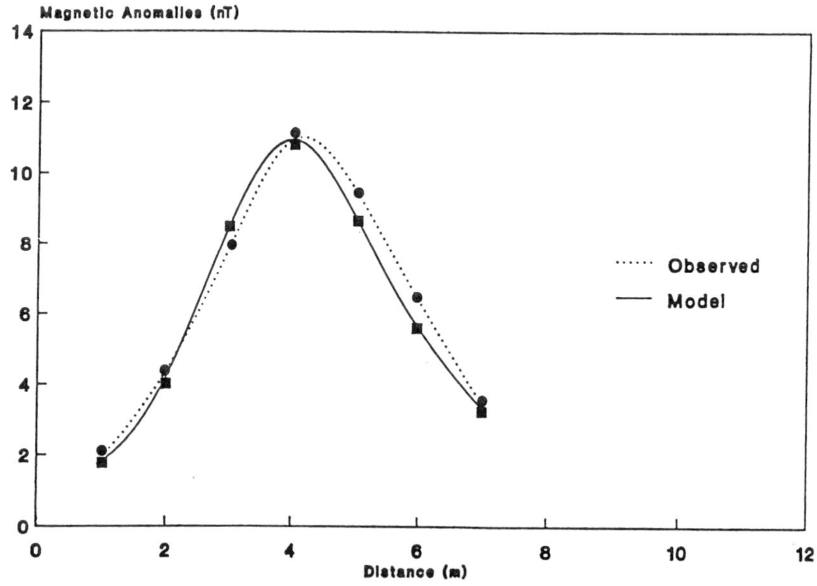

Fig. 21. Endrőd 19b: Magnetic anomaly A2 and the proposed model.

The situation is similar at Endrőd 19a and Endrőd 19b. However, aiming at identifying class 1 anomalies, of greater interest for the present research, the data matrixes were treated mathematically (least-square interpolation), assuming the class 2 disturbances to be regionals. The map of *Fig. 22* is a computerised representation of the residual matrix.

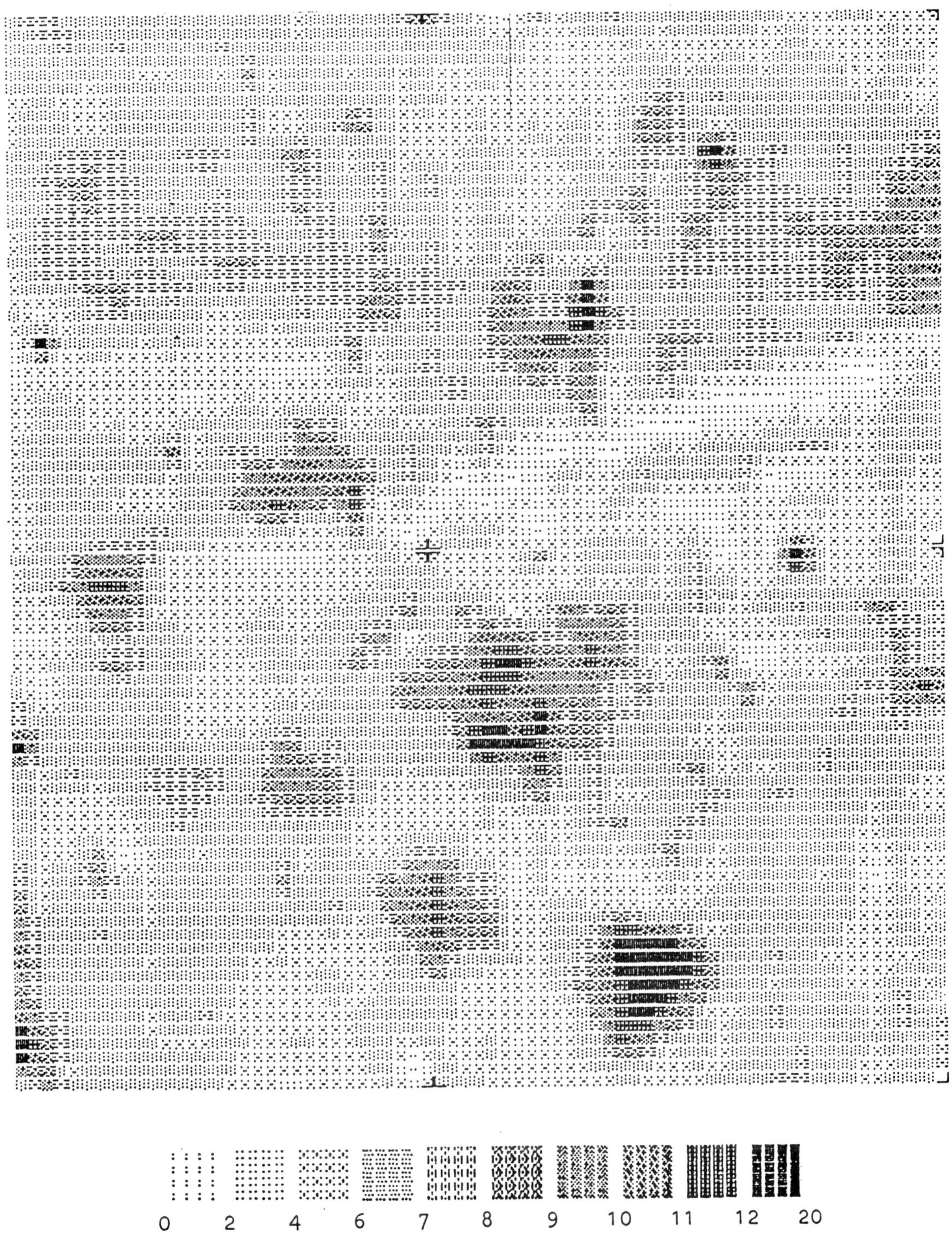

Fig. 22. Endrőd 19a: Map of residuals with some class of magnetic anomalies.

Core Drillings

Within the integrated geoarchaeological prospecting system, the study of stratigraphic samples obtained through core drillings is very important for the reconstruction of the ancient milieu and for a better understanding of how the sites were formed (Stein 1986). It is furthermore a direct, almost non-destructive test for hypotheses formulated on the basis of geophysical data (Cucarzi 1986).

During the 1988 campaign, this procedure was applied only at the site of Endrőd 124. The results are briefly described below.[4]

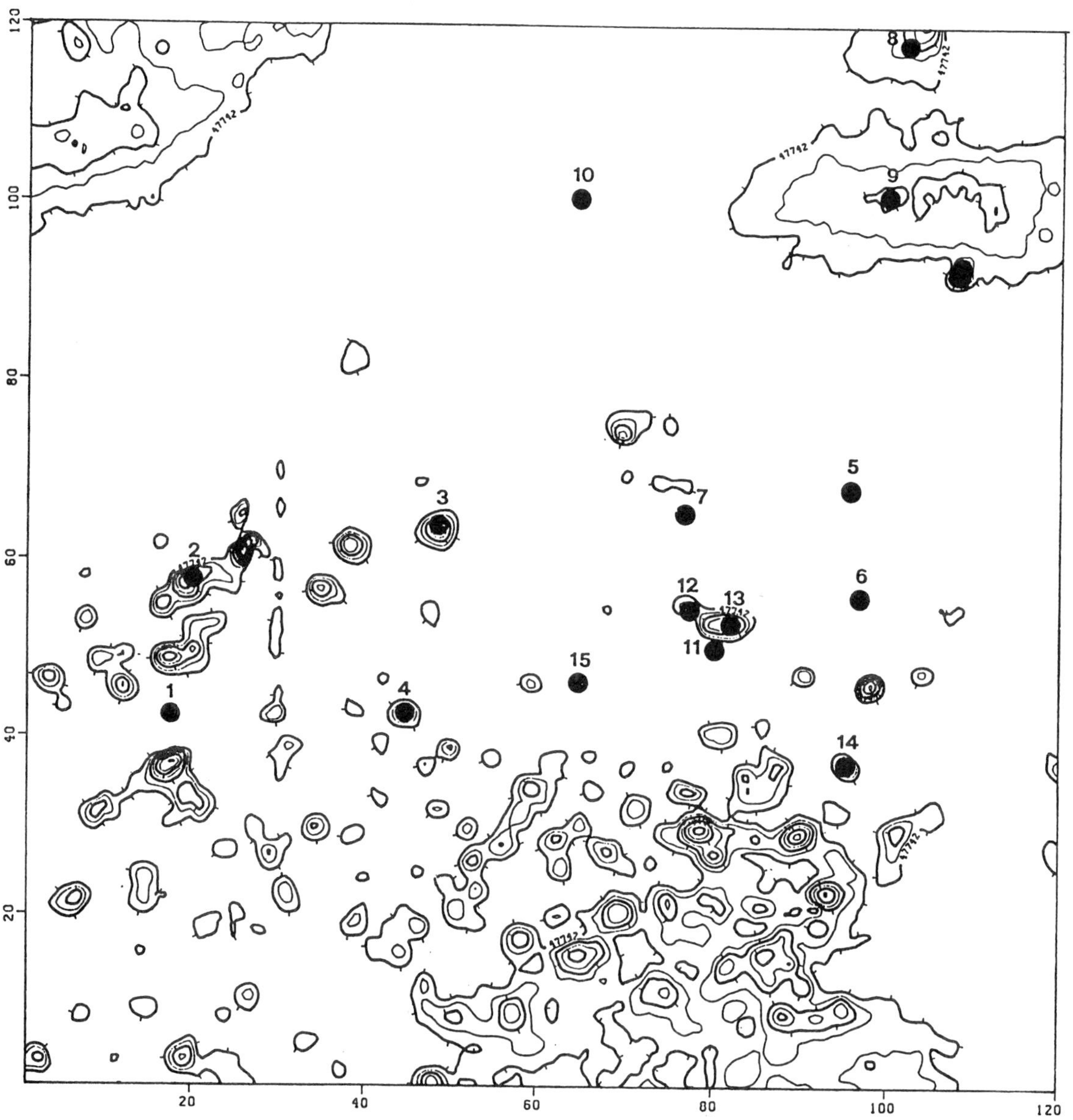

Fig. 23. Endrőd 124: The locations of core-drillings in comparison to the magnetic anomalies.

[4] The result of the stratigraphic studies will be described in a later report, by the person responsible for this sector, Dr. A. Rivolta.

In addition to the study of exposed section for the geomorphological description of the Microregion on a greater scale (for which Prof. M. Cremaschi is responsible) on an area explored by geomagnetic methods, 15 core drillings were made, some of which were placed directly over magnetic anomalies (*Fig. 23*).

The area studied is formed by a slightly undulating plain, with palaeoalveal remnants which are slightly hollow compared to ground level. The undulation may be interpreted as ancient river beds formed by sediments with a sandy-muddy texture.

The examined area's stratigraphic sequence consists of only two major pedostratigraphic units, with a gradual, lateral passage: chernozem and vertisuoli.

In addition to the specific geomorphological considerations which will be described in a specific further report by A. Rivolta, traces of ancient anthropic deposits are worth mentioning found in core drillings 2, 3, 4, 5, 12, 13 and 14 and holes of huge dimensions, which in some cases reach depths of approximately three metres.

The fill of these holes consists of material of anthropic origin, such as charcoal, ash, burnt slag perhaps from furnaces and often in centimetre-thick layers. They deepen from the bottom of the chernozem, pointing to the area having been inhabited during a period in which the soil had already begun its evolution into one with characteristics similar to the present.

Fig. 24 shows a simplified, computerised representation of the 15 logs.

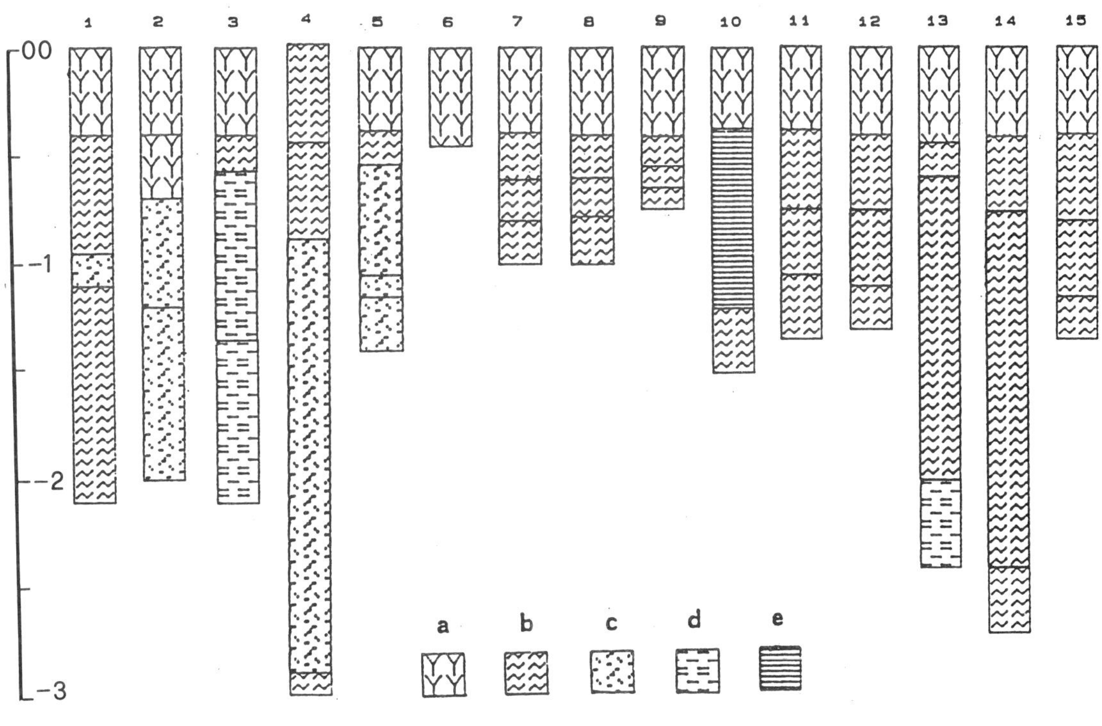

Fig. 24. Endrőd 124: Logs the 14 core-drillings taken at Endrőd 124. The symbol "a" represents the agricultural layer, "b" silt, "c" the anthropic layer, "d" chernozem, and "e" clay.

Interpretation of Data

On the basis of the geophysical data and the results of the drillings, some interpretational suppositions may be infered, regarding the area, so far explored.

Endrőd 124. Taking into account these observations and looking to the geomagnetic map of *Fig. 11*, it is possible to formulate some hypotheses about the horizontal extent of Endrőd 124.

The greater concentration of class 1 anomalies lie in the central and southern part of the explored area, where it is possible to recognize 70-80 anomalies. These types of anomalies are less frequent towards the north. It therefore might reasonably be expected that the northern limits of Endrőd 124 settlement may be found in this area.

In the northern sector of the investigated area there are some class 2 anomalies, which as has already been suggested, seem to be related to natural events.

Under the conditions of this campaign, we have not touched Endrőd 125, as on the contrary we made our assumptions based on the map generated by the collected surface data. Sites Endrőd 124 and Endrőd 125 seem to be divided by a zone in which anthropic structures are missing.

Endrőd 19. Class 2 anomalies are likely to be caused by palaeochannels or old riverbeds. In sector "a", the class 1 anomalies are spread over the whole of the explored area and hence no site borders can be identified (*Fig. 22*).

In sector "b", on the contrary class 1 anomalies are concentrated in the central northern part; the class 2 anomaly in the southern part seems to represent the site's border on this side (*Fig. 4*).

The intensity of class 1 anomalies, and above all the magnetic susceptibility contrast values of the models which represent them, are remarkably much lower than those in Endrőd 124. Considering that site Endrőd 124 is mainly from the Sarmatian period, whilst Endrőd 19 has Bronze Age and Le Tène components, it may be suggested that there is a link between the type of magnetic anomaly produced by anthropic structures and the archaeological period they belong to.

Örménykút 84. The area which could be explored was extremely limited due to reasons previously described. No class 2 anomalies were found, whilst some class 1 anomaly did appear (*Fig. 1*). The excavation carried out in that area by Dr. Cs. Bálint, showed that it was caused by a fire place. The intensity of the anomaly is comparatively low, but the closest model indicates a susceptibility contrast of 270 E-6 emu, which is definitely higher than those of Endrőd 19 and comparatively higher than those of Endrőd 124. The fact that Örménykút 84 is mainly an Avar site and that it produced an anomaly with different characteristics from both Endrőd 19 and Endrőd 124, may confirm the hypothesis of an association between structures from a certain period and a certain type of anomaly.

Gyoma 133. No class 2 anomalies were found in the Gyoma 133 sector although there were some class 1 anomalies with completely different characteristics from those at previously described sites, particulary because they have a higher gradient (*Fig. 2*).

Conclusions

During these experimental campaigns, it seemed evident that a single method of investigation, even if applied in a very detailed manner, cannot provide sufficiently satisfactory results to formulate interpretative hypotheses on a region's settlement patterns or characteristics.

The photointerpretation shows traces which may be interpreted as buried anthropic structures, but whithout support from other investigative methods, a sufficiently reliable interpretation cannot be guaranteed. Cross checkings with geophysical stratigraphical soundings reduce the margin of interpretative error and, at the same time, yield large quantities of data of varying levels of detail.

Procedures which provide very detailed information over space and time are not sufficient when it comes to the territory, as they do not consider location from that point of view.

Therefore, only an integrated system can yield properly reliable information on a multi-scale level on a territory if it is treated as a cultural deposit.

The integrated system, however, is not intended as a substitute for archaeological excavation, but rather it is designed to establish procedures which may give a series of data to use when planning an archaeological excavation, thereby reducing area which needs to be excavated.

Eventually, the results of the excavations should confirm the hypotheses and, at the same time, with a minimal destructive intervention, supply new data to the general interpretation process, hence contributing to the historical understanding of the territory.

Future prospect

The results obtained so far show the reliability of this approach in the general context of the Microregion. It is therefore, considered necessary to proceed from the phase of experimentation and verification to a more systematic application of the method over a greater area.

Already during numerous discussions with the Hungarian archaeologists responsible for the various sectors of the Microregion project, it has been pointed out that the area between Endrőd 170 and Endrőd 18 (see *Fig. 3*) is a sector of particular interest due to the many settlements from different cultural periods near the palaeoalveal system of the Körös river.

Therefore, explorations could be extended using these experimental methods, thus, activating those research sectors which were marked down for geoarchaeological prospecting but which have not yet been worked on.

Briefly, aerial photographs of the area taken at various altitudes will be used to reconstruct the palaeoalveal systems of the Körös river, in order to correlate old river beds with ancient settlements. Geophysical methods will be used to locate buried anthropic structures.

On the basis of this data and the data obtained by the analysis of the surface research already carried out by the team of experts from the Archaeological Institute, systematic geophysical prospecting could be commenced, in order to better establish its limits, locate structures, and register – where possible – the main nucleus of the settlement.

The research would then be continued on the basis of the ability to recognise, through the wave characteristics of geophysical anomalies, structures belonging to different cultural periods.

Premises exist in this field as well. What appears to determine the wave characteristics of magnetic anomalies is the intensity of magnetisation, which was schematically represented in terms of magnetic susceptibility. This may be one solution as, all conditions being equal, the increased magnetic susceptibility value may be due to a different level of pyro-technical evolution.

The core drilling program will provide the information required for controls as well as for a further correlation to be made between geophysical anomalies and anomalies caused by buried structures. It will also say something about the stratigraphic sequence in order to temporally correlate the formation of the site with the fluvial morphology of the Körös river.

Sedimentological and palaeobotanical analyses of stratigraphic samples collected within this system will provide the data required for a palaeoenvironmental reconstruction of the Microregion.

References

Agache, R., 1970. Détection aérienne de vestiges protohistoriques gallo-romains et médiévaux dans le bassin de la Somme et ses abords, *Bulletin de la Société de préhistoire* 7.

Chevallier, R., 1964. *L'avion à la découverte du passé*, Artheme Fayard, Paris.

Colwell, R. N., 1983. *Manual of Remote Sensing*, 2nd ed., American Society of Photogrammetry, Falls Church , Virginia.

Cucarzi, M., 1986. A Methodology for Geophysical Prospecting in Archaeology. An example: Mohenjo-Daro, *South Asian Archaeology* 1983, 23, 279-295.

Dassie, J., 1976. *Manuel d'archaéologie aérienne*, Editions Technip, Paris.

Ebert, J. I., 1984. Remote Sensing applications in archaeology, *Advances in Archaeological Method and Theory*, 7, 293-351.

Fassbinder, J. W. E., Stanjek, H. and Vali, H., 1990. Occurence of magnetic bacteria in soil, *Nature* 343, 6254, 161-163. Macmillan Magazines.

Fischer, W. A., 1975. History of remote sensing, *Manual of Remote Sensing*, 27-50.

Genito, B. et al. (in press), Preliminary Report of excavations on Gyoma 133, *East and West*.

Jankovich-B., D., Makkay, J. and Szőke, B. M., 1989. Archaeological Topography of Hungary: *MRT 8*, Budapest.

Jankovich-B., D. and Pattantyús-Á., M., 1990. Archaeological and geophysical survey of a multi-aged site Endrőd-1970, SE-Hungary. *Prospezioni Archeologiche*, 1/1990. 123-132. Rome.

Le Borgne, E., 1955. Susceptibilité magnétique anomale du sol superficiel, *Annales de Géophysique*, 399-419.

Le Borgne, E., 1960. Influence du feu sur les propriétés magnétique du sol et sur celles du schiste et du granite, *Annales de Géophysique*, 16/2, 159-195.

Ray, R. E. and Fischer, W. A., 1957. Geology from the air, *Science*, 126, 725-735.

Seguin, M. K. and Samson, S., 1980. Application archéologique des techniques de resistivité électrique des sites amérindiens, lac de la Hutte Sauvage (Nouveau-Quebec), *Geoexploration*, 18, 15-32.

Stein, J. K., 1986. Coring Archaeological Sites, *American Antiquity*, 51 (3), 505-527.

Stoner, E. R. and Baumgartner, M. E., 1981. Characteristic variations in the reflectance of surface soils, *Soils Science Society of American Journal*, 45, 1161-1165.

Strandberg, C. H., 1967. Photoarchaeology, *Photogrammetric Engineering*, 33, 1152-1157.

Tite, M. and Mullins, C., 1971. Enhancement of the magnetic susceptibility of soils on archaeological sites, *Archaeometry*, 13, 209-219.

Weymouth, J. W. and Lessard, Y.A., 1986. Simulation Studies of Diurnal Corrections for Magnetic Prospection, *Prospezioni Archeologiche*, 10, 37-47.

Results of geophysical surveying in the investigation of archaeological sites in the Gyomaendrőd region, Hungary

BALÁZS SZÉKELY - ORSOLYA MAGYARI
PÉTER STEINBACH - TAMÁS TÓTH

I. Introduction

Excavation as a direct method of archaeology – although not always – can be useful, but the result is uncertain if the selection of the site is not correct. That is why science has always searched for new patterns which provide indirect information on underground data. In these cases, geophysics offers a variety of methods such as:

– resistivity measurements
– ground radar tomography
– seismic sounding
– magnetic prospecting
– thermal exploration
– electromagnetic detection
– gravity surveys.

Compared with industrial application, archaeological objects are usually smaller and lie closer to the surface, that is why a special geophysical approach is needed in archaeological site surveying.

In our paper we will outline two methods used here: magnetometry and resistivity measurements and we shall summarize their physical basis as well.

The Geophysical Department of Eötvös University of Budapest has been taking part in archaeological research since 1987 in cooperation with the Archaeological Institute of the Hungarian Academy of Sciences. The sites we worked at are in the neighbourhood of Gyomaendrőd (south-eastern part of Hungary): *Endrőd 6, Gyoma 133, Örménykút-52* and *Örménykút-3*. The location of these areas can be seen in *Fig. 1*.

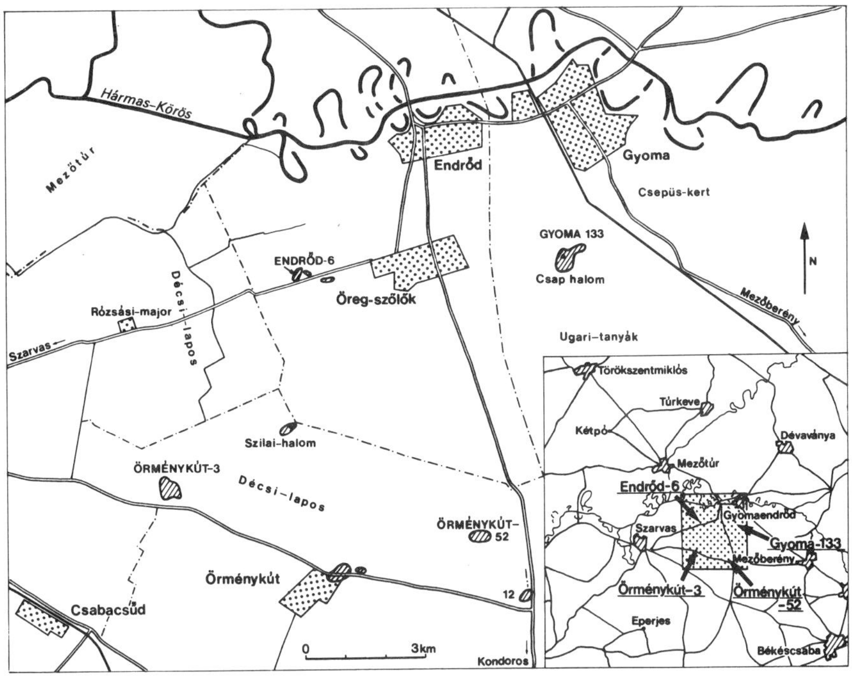

Fig. 1. Locations of the sites surveyed in Hungary.

These areas are marked by sherds found in the ploughed land. The findspots are unsuitable for traditional excavation because of their large size. On the basis of preliminary information it was expected that by using quite accurate and dense geophysical measurements we could find anomalies characteristic of archaeological objects. Geophysical measurement has a great advantage over excavation and boring because it supplies useful information about the location and size of the objects sought without causing any damage to them and, last but not least, they are cheaper than excavation. In the last few years the tendency to keep cultural remains at their original place becomes stronger (Scollar et al. 1986, Weymouth 1986). So future generations have a chance to study them where they were found. Geophysical measurements are excellent for this purpose.

In our case the measured area will have been uncovered soon. This will give valuable information because the archaeological objects unlike to geological bodies (e.g. ore deposits), are totally excavated, so we will know their exact geometrical shape and material. This helps us to correct our hypothetical models and our further measurements as well.

II. Physical Basis of Magnetic Measurements

1. The Magnetic Field

The phenomenon of magnetism was already recognised by ancient scientists. Nowadays, it is well-known that the Earth is surrounded by a magnetic field made up of two parts:

- The magnetic field originating in the interior of the Earth which accounts for 95% of the whole field.
- The remaining 5% is of external origin.

The internal field can be explained by magneto-hydrodynamic processes in the inner core. The primary field produced by them can be approximated by the field of a magnetic dipole. This dipole is approximately geocentric and its axis makes an angle of 11° with the axis of rotation of the Earth. Consequently, the magnetic and the geographic poles differ by this amount as do the corresponding equators. The north pole is of south polarity while the south pole is of north polarity. This dipole produces an easily calculated magnetic field which depends on geographical coordinates and which regularly changes over the surface of the Earth. This dipole-field can be seen in *Fig. 2*.

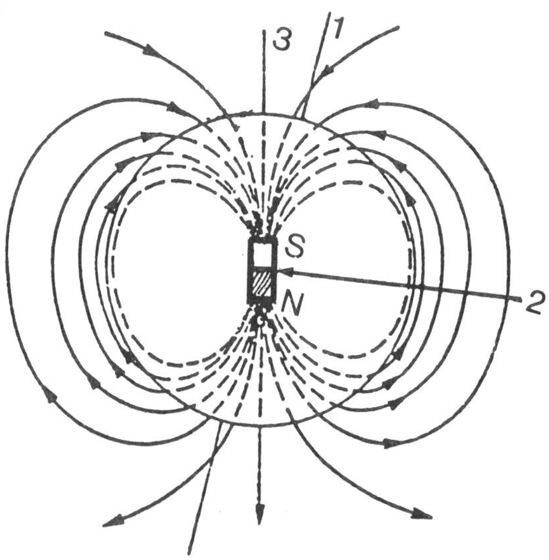

Fig. 2. The dipole field of the Earth.
1: rotation axis 2: hypothetical dipole 3: geomagnetic axis.

Experience shows that the regional magnetic field has an irregular component which is caused by different objects beneath the surface (e.g. iron ore deposits). It is this irregular component or local deviation that is interesting for investigation.

The magnetic field can be represented by a vector (**T**, and its magnitude *T*) at a given point on the Earth's surface whose direction coincides with that of the magnetic field at the given point and its magnitude is directly proportional to the strength of the field. This vector can be described in the geomagnetic system of coordinates in the usual way. In that system the x axis is directed to the north, y to the east and z vertically down. In this coordinate system the total field **T** can be described as follows:

The projection of **T** on the x-y plane is the horizontal component (**H**) and that on the z axis is the vertical component (**Z**). The angle made by **H** and the x axis is the declination (D) and the angle between **H** and **T** is the inclination (I) (see *Fig. 3*).

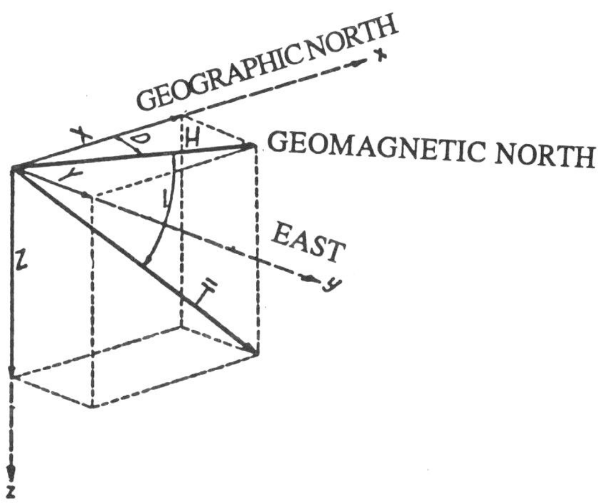

Fig. 3. The elements of the geomagnetic coordinate system. (See explanation in the text.)

At any point on the surface of the Earth the magnetic field is defined unambiguously by D, I and T. D and I can be measured with a freely rotating magnetic needle suspended in its centre of gravity. The loci with I=90° are the magnetic poles and those with I=0° define the magnetic equator.

With our instrument we measured the magnitude of the total field (T). The distribution of the total field (T) on the surface of the Earth is given in *Fig. 4* (after Márton 1975) as an isoline map.

The magnetic field of the Earth is a function of time as well. As a result, one measurement of the magnetic field is valid only for the given time. Therefore, it is also necessary to pay attention to the time variation of the magnetic field. This variation is quite complicated and has several components in different frequencies. In our case, we must pay attention to the frequencies corresponding to daily or shorter periods. These are caused by currents flowing in the upper atmosphere (ionosphere), affected by the Sun.

The intensity of the field decreases after sunrise, reaches its minimum at noon and increases during the afternoon and during the evening. The amplitude of this variation is 5 nT in winter but can rise to 50 nT in summer. Examples of these daily variations measured on two different days can be seen in *Fig. 5*.

As the variations are irregular, they cannot be easily corrected for unless they are measured during the whole survey at a base point on or close to the survey area. Another way of avoiding the effect of the variations is to run the measurements over a relatively small network quickly. A reference instrument still has to be used when the solar activity is strong. During magnetic

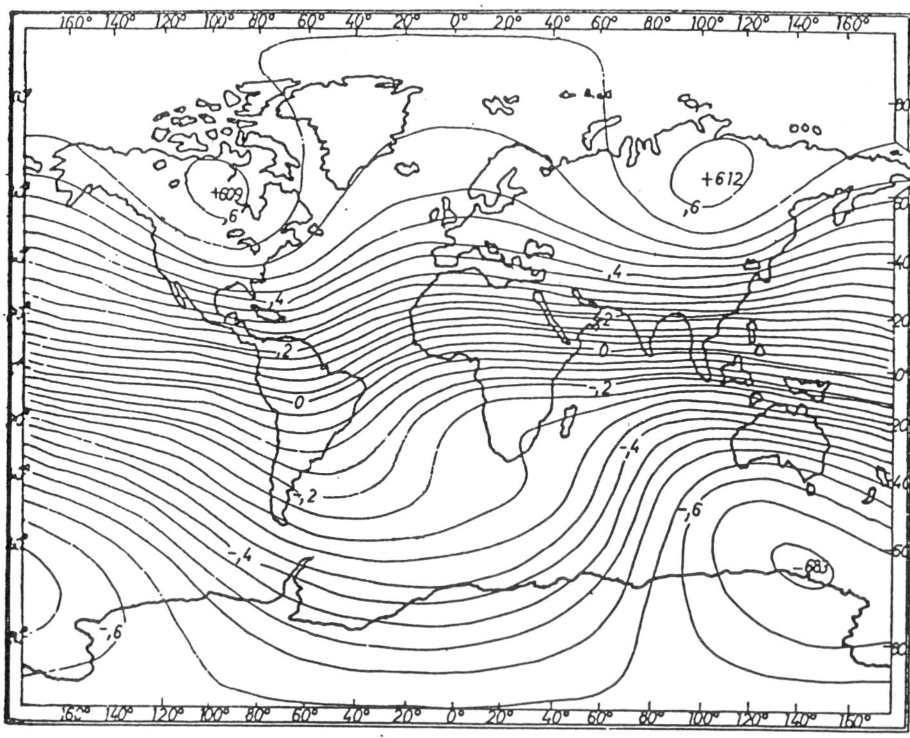

Fig. 4. The total component (T) of the geomagnetic field for 1960.0. Isoline distance is 2000 nT.

storms the variations can reach 1000 nT or so. The probability of the occurrence of these storms is connected with the eleven-year-period of the sunspot cycle. The last maximum was in 1980 and the years of the survey (1989-90) were the years of the new maximum. During large storms fieldwork should be suspended. With our measurements we applied corrections using data of the base station (see detailed description later).

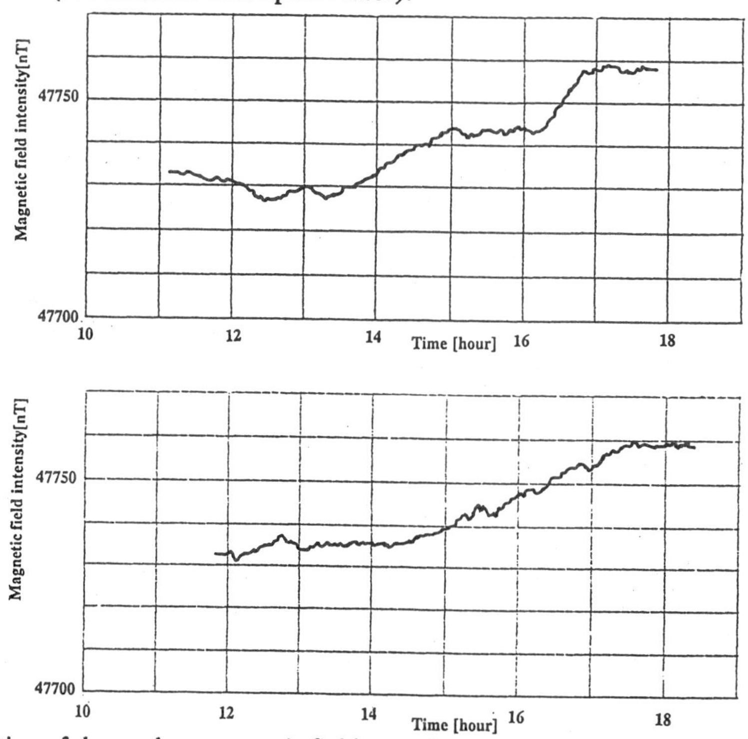

Fig. 5. Variation of the total geomagnetic field near Gyomaendrőd for two days (3 and 6 April 1989).

2. *Magnetic Research in Archaeology*

Magnetometry is used mainly to search for iron deposits and is one of the oldest geophysical survey methods. Smaller objects can also be detected by magnetometry. For this purpose a sensitive instrument is needed which is realized by the so called proton precession magnetometer. This instrument is sensitive enough, portable and fast. One of the current applications of the proton magnetometer is the search for archaeological objects. In the sixties the magnetometer was successfully used for detecting ovens in England. The physical basis of this is the *remanent magnetism* of the objects. There are different kinds of remanent magnetism:

- *isothermal remanent magnetisation* if the temperature was constant during the magnetism process
- *thermoremanent magnetisation* if temperature decreased from a higher temperature to a lower temperature during the magnetism process
- *depositional remanent magnetisation* if the magnetizable particles of the deposit are oriented by the external geomagnetic field during sedimentation.

Characteristic of thermoremanent magnetisaation is that its direction is parallel to the direction of the magnetizing field during cooling and the intensity is directly proportional to field intensity. This magnetization is stable against time.

Not only burnt objects, like ovens, kilns etc. cause a magnetic signal (anomaly) that can be detected by the magnetometer, but other features too, e.g. buried ditches, wells etc. (Aitken 1974). The magnetic anomaly caused e.g. by a buried hole comes from the difference in susceptibility between the fill and the surrounding soil. In favourable conditions roads and foundations can also be detected using this method. The greatest problem is posed by modern iron objects in the vicinity of the measurements which make the analysis of data more complicated. An environment of igneous rocks renders the magnetometric method useless because igneous rocks have strong thermoremanent magnetisation which overprints the weaker anomalies of the archaeological objects. Our surveys were run over surfaces with several hundred metres of silt and clay underneath.

3. *The Instrument Used in Magnetic Research*

This measurement requires a sensitive instrument. A proton precession magnetometer is suitable for these purposes. We used the G-816, G-826, and G-856 magnetometers made by *Geometrics*. The principle of the instrument is based on the gyroscopic behaviour of the nucleus. Hydrogen has the simplest nucleus consisting of a single proton. The behaviour of the proton in a magnetic field can be modelled by a gyroscope. This model-gyroscope has a magnetic dipole aligned parallel to the rotational axis. This dipole tends to stand parallel to the magnetic field of the Earth. Because it rotates quickly however, the gyroscope-effect forces it to precess. The model is similar to that produced by a symmetrical heavy gyroscope. The frequency of precession is directly proportional to the intensity of the magnetic field (T). This connection makes possible the measurement of the field intensity.

The probe of the instrument consists of a half litre bottle of water (as a supplier of protons) inside a coil of about 1000 turns. The measurement consist of two parts: polarization (2 sec) and counting (0.5 sec). During polarization, a current of about 1 A flows in the coil producing a magnetic field of 1000 A/m, which aligns most of the protons' magnetic dipoles parallel to the field. Counting begins with the switching on of the current. The protons start to precess around the geomagnetic field. The precessing protons produce an alternating field which in turn, induces a voltage of approximately 1 μV in the coil. The signal is amplified and filtered, and its frequency measured for field intensity. Thus, the magnitude of the field can be measured indirectly.

III. Description of the Magnetic Measurements, Data Processing

In our work we proceeded in the following steps:

1. Site selection and geodetic marking.
2. Magnetic survey.
3. Data processing and visualization.
4. Calculating model parameters for the measured data.
5. Interpretation.

1. Site Selection and Geodetic Marking

Precise geodetic surveying of the measured area is very important for the exact location of the anomalies (finds). To avoid any later inconvenience arising from a possible discrepancy between the base map and the magnetic map, we fixed, with additional surveying, certain reference points with respect to the surrounding objects (e.g. kilometre stone, water tower).

After marking the area we performed several magnetic measurements along different profiles to determine whether there were anomalies at the site. We found some anomalies and so managed to find the site. The next step was to peg out a network of squares of 20 by 20 m using a theodolite and a measuring tape, and a base point was selected to serve as the reference point. At site *Örménykút-3* there was a mound in the measured area. To avoid distortion the distances were measured horizontally. In some steps it was necessary to reduce the size of the network to 10 by 10 m in the case of steeply sloping surfaces. Taking the heights by levelling we constructed a topographic map of the mound. Its isoline representation can be seen in *Fig. 6.*

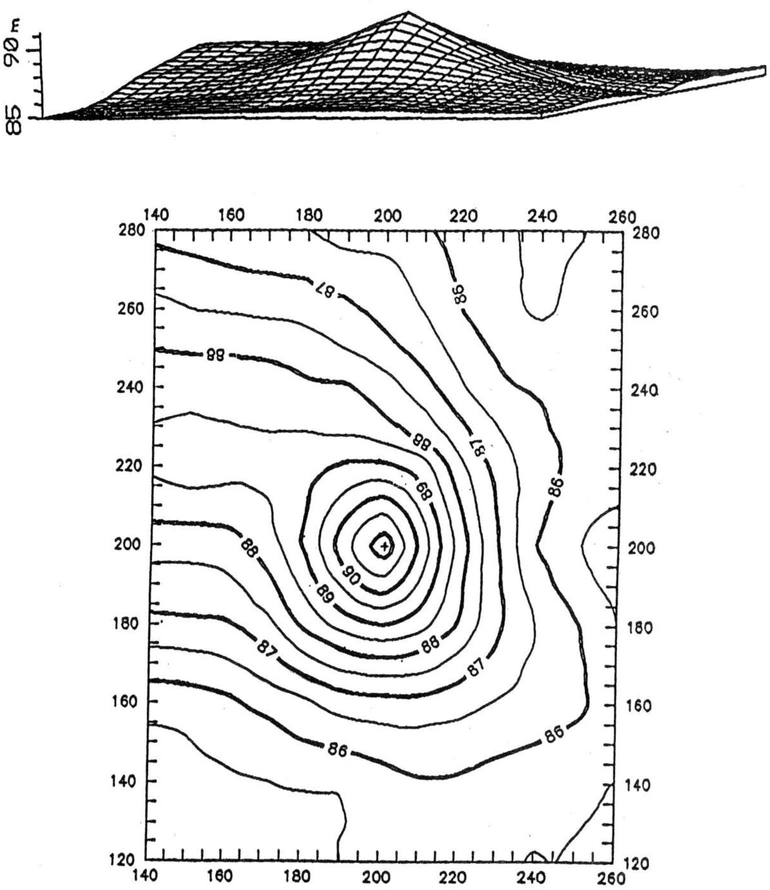

Fig. 6. The relief of the archaeological site Örménykút-3. All data are given in metres.

In many respects it was advantageous to use the 20 by 20 m network, for using smaller squares would have made the field measurements and the data-processing more difficult. From the point of view of registering the data on a map measurements arranged within a single large square would be the best. On the other hand, the use of a single big square is disadvantageous because of the time variation of the magnetic field and the limited capacity of a battery set (1500-2000 readings) that limits the size of the area that can be measured at one go. While the typical magnitude of an anomaly originating from archaeological objects is 10-20 nT, the running down of the batteries may cause a difference of 5-10 nT in the readings at the same measurement spot, which is too large and not permissible. Our experience has shown that a network consisting of 20-metres squares is the optimum for archaeological purposes.

2. Measuring

The magnetic measurements were taken using a sensor height of 70 cm. This has been found to be the best elevation for the sensor, because smaller heights result in noisy measurements and the resolution of the data is not sufficient at greater heights. We also found magnetic cleanliness to be an important factor so we took care of it. The batteries were peeled because we found even this quantity of metal in the magnetometer to be very disturbing.

Usually, two persons ran the survey, one holding the sensor and the other taking the readings and recording or one holding the sensor and taking the readings while the other did the recording.

A 20-metre square was subdivided by a one-metre grid and the measurements were taken at grid points with a total of 441 readings for each 20-metre square including the sides. The measurements along the sides were indispensable to allow neighbouring squares to be matched. The one-metre grid distance was chosen because the horizontal extension of an archaeological anomaly is usually not greater than 10 metres and one metre seemed to be a good compromise in the 0.5 to 2 m range that is used in the search for archaeological objects. In some cases, mainly for modelling purposes, a 0.5 m grid was used.

Preliminary data processing was carried out in tandem with the magnetic measurements in order to see the trend of the anomalies to be traced, that is, to separate areas with or without magnetic anomalies, respectively.

With the combination of measurement and data processing, effort and time could be saved since the survey was not run over areas free of magnetic signal. Such was the case with the *Örménykút-52* area. Here, we searched for the remains of a settlement. The supposed limits of the settlement can be seen at the western and eastern part of the map in *Fig. 7*.

As previously mentioned, a reference instrument (basis instrument) was used which measured the intensity of the magnetic field at a fixed point at regular intervals. The basis instrument, a programmable G-856 magnetometer, was set up on an anomaly-free area, far from metallic objects and the measurements themselves. It took readings every two minutes, or every minute if the field was disturbed. The base correction will be described later.

The sensitivity of the magnetometers is 1 nT making them suitable for detecting anomalies having 10-20 nT amplitude. When a difference larger than 10 nT was found between adjacent points, the measurements were repeated for these points to avoid gross errors.

The consistency of the trends in the magnetic data along adjacent sections (lines) was also checked. It was found to be very important that the measurements were taken without interruption because stops invariably resulted in errors. One of the advantages of the preliminary data processing was that it indicated if an error had occurred when the measurements were taken and enabled immediate correction.

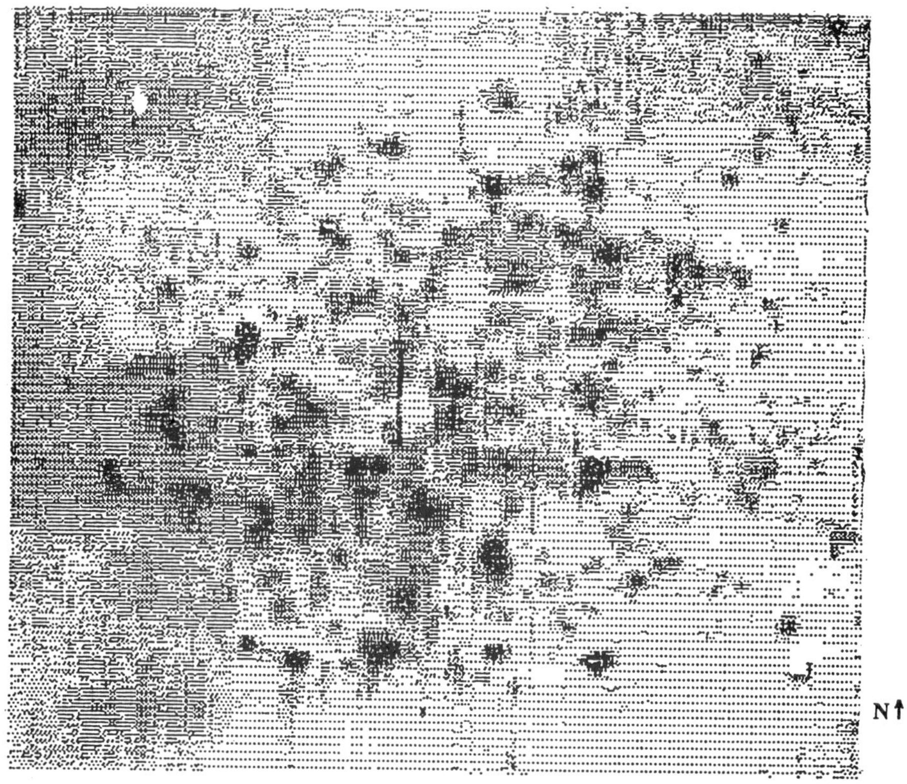

Fig. 7. Residual dot density map of the total magnetic field measured at the archaeological site Örménykút-52. The dot density scale used is equidistant, the lightest shade is for -20 nT, the darkest is for +40 nT.

3. Data Processing

Data processing consisted of an on-the-spot stage and a laboratory stage. With the first, we used the portable *Commodore-128* machines which were suitable for processing a 20-metre square since other portable machines had not been available, while in the laboratory IBM PC-s having larger capacity were used. Now for the field work a laptop machine is used.

The main steps in the data processing were:

(a) Feeding the data into the computer

(b) Base-line correction, calculation of the residuals

(c) Square fitting, zero level calculation

(d) Visualization

(e) Model computation (if the data were appropriate for it).

Several program codes were written for these computations which were executed on C-128 machines (steps (a)-(c)) and on IBM PC-s ((d) and (e)).

(a) *Input data*

The data of each square were stored in separate sequential files named with the geographic coordinate of the northwestern corner of the square. The base line data were input directly using an RS-232 port. Data recorded manually were keyed in directly.

(b) *Base-line correction*

Residuals were computed on the basis of the formula

$$P(i,j)_{res} = P(i,j,t)_{measured} - B(t)$$

where $P(i,j)_{res}$ is the residual value for the grid point with coordinates (i,j); $P(i,j,t)_{measured}$ is the measured magnetic field intensity for the same point at time t and $B(t)$ is the base line value at the time of the measurement at point (i,j).

(c) *Square fitting and zero level calculation*

This was the most difficult part of the data-processing. The different squares were measured by different instruments on different days when the circumstances differed. Thus, there were differences between the average residual values for the adjacent squares. Since the magnetic field is represented by a vector, as discussed above, these changes occur both in its intensity (*T*) and orientation (*I,D*). While the total field in Hungary is typically 48000 nT, these changes are approximately up to 10 nT. That is why the orientation of the vector does not vary significantly, so in theory adding a constant value to intensity data in a square should solve this problem. It is for this reason that the common lines were measured twice. In ideal cases, with the help of these measurements, the squares can be fitted together by adding a constant value to the residual data of a square.

In practice, however, this is rarely feasible because of measuring errors. These usually make it impossible to fit even four squares without a closing error. Our method was as follows:

Let us characterize each square with the rectangular coordinates of its north-western corner and begin with the north-western square of site *Örménykút-52* chosen as an example. Assume that its coordinate is (-100,100). Fit its eastern neighbour, square (-100,120), to it then continue this process on the northern border with squares (-100,140), (-100,160) etc. After that, fit the squares along the border (squares (-80,100), (-60,100) etc.).

In this way the squares making up the northern and western borders are fitted together. Now fit the inner squares to the bordering squares. Begin with the (-80,120) square. That must be fitted in two directions: to square (-100,120) to the north and to square (-80,100) to the west. Usually, two different constants are arrived at and we take the arithmetic mean for the fitting constant. Continue on in this way.

On completion of this procedure, the initial magnetic map which looks somewhat chequered before square fitting becomes more homogeneous. One problem still remains however. It arises from the fact that the magnetic data measured over the whole area are referred to those in the starting square. If they are, on average, positive along the borders then the average magnetic level of the area should also be positive i.e. positive anomalies would be stressed and the negative ones suppressed.

By calculating the average magnetic value for the whole area and subtracting it from each value the distortion can be eliminated.

(d) *Visualization*

can be carried out using
- an isoline map
- a dot density map
- a coloured map

As Scollar et al. 1986 notice, in several cases, where the fine structure of the archaeological site has to be studied, the isoline maps of the magnetic field are not adequate. Our experience was the same, therefore the isoline maps were used only on the first few occasions. The other methods appear to have more perspective. For the model calculations this type of visualization was acceptable, however. An isoline map can be seen in *Fig. 8*, with a part of site *Gyoma 133*.

Later dot density and coloured maps were used. Dot density maps were used on the spot and these contain useful information for the archaeologists as well. These maps were prepared on two different scales: 1:300 and 1:100. The 1:300 maps provide a general outline while the 1:100 maps contain sufficient detail to be useful for the excavating archaeologist. *Fig. 9.* shows magnetic anomaly maps of the *Örménykút-52* area.

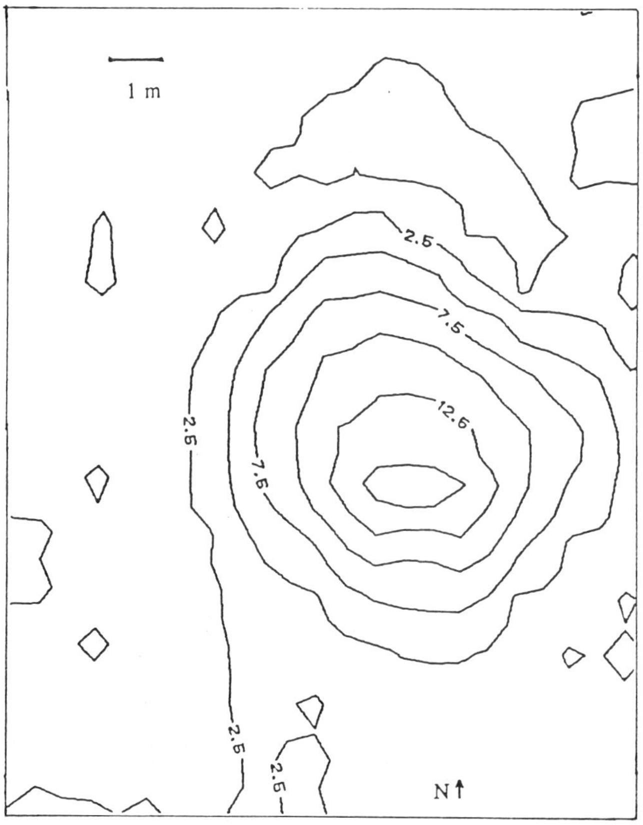

Fig. 8. Residual isoline map of the total magnetic field measured at Gyoma 133. Isoline distance is 2.5 nT.

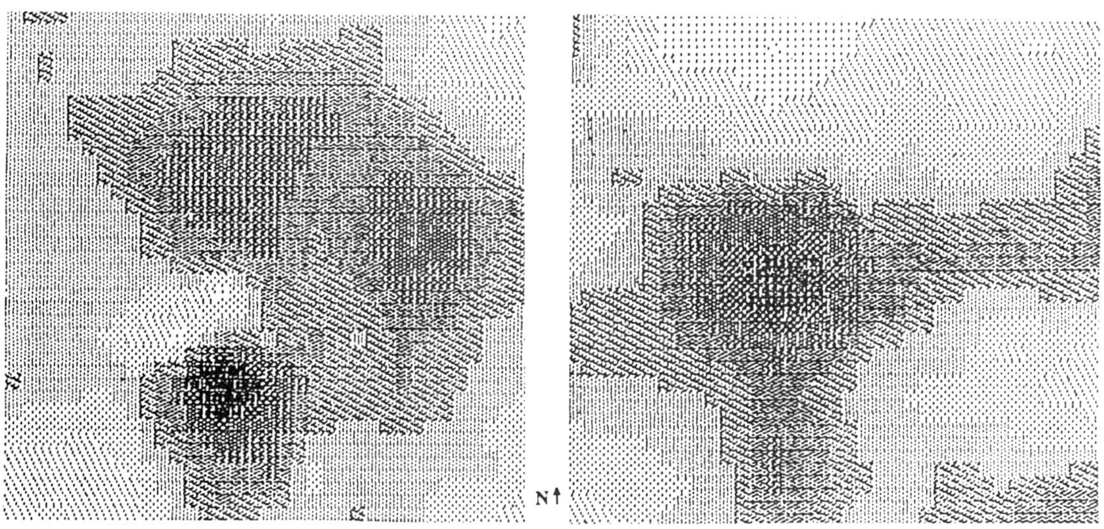

Fig. 9. Two typical magnetic anomalies of the dot density map of Örménykút-52.

Maps in colour were produced

1. by taking photos of the colour monitor displaying the maps of each 20-metre square one by one and joining the photos to form a map. It was necessary to show only a relatively small area in one step to avoid projection distortion.

2. by some additional data processing and using a colour printer (*HP PaintJet*) facility. Printing limitations prevented us from reproducing a colour map here.

(e) *Model computation*

Magnetic maps offer only qualitative information to the archaeologists, and the interpretation of anomalies is difficult if there are no quantitative data about the originating bodies. It is highly important to get some really adequate information about the extent and depth of archaeological objects. That is why we underline the importance of model computations (see *Fig. 18)*.

The method used is described elsewhere (Székely et al. 1991).

IV. Geoelectrical Measurements and Instruments

In the case of not too significant magnetic anomalies, the interpretation of the results is sometimes very difficult. It cannot be decided exactly whether the anomaly is caused by a real archaeological object or by something else. In such cases, electrical measurements, although time consuming, may be helpful (Pattantyús-Á., M. 1986, Young and Droege 1986). If the anomalies are situated in the same place both on the magnetic and the electrical map and their extents are similar, we can be sure that it is caused by a real archaeological object. In other cases, the geoelectrical survey can still be carried out successfully (Clark 1986, Frohlich and Lancaster 1986, Hesse et al. 1986, Tabbagh 1986).

1. Principles

According to Ohm's law, the electric current in a conductor is proportional to the voltage that generates the current. The factor of proportionality is the resistance which itself is proportional to the length and inversely proportional to the cross-section of the conductor. The reciprocal of the resistivity is the conductivity, the measure of the ease with which current flows through the conductor. Rocks have the ability to conduct electric current, especially those which are permeable to water. Water usually contains some dissolved salts, the ions of which are able to move under the effect of an electrical field, and thus, to conduct electrical current. The resistivity (or conductivity) of the rocks saturated with salt water varies with their porosity and water content and quality. E.g., the best conductors are rocks with high permeability (porosity) saturated with salt water, while conductivity decreases with decreasing permeability and the salinity of the interstitial water. The conductivity of the soil partially depends on the same factors. Loose, dry soils are highly resistant and their resistivity decreases with increasing water content. Another factor is the clay content, as clay itself is highly conductive.

What makes the electrical method useful in archaeological studies is that human habitation usually leaves the kinds of imprints in the soil which result in resistivity anomalies. These can be detected by geoelectric surveying under favourable conditions. The best objects for geoelectric research are remnants of roads, footpaths, foundations and floors of various buildings etc., if their extent is sufficiently large so that they can be resolved by this method.

2. *Measurements*

In geoelectrical prospecting, a DC current is sent through two current electrodes (*A* and *B*) into the ground. The electrical voltage resulting from the current is measured between the measuring electrodes *M* and *N*. Several electrode arrangements are used:

- The Wenner arrangement consists of four, equally spaced electrodes in a line.
- The Schlumberger arrangement is also a four electrode one in which the measuring electrodes are in the middle of the layout and situated very close to each other.
- In the dipole-dipole arrangement, electrodes *AB* and *MN* each constitute a dipole which can be laid out in different relative positions. During our work we used the so-called axial arrangement.

The measurements were carried out over a 25-metre square with an instrument of type *D-10 R* around point (-140,160) in the area of *Örménykút-52*. We chose this area because of a magnetic anomaly which had been detected here previous to the electrical measurement.

The measurements were taken using the Wenner arrangement at two different *MN* distances, 1 and 2 m. *MN*=1 m is appropriate for detecting objects at a depth of 1.5 m and *MN*=2 m is needed for objects at a depth of 2 m. Since the arrangement was east-west oriented and had no cylinder symmetry, the maps have been transformed by a two-dimensional filtering via a 5 x 3 matrix with weighting factors decreasing as $1/(r^2+1)$, where r is the distance from the centre. The results obtained for depths of 1.5 m and 2 m can be seen in *Fig. 10* and *Fig. 11* in units of 0.1 Ωm, respectively. Contrary to magnetometry, isoline representation of the geoelectric data is adequate. That is why it was chosen for demonstration.

In the figures, the negative anomaly situated in the north-eastern corner which is important for us, is conspicuous. This coincides with the magnetic anomaly. Consequently, the two anomalies must have been caused by the same object.

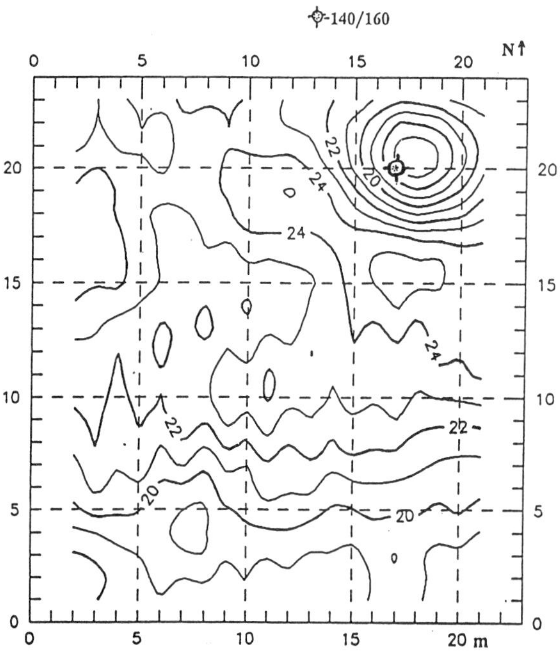

Fig. 10. Filtered resistivity map of the measured part of archaeological site Örménykút-52. Effective depth: 1.5 m

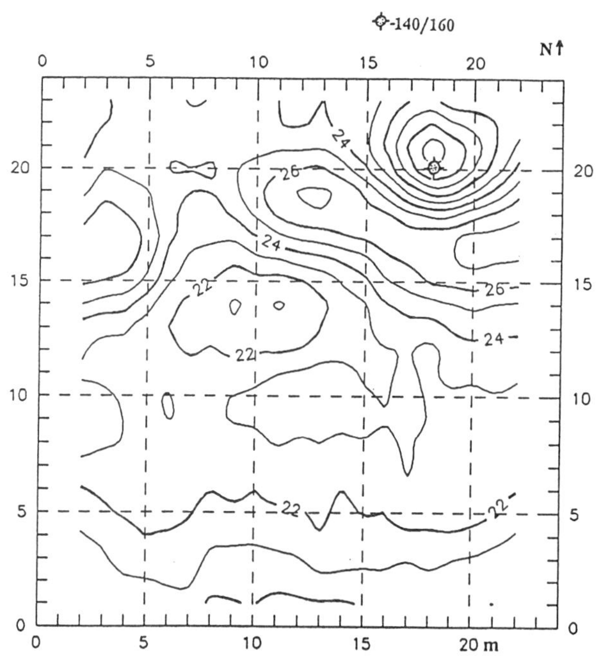

Fig. 11. Filtered resistivity map of the measured part of archaeological site Örménykút-52. Effective depth: 2 m

V. Interpretation

The result of a magnetic survey is a magnetic anomaly map. It contains all the information that can be used to draw conclusions about the place, size and shape of the objects lying beneath the surface.

Before the interpretation of the anomalies we show anomalies caused by simple buried objects. The first example is a magnetic dipole at depth *d*. The shape of the magnetic anomaly in a north-south section on the surface is shown in *Fig. 12* (after Drahos et al. 1979).

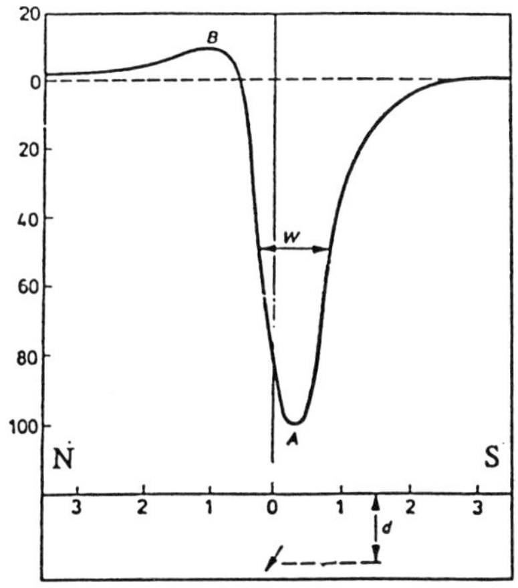

Fig. 12. North (at the left)-south section of anomaly of a small dipole at depth d. The horizontal axis is scaled in d units. Note that the position of both the maximum (B) and the minimum (A) is different from the position of the buried object.

By varying the dip of the magnetic dipole, we get the anomaly forms shown in *Fig. 13*. Note that the maximum of the anomaly is displaced to the south with respect to the source (on the northern hemisphere) and also that it decreases to below 2% of the maximum value at a distance of 3 *d* from the surface. When the depth of the source object is at a triple distance, the anomaly no longer appears significant.

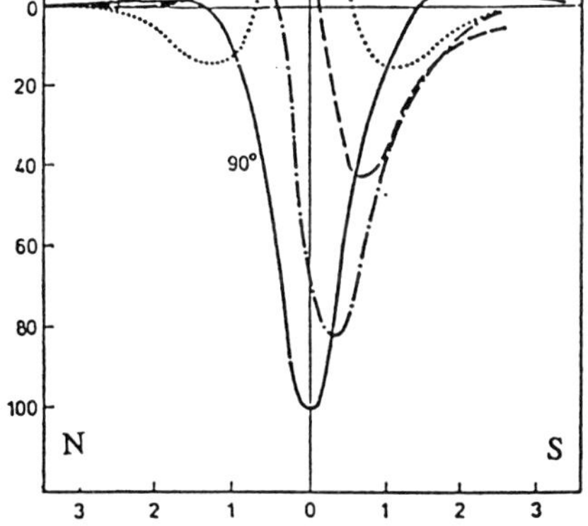

Fig. 13. The influence of the inclination on the shape of the anomaly. (North-South section, north is to the left, the axes are scaled in arbitrary units.)

If we examine small objects, the magnetic dipole is a good approximation. For larger sources, the width of anomaly is greater. Measuring in the field, we usually find anomalies which differ from the ideal one shown in *Fig. 12.* This may be for example because of the effect of neighbouring objects. Magnetic anomalies can be caused e.g. by magnetic materials (ores, metals, pots and pans etc.), but an anomaly can be also as a result of a ditch, if the magnetic susceptibility of the filling soil is different from that of the surrounding soil.

The shape of an idealized anomaly caused by a filled ditch running in an East-West direction is shown in *Fig. 14.* with a North-South profile.

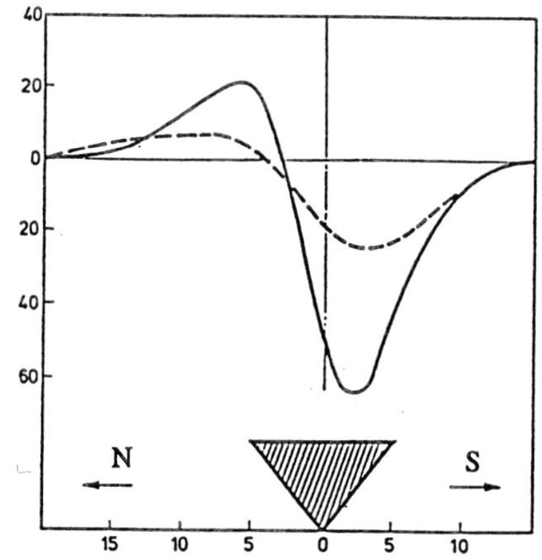

Fig. 14. The hypothetical anomaly of a ditch.
(North-South section, north is to the left, the axes are scaled in arbitrary units.)

With the help of these typical anomaly shapes, the sources of anomalies can be guessed from the results of the measurements. To verify the theory, we have measured the total magnetic field of an iron plate measuring 75 x 60 cm. The results can be seen in *Fig. 15.* Because of the low disturbing influence of the surroundings, the shape of the anomaly is almost ideal.

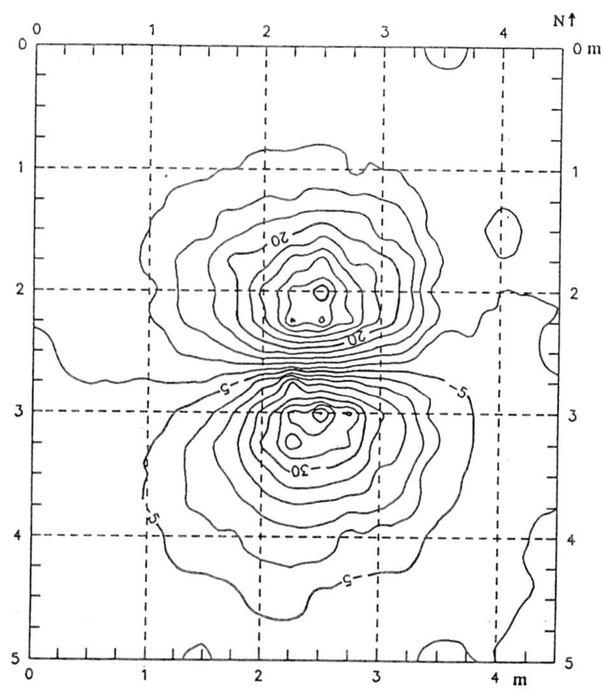

Fig. 15. The measured magnetic field of an iron plate.
(Extension data are in metres, the isolines are scaled in nanotesla, isoline distance is 5 nT.)

One of the most important questions is the position of the object causing the anomaly. The horizontal position can approximately be determined with the help of the magnetic map. As was mentioned earlier, the source of the anomaly must be somewhere between the negative and the positive peaks but a little closer to the positive one. It is much more complicated to find the depth or to determine the thickness of the object. Sophisticated computer programs are available for the identification of the parameters of different source objects.

Another important question is what kind of object causes the anomaly. The results of former excavation were a great help in the analysis of the magnetic maps.

The *Gyoma 133* site was excavated just after the measurements were taken and we received the results of the excavation. Those excavations found remains from the late Sarmatian period (approximately 4th century A.D.), including remnants of workshops, fireplaces, wells, houses and filled pits. Thus, we could compare our magnetic map with the results of the excavation. Thus, we would expect similar objects from similar sites (e.g. at *Örménykút-52*).

The result of the comparison was very encouraging. It is possible to say that every typical magnetic anomaly had an archaeological source. Four examples can be seen from this site in *Fig. 16.* The figure shows magnetic anomalies caused by various archaeological features excavated at *Gyoma 133.*

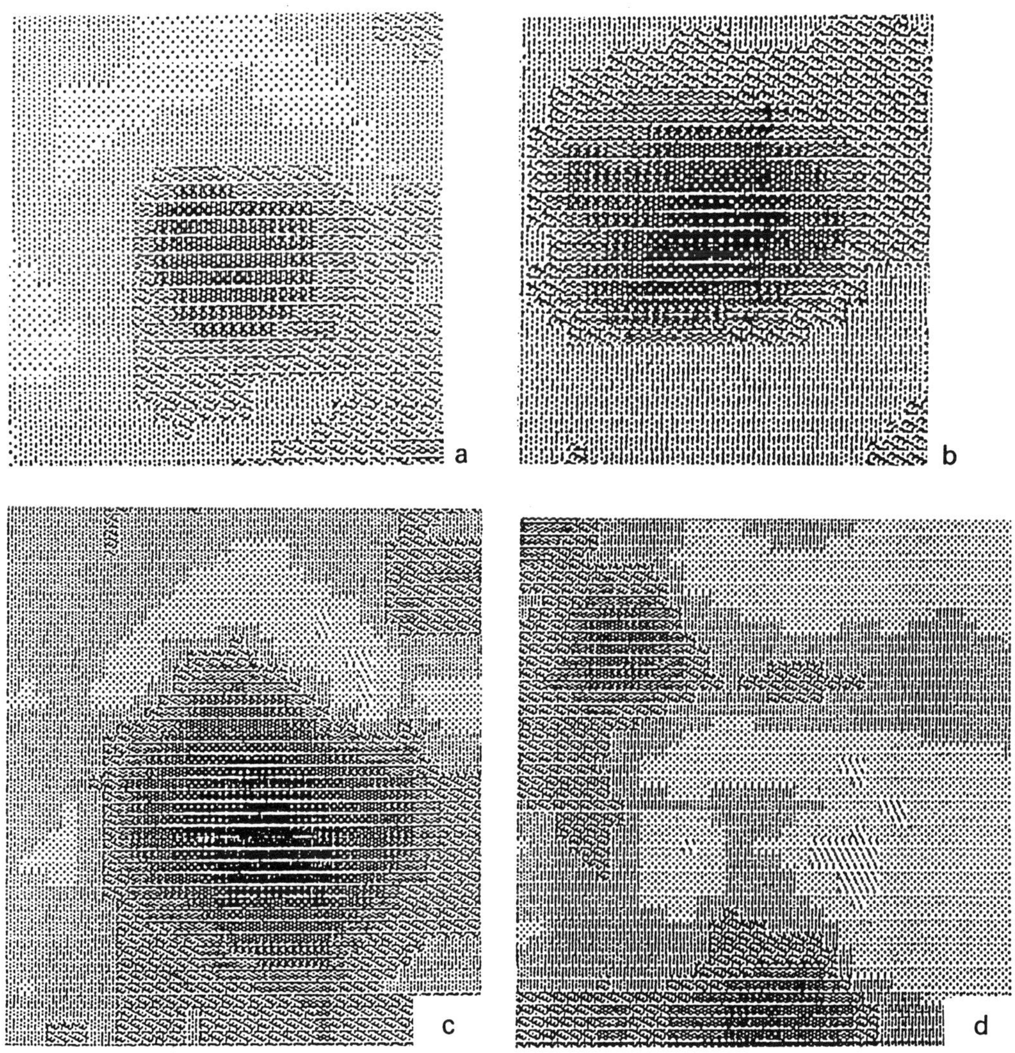

Fig. 16. Four typical anomalies of the site Gyoma 133.
(a) Well (b) Pit (c) Workshop (d) Stove

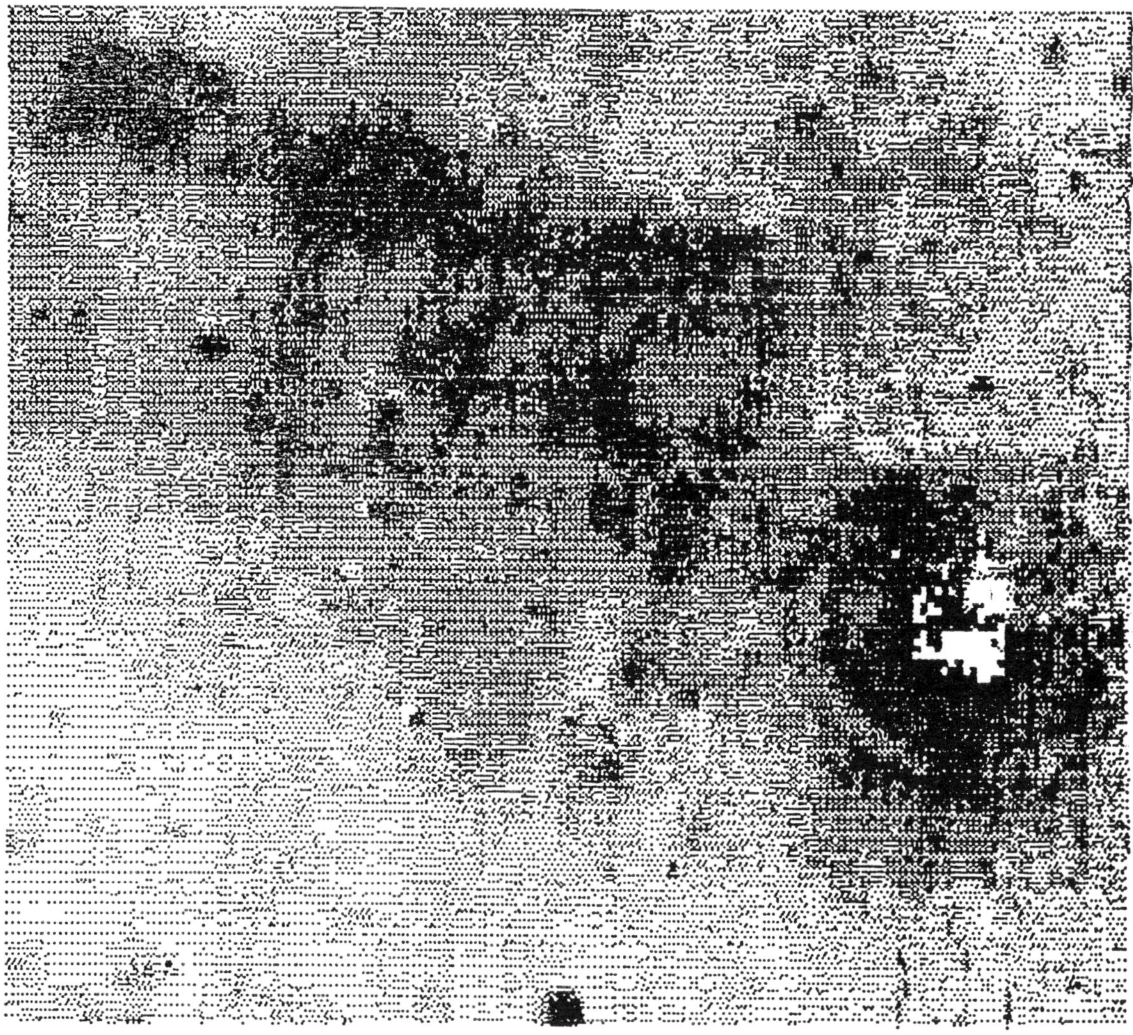

Fig. 17. Residual dot density map of the total magnetic field measured at the archaeological site Örménykút-3. The used dot density scale is equidistant, the lightest shade is for -20 nT, the darkest is for +40 nT.

Anomalies similar to those on the map of site *Gyoma 133* can be seen on the magnetic map of *Örménykút-52* (see *Fig. 7*). These are distinct and typical anomalies so we can conclude that these anomalies are caused by similar archaeological objects. After the positive result of the magnetic and resistivity survey we drilled into the soil down to depths of 2.5 m at point (-140/160) (see *Fig. 10* and *Fig. 11*). Beneath the anomaly we found a layer with the remnant of a fireplace at a depth of 2 m.

The interpretation of the site *Örménykút-3* is difficult. There was a mound at the point (200,200) with a boundary stone made of iron concrete on the top of it. One can see the effect of this stone on the magnetic map of the site (*Fig. 17*). We draw attention to the two circular shaped anomalies around the top of the mound. They could not have been caused by the mound or by the stone so we suppose the presence of the remains of walls or ditches as causative bodies for these anomalies. In the northwestern part of the site, one can see a low amplitude anomaly having relatively large extent lying in a southwest-northeast direction.

The anomaly is situated on higher ground. Considering that this land is ploughed frequently and that we found fragments of fireplaces on the ploughed surface during the survey, we suppose that this wide but not too large anomaly is caused by the remnants of a settlement ow flattened and dispersed, while the source of the typical dipole-shaped anomaly over low ground at the bottom centre happened to remain undisturbed (perhaps it is deeper). This clear-shaped anomaly was measured using four different sensor heights, so a large number of data is available for the model computations. That is why several runs of the parameter estimation procedure were carried out to recover the parameters of the buried object (*Fig. 18*) (Székely et al. 1991).

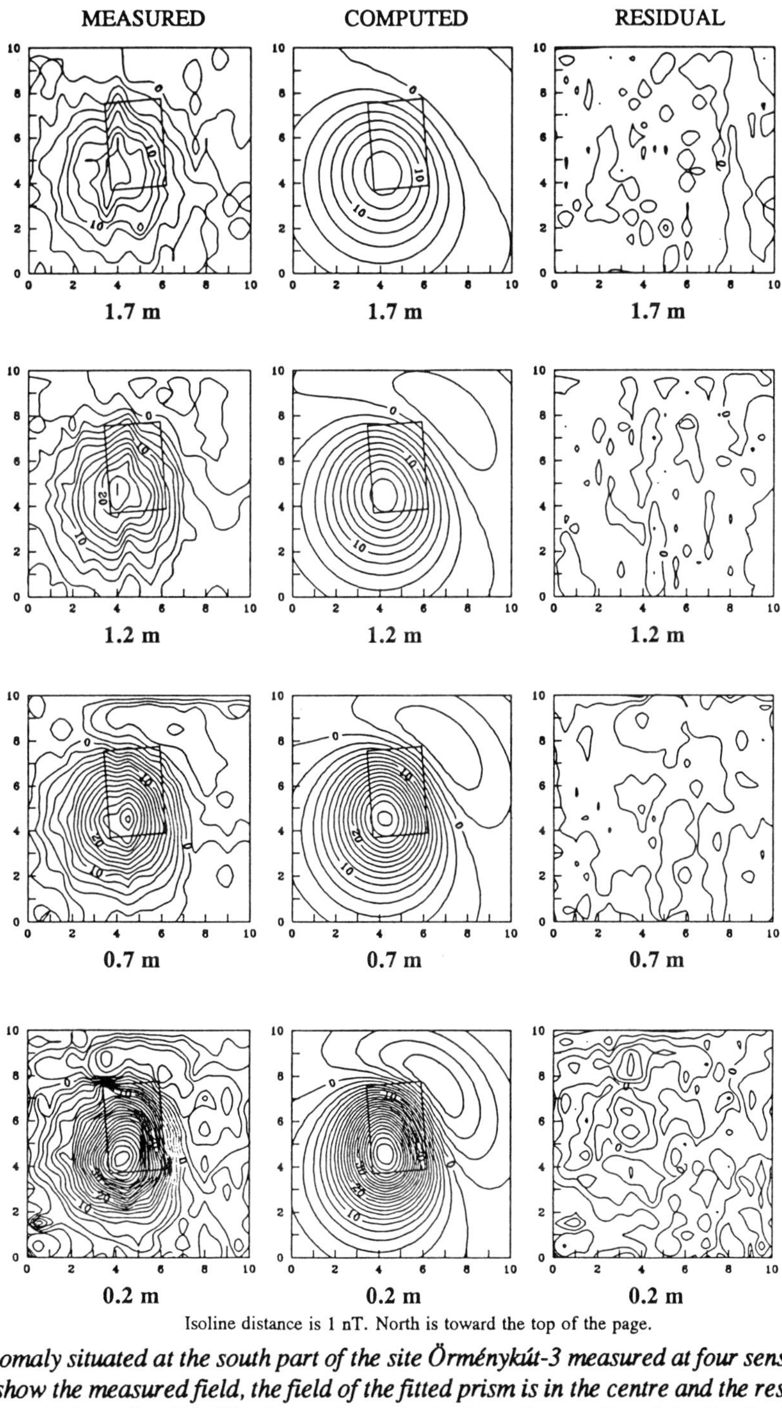

Isoline distance is 1 nT. North is toward the top of the page.

Fig. 18. The anomaly situated at the south part of the site Örménykút-3 measured at four sensor heights. The left graph show the measured field, the field of the fitted prism is in the centre and the residual ($T_{measured}$ - $T_{computed}$) is on the right. The frame shows the horizontal extension of the fitted prism.

VI. Conclusions

The paper summarizes the field work, data reduction and interpretation of magnetic and electric measurement conducted over selected archaeological sites in Hungary. It has been shown that not only the magnetic and resistivity map can be useful for archaeological purposes, but the computation of parameters for the clear-shaped anomalies can result in a quantitative approach to archaeological objects. Our future intention is to apply and compute models for the resistivity data too, as it was presented by Pattantyús-Á., M. 1986.

Acknowledgements

We are grateful to the Geophysics Department, Eötvös University, Budapest, for organizing, financing and supervising our work. We thank Prof. A. Meskó, Head of the Department, for initiating and continually supporting these activities. Special thanks are due to our supervisor, S. Puszta for his close cooperation. The interpretation of some anomalies was a cooperative work with D. Jankovich-B., A. Vaday and M. Vicze. Some of the computer codes used were made by A. Abbas and Sz. Pásztor. Finally, we would like to thank our colleagues who participated in the field work as well as Dr. Gy. Tarcsai and Prof. P. Márton for their help in the preparation of the manuscript.

References

Aitken, M. J., 1974. *Physics and archaeology,* Oxford.

Clark, A. J., 1986. Archaeological geophysics in Britain, *Geophysics,* 51, No. 7, 1404-1414.

Drahos, D., Kis, K., Meskó, A. and Salát, P., 1979. *Bevezetés a gyakorlati geofizikába* (Introduction to practical geophysics), Budapest.

Frohlich, B. and Lancaster, W. J., 1986. Electromagnetic surveys in current Middle Eastern archaeology: Application and evaluation, *Geophysics,* 51, No. 7, 1414-1425.

Gibson, T. H., 1986. Magnetic prospection of prehistoric sites in Western Canada, *Geophysics,* 51, No. 3, 553-560.

Hesse, A., Jolivet, A. and Tabbagh A., 1986. New prospects in shallow depth electrical surveying for archaeological and pedological applications, *Geophysics,* 51, No 3, 585-594.

Márton, P., 1975. *Általános geofizika: Földmágnesség* (General geophysics: The magnetic field of the Earth), Budapest.

Pattantyús-Á., M., 1986. Geophysical results in archaeology in Hungary, *Geophysics,* 51, No. 3, 561-567.

Scollar, I., Weidner, B. and Segeth, K., 1986. Display of archaeological magnetic data, *Geophysics,* 51, No. 3, 623-633.

Székely, B., Magyari, O., Steinbach, P., Tóth, T. and Tarcsai, Gy., 1991. Inversion of total magnetic anomalies measured over archaeological sites using the vertical prism model. In: *Theory and Practice of Applied Geophysics,* Special monograph on Geophysical data inversion in archaeological site investigation, Wiesbaden (in press).

Tabbagh, A., 1986. Applications and advantages of the Shigram electromagnetic method for archaeological prospecting, *Geophysics,* 51, No 3, 576-584.

Weymouth, J. W., 1986. Archaeological site surveying program at the University of Nebraska, *Geophysics,* 51, No. 3, 538-552.

Young, Ch. T. and Droege, D. R., 1986. Archaeological applications of resistivity and magnetic methods at Fort Wilkins State Park, Michigan, *Geophysics,* 51, No 3, 568-575.

The combined use of archaeometric methods preceding the excavation of archaeological sites

ERZSÉBET JEREM - ZSUZSA KISS
MIKLÓS PATTANTYÚS-Á. -ANDRÁS VARGA

1. Introduction

A variety of methods applicable both individually and in combination is available to specialists for the location and pre-excavation testing of archaeological sites.[1]

Within the framework of the settlement research programme on the Great Hungarian Plain (Békés County, Microregion Research Project), traditional surveying methods, especially field survey and topographic preparatory work were supplemented by additional archaeometric studies.

A joint English-Hungarian project with similar research aims was carried out in the environs of Szeghalom (Békés County) between 1979 and 1981. This project was essentially a survey and had narrower limits in terms of the material covered (both spatial and temporal) and in its range of methods than the later fieldwork which is the subject of this report (Sherratt 1982 and 1983 a, b).

Two considerations were of paramount importance during this work. First, the studies pursued should never be an aim in themselves, even if they are being applied in a new context or in new combinations for the first time.[2] They had to be directed towards deciding whether the site in question was appropriate for excavation and if so, which recovery method should be employed.

The other aim was to gather particular highly specified information from a small, well defined area (perhaps comprising several sites) suitable for excavation which would not have been revealed as archaeologically significant by traditional methodology. Such data represent an indispensable starting point in the excavation of settlements.[3]

2. Factors considered in the selection of the excavation area

The settlement research project covered an area of ca. 42 square kilometres of the Szarvas district in Békés County. Our inconclusive preliminary field surveys in the spring and autumn and our first summer trial excavation[4] indicated that we had to concentrate on a smaller area in order to determine the sites for future excavation.[5]

---

[1] A few examples of archaeological handbooks on prospecting and excavation techniques include: Aitken 1974, Clark 1990, Hrouda 1978, Mommsen 1986, Parkes 1986, Rottländer 1983, Barker 1982 and 1986, Brown 1975, Gersbach 1988 and 1989, Kossack et al. 1980, Joukowsky 1980, Webster 1974.

[2] E.g. soil core sampling on mounds and the multi-faceted palaeobiological study of such cores.

[3] More precise information about palaeohydrographic and soil conditions and natural geographic as well as environmental factors.

[4] We commenced fieldwalking in 1986, and continued every year until 1990. A short interim report on the 1986 excavation is published in *RFüz* 40 (1987) 13. and *ArchÉrt* 114-115 (1987-1988) 261.

[5] It should be mentioned that the need for research within a smaller regional unit, so-called Microregion, was recognized while the volumes of the Archaeological Topography of Hungary were in preparation. *MRT* 8, 14.

We therefore employed a geomorphological survey to identify units which were potential settlement sites, on which we could concentrate our future activities (Crumley et al. 1989, Davidson and Shackley 1976, Hodder and Orton 1976, Gladfelter 1981, Macready and Thompson 1985, Waldhauser 1976, 1984, 1986).

The compilation of an inventory of mounds was set as a longer term goal, since kurgans are not only among the most significant prehistoric sites on the Great Hungarian Plain (Ecsedy 1979), they are also continuously destroyed by land cultivation. This project was not merely intended as an inventory of these features. Soil soundings were also used to determine whether the mounds traditionally regarded as kurgans were in fact artificially created burial places.[6] Another important task was clarifying the relationship between the mounds we had identified in this way and the settlements in their immediate proximity; even if such associations could not always be established without excavation.[7]

Last but not least, we carefully selected Iron Age sites whose date – based on surface finds – was reliable[8] and settlements which were not damaged by subsequent, more recent (predominantly Sarmatian) habitation.[9]

Thus, our activity was concentrated on two areas with mounds as their focal points. One of these lay south to southeast of the village of Gyoma encorporating the environs of Rigó-mound (Gyoma 117) and Egei-mound (Gyoma 121) (*Fig. 1*). The other area is located south of Endrőd including the immediate proximity of Papp-mound (Endrőd 18) and the region between Endrőd 18 and Sós-mound (Endrőd 33)[10](*Fig. 2*).

3. Fieldwork methods

Following the guide-lines described above, the selection of research methods was influenced by several considerations. A need for the palaeoecological reconstruction became clear right from the start of the project; both for the overall area of the study and for the individual sites to be investigated subsequently in a systematic manner. (Becker et al. 1985, Butzer 1982, Dincauze 1987, Fieller et al. 1985, Jerem et al. 1984-85, Jerem 1986, Kossack et al. 1984) At the same time, it seemed necessary to complement the results of traditional topographic survey with large scale prospection methods in order to determine settlement size and structure with special attention to concentrations of features within the settlement.[11] Following the same principles, data were sought to define the microstratigraphy of large mounds. Without this information, the required mode of excavation could only be guessed at.[12]

---

[6] This will probably be a long term programme since it needs considerable commitments in terms of both time and finances.

[7] Intensive and complex studies carried out in the proximity of mounds were aimed at such a goal, as will be shown later, especially, in the case of the site Endrőd 18.

[8] Refer to the relevant section of *MRT* 8, 15 and 26 p. concerned with the problems of dating material from the Late Bronz Age – Scythian period and the Scythian – Celtic transition period. The increased quantity of pottery now available for study has allowed us to show that some of the previous sites attributions were incorrect.

[9] This fact justified surveying and excavation at sites Gyoma 264 and Endrőd 20.

[10] Some other more distant kurgans, e.g. Szarvas 46, Borgulya-mound, were assessed within the framework of the mound research programme. These, however, will not be discussed in detail here.

[11] See the uncertain extent (e.g. Gyoma 238) and unreliable size estimates (e.g. Gyoma 116) of these sites. Similar difficulties are discussed in relevant chapters of each volumes of the Archaeological Topography of Hungary.

[12] Previously, soil core sampling had only been used for testing tell settlements and horizontal stratigraphy of sites in Hungary (Varga 1989).

Fig. 1. Low-level aerial photographs showing the hydrology and floodplain around the burial mound Endrőd 18 (Papp-mound) and the settlement sites Endrőd 19, 20, 124, 125, 126.

1

3

2

4

5

Fig. 2. Low-level aerial photographs showing sites 1-2: Gyoma 116, 117 (Rigó-mound), 3: Gyoma 121 (Egei-mound) and 264, 4: Endrőd 33 (Sós-mound) and 50. 5: Gyoma 133 and 238.

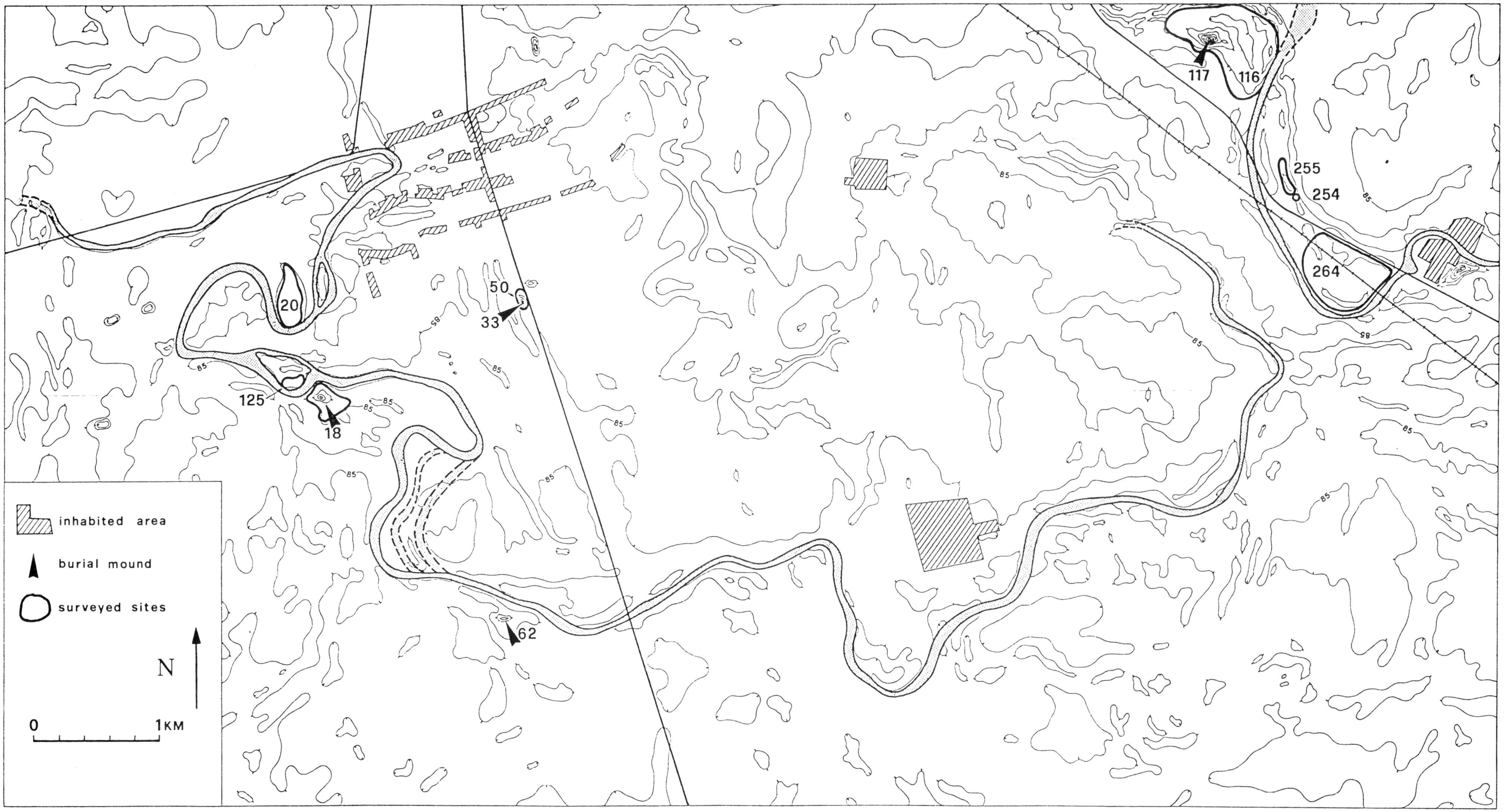

Fig. 3. Contour map of the reconstructed ancient river courses (palaeochannels).

The numbered sites are those which have been investigated in the surrounding area.

The preparation of a palaeohydrographic map was deemed similarly important. This work needs to be further refined, especially from a chronological point of view.[13]

Within the framework of this composite programme the following methods were used:

3.1 Aerial reconnaissance
3.2 The analysis of older and more recent maps
3.3 Surface artifact distribution
3.4 Magnetic survey
3.5 Sub-surface site testing by soil sounding, e.g. core drilling
3.6 Test excavation

3.1 Aerial reconnaissance

Aerial photographs were taken on three occasions within the Békés County Microregion Research Project.[14] The first series of pictures were taken from an A2 type airplane using a fixed camera with a nearly vertical axis to cover the area. A Hasselblad camera with 80 mm basic lens was used at a height of 800 m. Pictures of a scale 1:10,000 were also taken on 6 x 6 cm Kodak Aerocolor negative film. At the height of 1200 m, a 50 mm lens was used and pictures of a scale 1:25,000 were taken on Kodak Infracolor film. [15]

The second flight took place on the first of April 1989 on board a K26 helicopter at a height of 50-250 m.[16] One set of shots was taken with an 80 mm Mamia 645 camera using 4.5 x 6 cm Fuji HR 100 negative film. The other pictures were taken with a Practica MTL 3 camera equipped with a Flektogon lens using Ectachrom HC 100 positive slide film. These pictures were taken during the late afternoon to make best use of the slanting rays of the setting sun which threw the landscape features into high relief (Nikitsch 1987a, b and 1989, Riley 1987, Wilson 1982). Information from these photographs was of immense significance for the preparation of the palaeohydrographic map.

3.2 Map analysis, palaeohydrographic reconstruction

This section is not aimed at presenting a picture specific to the area. Rather it attempts a short but comprehensive description of palaeohydrological developments characteristic of the Körös region, as a typical part of the Great Hungarian Plain (Papp 1969, Pécsi et Sárfalvi 1977).

---

[13] A well defined area has been chosen in order that it should prove possible to reconstruct a detailed sequence of events in terms of changes to the ancient river system and to the depressions caused by flooding in low lying regions. It is hoped that these changes over time can be illuminated by means of core sampling and biological, especially palynological analyses of this material. At present, the number of these studies is too small to allow this kind of reconstruction.

[14] Financial as well as administrative difficulties related to obtaining permits limited us and influenced the choice of flying time in August 1988. Since then, a second and third series of pictures has been taken. During these flights, target oriented photography was carried out in order to gather data needed for mapping.

[15] In all three cases photography was carried out by Gy. Csáky, a scientific research worker at the Institute for Geodesy and Cartography in Budapest. We intend to conduct low-level aerial photography over the whole area of the Microregion Research Project. A complete series of pictures could serve as an input for the computer aided compilation of maps. The results of that work will be published as a separate study (Jerem and Csáky, forthcoming).

[16] Not only did the helicopter prove an ideal tool in terms of choosing the flying altitude, but it could also be held stationary above the feature to be photographed, thus contributing to an even and high quality picture.

The major stages in the formation of the present hydrography may be summarized as follows: The water system of the "Ancient Danube" was shifted eastward by the gradual filling of the Dráva depression and the increased rate of sinking of the basin represented by the Great Hungarian Plain.

In the transitional period between the Upper Pliocene and Oldest Pleistocene, the new flow of the "Ancient Danube" was determined by further sinking activities southwest of the Budapest – Cegléd – Tiszaföldvár line, while water systems to the east flowed toward the Tisza depression.

The "Ancient Tisza", uniting the Szamos, Kraszna and Ér rivers, lay along the Ér – Berettyó line and extended into the plain's depression. It flowed into the "Ancient Maros" river after having followed the beds of the "Ancient Körös" and Aranka rivers (Lászlóffy 1982, 24-25). During the Quarternary, in the second half of the Pleistocene, the flow of the Tisza river was diverted by smaller depressions and elevations. The shape of the river bed known prior to regulation may have come about by the beginning of the Holocene.

The Lower Körös carried waters from the three branches of the Körös and the Berettyó rivers. The Szamos and Kraszna rivers which together formed the "Ancient Tisza" first flowed northward, then turned southwest as a united river subsequently collecting waters of the previously separate Bodrog, Sajó and Zagyva rivers. Following this, the "Ancient Tisza" turned to the south and combined with the Körös and Maros rivers. The Tisza and the above mentioned tributaries cut their beds into the Pleistocene alluvial deposit.[17] Once the geological formation of these rivers had ended, their riverbeds were subject only to limited change within a well defined area. This became one of the most typical traits of rivers on the Great Hungarian Plain. A variant of riverbed alterations is represented by periodic floodings. During these floods, several tens of kilometres were covered by water for various periods of time.[18] The other typical aspect is the formation of meanders. These oxbow-like curves are one of the most quickly changing of riverbed formations. The meanders of the Tisza river are uniquely regular in Europe. The majority of these meanders were cut during the large scale river regulation projects carried out during the last century (Pécsi 1969, Lászlóffy 1982, Mike 1991).

Similar processes took place in the Körös river region (Huszár 1822, [1985]). The speed and quantity of water decreased once meanders were cut through and sealed off. Since the remaining water could not be drained, the meanders turned into cutoffs. With the passing of time these lakes silted in and filled, thus contributing to the formation of newer meanders. These events have been documented by several soil core samples. This allows processes in the immediate proximity of the studied sites to be analysed.[19]

The palaeohydrographical map compiled here (*Fig. 3*) was produced by conflating information from maps published previously at various scales.[20] Our own aerial photographs taken at different times, field surveys and sub-surface soil core testing greatly supplemented the data available from the old maps.

---

[17] This late Pleistocene layer has been identified several times during core sampling.

[18] Aerial photographs clearly show deeper areas which were covered by water for longer periods of time and which were completely uninhabited particularly during periods of deteriorating prehistoric climate.

[19] The expansion of systematic soil core sampling in combination with biological evaluation and dating is also a long term plan.

[20] András Varga used the map of the Körös river made by Mátyás Huszár between 1818 and 1822, the 3rd military survey map of 1:75 000 scale, the 1:25 000 map from 1970 and the 1:10 000 scaled map edited in 1984 as a revised version of the one issued by the Institute of Geodesy and Cartography in 1979. The orthography of geographical names and the digitized version of the general map were both based on the latter.

3.3 Surface distribution of artifacts

Study of the surface distribution of artifacts played a leading role in the systematic surveying of sites.[21] Traditional field survey methods and artifact collection for statistical purposes, aided by geophysical measurements were carried out within the framework of this research activity, (see also Jankovich et al. in this volume).

3.3.1 In this case, traditional field survey served the following purposes:

(a) Since preparatory work for Volume 8 of the Archaeological Topography of Hungary was finished some ten years ago, it was considered necessary to ascertain possible changes at the sites of chief concern here. Such changes include new building work, land cultivation and all other disturbances which may affect structures below the surface.

(b) Data obtained during the analyses carried out previously had to be refined as much as possible before starting actual excavations. This was because the 198 sites identified as Scythian in the volume of the Archaeological Topography of Hungary appeared to have been overestimated once Iron Age material from the area had been re-examined. The majority of catalogued artefacts were either sherds with no particular character or not Iron Age at all. In the case of Celtic settlements, confusion of Celtic with Sarmatian sherds was common.[22] An additional serious problem was posed by sites, the extent of whose area had not been clearly defined or which had been significantly overestimated. Such sites often appeared actually to have incorporated several separate small units rather than being a single stratified settlement.[23]

3.3.2 Collection of artifacts:

Another form of surface sampling involved the collection of all artifacts from the areas covered by magnetic survey for the purposes of statistical evaluation. Here it should be noted that animal bones and remains of plastering were not included in this analysis since it could seldom be confirmed that they were found 'in situ' due to their small numbers. Both frequencies and size categories of ceramics were taken into consideration. Classification by size was of help in interpreting these phenomena: Size parameter is a significant factor in terms of assessing the sherd's origin, whether they have come from the same vessel or have been widely distributed across the site (Jerem and Somogyi 1992).

3.4 Magnetic survey

Magnetic survey is the most frequently used and most efficient of all geophysical methods employed in the prospection of archaeological sites (Aitken 1974, Clark 1990). Geophysical methods implemented within the framework of the Microregion Research Project also concentrated on the measurement of the natural magnetic field complemented in some cases by the analysis of soil resistivity (Jankovich and Pattantyús 1990, Jerem, Pattantyús and Varga 1990).

---

[21] This clearly shows the value of topographic research despite uncertainties concerning interpretation. On possibilities of evaluating and understanding the data see Dunnel and Dancey 1983, Haselgrove et al. 1985, Macready and Thompson 1985, Shennan 1985 and especially Schofield (ed.) 1991.

[22] Our first activity comprised the repeated, piece-by-piece analysis of the material collected during earlier fieldwalkings and which had been identified to belong to the Iron Age.

[23] The most obvious examples of this phenomenon in the studied area are Gyoma 116, 133 and 238. *MRT* 8, 213-214, 227-230, 261.

Investigations were carried out by a G-816 type magnetometer measuring the total magnetic field at a sensor height of 0.6 m, following a 1 by 1 metre grid in all areas studied. Base measurements were taken in the corner of each square unit for diurnal corrections. Most such adjusted data were plotted in the form of an isometric map using a computer.

A number of mathematical filtering procedures were used during the computer analysis (Kulhanek 1976). These were selected in accordance with the properties of individual data sets. In the case of data displaying great diversity, "low-pass filter" was applied in order to emphasize the most important characteristics.[24] In the case of strong surface or regional effects, "high-pass filter" was used in order to give greater emphasis to the potentially supressed effects of features close to the surface (Pattantyús 1986).

Once surface noise has been filtered out, a summary magnetic map may be compiled by combining the various maps, which in most cases, clearly reveals the density and location of archaeological features. In other words, information is obtained relevant to internal settlement structure.

As mentioned previously, these measurements were accompanied by the surface collection of all possible artifacts to provide data for a statistical evaluation and chronological classification. Characteristics interpreted as archaeological phenomena in certain areas subjected to magnetic survey were further studied by analysing soil cores or carrying out test excavations. [25]

3.5 Sub-surface site testing or soil testing

Soil core samples obtained by drilling were primarily used in gathering information on natural and occupation layers as well as on the extent of the settlement. Statistically representative samples obtained in this way also provide material for other disciplines.[26] The type of features thus explored may be identified with great certainty while more recent disturbances may also be separated out. Another advantage is the information provided on the economic efficiency of excavation such as the quantity of earth that needs to be moved and an assessment of the manpower and machinery required.

The machinery used in this work consisted of a drilling machine equipped with a 40 mm diameter and 1 m long bit. It is formed in such a way that the stratigraphy of the retrieved samples may be laid out without risk of contaminating the layers.[27] This method also provides an opportunity to gather samples to depths as great as 15 m. Drilling is carried out without lubrication. A one metre long core segment is taken out in a single operation.[28]

A topographic plan based on adequately detailed geodetic measurements is indispensable in locating the best positions for soil core sampling since these are always determined by the character of individual sites. (See the different layout of test points according to each investigated mound.)

---

[24] This method of filtering is normally used in the measurement of soil resistivity in order to exclude electrode inhomogeneities.

[25] For detailed discussion and evaluation see individual sites.

[26] Soil sampling apart from the physical and chemical analysis of the soil is additionally useful for bioarchaeological investigations: molluscs, seeds, pollen and macro finds such as bone.

[27] The prototype of this boring equipment was built in Hungary following the example of the Swedish "Borro" model.

[28] For more detail about the methodology of this procedure: Jerem and Varga, forthcoming.

In the case of mounds and tell settlements, it is of fundamental importance to determine the original depths of occupation layers, where features are located and the precise nature of their fill. Such information should contribute to a clear understanding of the structure and sequence of the occupation layers[29] (Harris 1989).

Using this method, exact information may be obtained on the area to be investigated prior to excavation (Stein 1986, Becker et al. 1985, Schlichterle 1990). Decisions can also be made about the location, size and techniques of this excavation. Since the complete recovery of mounds and tell settlements is rarely possible, soil core sampling can also complement the excavation itself.[30]

Finally, it should be mentioned that this method of soil sampling is particularly appropriate for testing the validity of aerial photographs and maps made during the investigation of ancient riverbeds. (*Fig. 4*) Such features may not be visible or may be barely apparent on the modern surface due to natural processes of filling and erosion and intensive agricultural activity. Soil stratigraphy also provides reliable information on the location of abandoned stream- and riverbeds, cutoffs, lakes and frequently flooded areas.

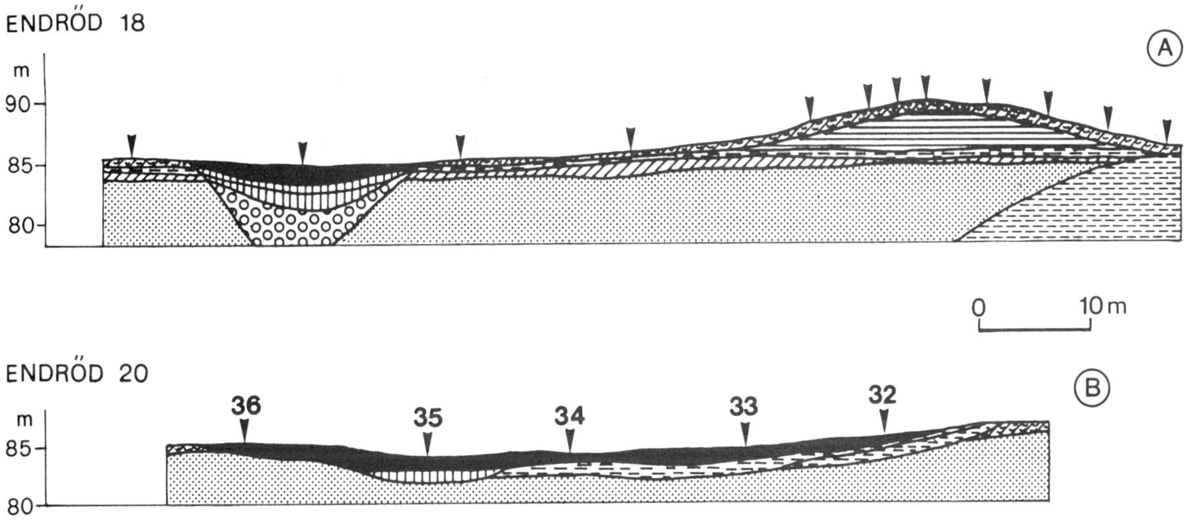

Fig. 4. Geological sections of the old river-beds constructed from core samples.

3.6 Test excavation

Although the previously described methods yield ample information on individual sites, these data can only be verified by means of excavations.[31] At this point, appropriately planned test excavations were carried out at a single site, Gyoma 264. That excavation was justified in part by the character of surface phenomena as well as their chronological range (Jerem and Kiss 1992).

---

29 The precise layout of the bores will be determined by, among other factors, the nature of the soil, its formation and the character of the settlement itself. This form of investigation was conducted by A.Varga at the following sites in Hungary: Tápé-Lebő, Szegvár-Tűzköves, Deszk-Vénó, Hódmezővásárhely-Gorzsa and Kökénydomb, Sopron-Krautacker, Berettyóúfalu-Herpály, Gerla-Nagyalmás, Galgahévíz-Szentandráspart, Murony-Földvár, Tiszaug-Kéménytető, Öcsöd, Polgár-Csőszhalom, Túrkeve-Terehalom, Bogács and Borsod-Földvár.

30 E.g. Berettyóúfalu-Herpály, Öcsöd, Polgár-Csőszhalom etc.

31 A variety of archaeometrical techniques can provide important information with which to supplement data retrieved in excavation. The important point is that neither technique can replace the other; they are essentially complementary and in some ways interdependent.

4. Evaluation of surface collection

In accordance with the principles outlined in the first section of this paper, surface collection concentrated primarily on sites in the proximity of mounds selected as focal points of research. These may be assumed to be Iron Age settlements using data from the archaeological topography.[32]

4.1 Gyoma 117, Rigó-mound and its environs

The artifacts collected on the western side of this mound were predominantly Sarmatian and Árpád Period, while late Migration Period material was found in the southwestern sector. No traces of an Iron Age settlement could be found. Several fieldwalkings were carried out at the site of Gyoma 116 in a semicircle radiating from the mound. There was a concentration of occupation debris mostly to the east and southeast of this site, especially in the proximity of the Csapó farmstead. The geophysical surveys conducted in this area will be described below. Migration Period and Árpád Period finds dominated over the whole area, while prehistoric periods were represented by a much smaller number of sherds, mainly of undiagnostic type.

A relatively large number of La Tène C period sherds were found on the elevation extending east of the mound. This elevation was formerly a terrace of the Körös river. Despite the presence of these finds no Celtic settlement could be located by geophysical survey.

4.2 Gyoma 121, Egei-mound and its environs

A number of sites marked as belonging to the Iron Age are present in the neighborhood of this well known mound (*MRT* 8, 220-222, Pl. 31,1-8 and 32, Maráz 1977, 53, 55, 61 and Fig. 5., Maráz 1981, 97 ff.). They were probably scattered along the bank of a tributary the Körös river. Two walks covered the site of Gyoma 256 (Pólus-mound) which had been described as a Scythian Period site in the Archaeological Topography of Hungary (*MRT* 8, 266). The few and diagnostic sherds gathered did not confirm this dating, although the location of this site fits into the pattern of Iron Age settlements.

The most significant site in this area, and the one also subject to the most detailed analysis in our project, is Gyoma 264[33](*MRT* 8, 267-268). A considerable quantity of material had already been gathered during the first field surveys across the settlement. These covered approximately two hectares and were enclosed in a semicircle by the contemporary (i.e. Iron Age) riverbed. Finds from six periods could be distinguished when the sherds were studied, what suggested that we are dealing with a multi-period settlement.[34] It could also be observed that the highest frequency of surface finds was concentrated mostly on the south-southwestern sector[35] and on the bank of the riverbed closing off the settlement to the east.[36] The above

---

[32] Two periods of fieldwalking were carried out every year (during the spring and autumn) between 1986 and 1990. As a result of this fieldwork the most promising sites (Gyoma 116, 117 and 264 and the environment of Endrőd 18) were surveyed several times.

[33] The kind help of D. Jankovich-B. who directed our attention to this site and supported excavation must be acknowledged here.

[34] Copper Age, Late Bronze Age (Gáva culture), Scythian, Celtic Period, Sarmatian and Late Avar Period.

[35] Surface collection within the distinct greyish spots by the railroad tracks, visible on the aerial photographs as well, yielded a number of sherds some of which could be fitted together.

[36] Additional fieldwalkings were carried out within the ancient riverbed at the sites of Gyoma 259 and 260 located in the immediate proximity of the settlement Gyoma 264. It is possible that these two sites were part of Gyoma 264. The fact that they yielded much fewer sherds is due to modern disturbance in the area.

observations, combined with the fact that soil improvement activities are threatening archaeological features, lead to the choice of this site for a series of complex investigations.

Since the majority of other sites previously observed in the proximity of Egei-mound (*MRT* 8, 220-226, Maráz 1977, 51, 53, 55, 61, Maráz 1981, Fig. 3, 4, 6, Pl. II, III, IV, VI and VIII) had been destroyed by construction activities, field surveys were extended to cover the loess elevation east of the mound.[37] This spot had already proved attractive for settlement during prehistoric times due to the proximity of water and its protected position. Although this area fell somewhat beyond the border originally drawn for the Microregion Research Project, its presence in our survey was justified not only because it is part of the mound's immediate environment but also because of the proximity of Scythian and Celtic cemeteries, reliably documented in the literature.[38](Szénánszky, *RFüz* 37, 1984, 11; 38, 1985, 8; 42, 1991, 9; Maráz 1981, Fig. 6, Sites 24, 43-46).

4.3 Endrőd 18, Papp-mound and its environs

Systematic artifact collections have been carried out in this area since the autumn of 1987 (*MRT* 8, 136. Especially Endrőd 124, 125, 126. *MRT* 8, 167, 168). The concentration of finds indicating a predominantly Sarmatian settlement is present in the area to the northeast (Endrőd 19) of the mound and to the southwest (Endrőd 124) at its foot.[39] The most promising site in this region was Endrőd 20[40](*MRT* 8, 137). This large area, originally surrounded by water, mostly yielded late Bronze Age early Iron Age (earliest Gáva culture) find material, accompanied by a smaller Celtic component. However, sherds, possibly of Sarmatian and Migration Period origin, were found sporadically as well.[41]

4.4 Endrőd 33, Sós-mound and Endrőd 50

No artifacts were found on the surface of this mound, which was covered by black locust trees (*MRT* 8, 140-141, Pl. 18,5). On the other hand, the site of Endrőd 50, located north of this mound, is worth mentioning because a great number of Neolithic sherds were retrieved from it (*MRT* 8, 149). These pottery fragments lay scattered in an approximately hundred metre long band. This material dates to the most recent phase of the Alföld Linear Pottery culture, immediately preceding the Szakálhát culture.[42]

---

[37] Sites Gyoma 125 and 128, *MRT* 8, 224-226. Refer to Fig.3. and also the general map enclosed at the back of the volume.

[38] Aerial photographs were taken in this area in spring 1990, and a new fieldwalking campaign was planned on the basis of these. This proved unsatisfactory due to the maize crop planted in the area.

[39] Endrőd 125, 126 and the environs of Endrőd 20 must also be mentioned here.

[40] Identified as an exclusively Scythian settlement.

[41] The chronological range of finds at this settlement is similar to that of Gyoma 264.

[42] We are grateful for the help of N. Kalicz in the precise dating of these sherds. See also Szénánszky 1979, Goldmann 1983.

5. Site prospection using complex methods

5.1 Gyoma 116 (Gyoma 117, Rigó-mound environs)

Geophysical measurements were taken in close association with the gathering of archaeological data for statistical analysis in three separate areas representing different archaeological periods at this remarkably large settlement (*Fig. 5*).

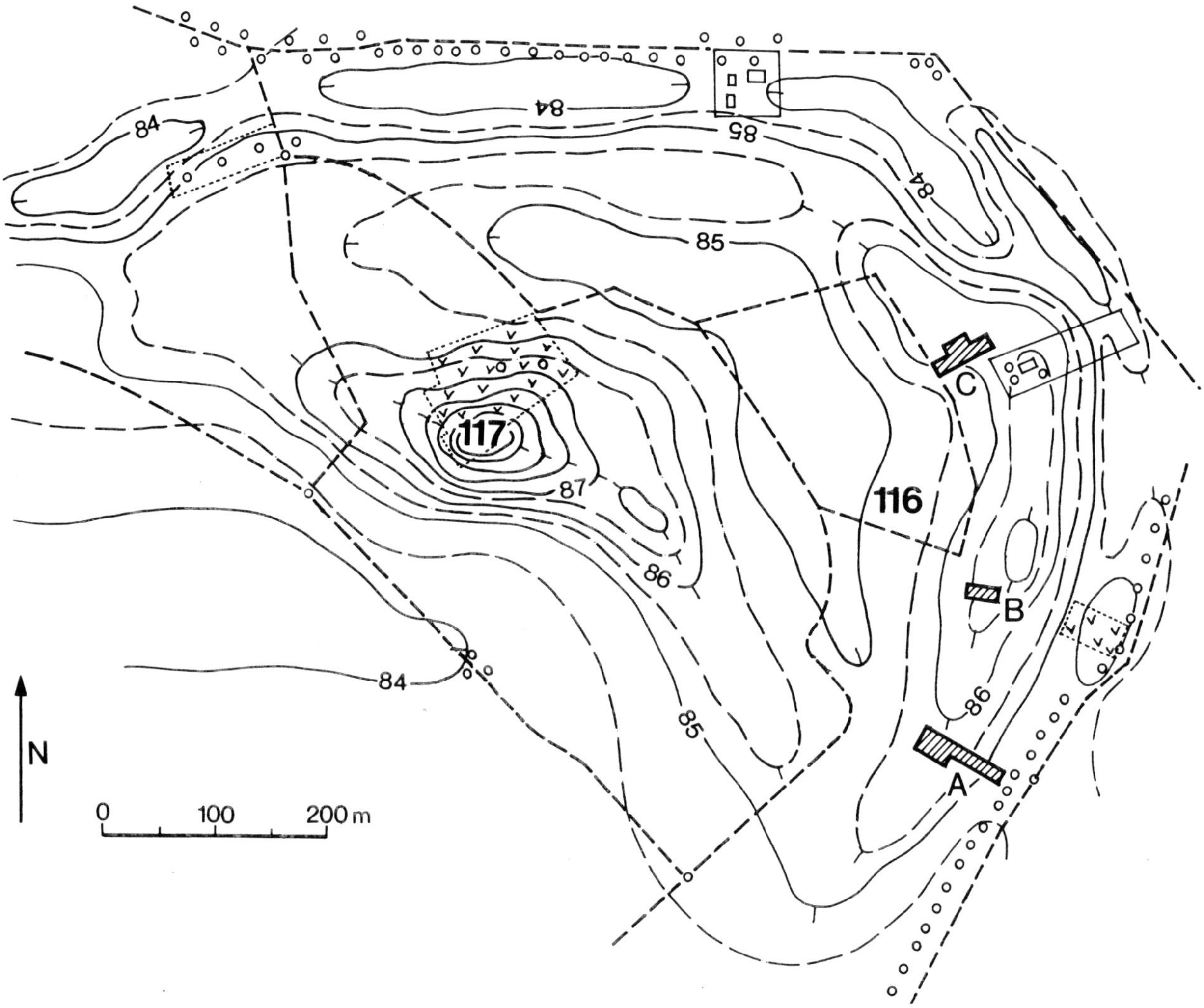

Fig. 5. Location map of archaeological sites Gyoma 116 and Gyoma 117 (Rigó-mound). A, B, C: sites of magnetic surveys (isomap of site "A" is shown on Fig. 6).

The elevation stretching along the former riverbed was suitable for settlement and seemed appropriate for archaeomagnetic measurements as well. Of the test areas, the one labelled "B" had no obvious archaeological features. Nor were significant differences detected within an area of 10 by 30 m. The surface distribution of sherds was rather sporadic. The fragmented, usually small, pieces could not be dated.

On the basis of the encouraging results obtained, from area "C" it was decided to carry out magnetic studies. Effective interpretation at this point, however, is hampered both from an archaeological and geophysical point of view, by the relatively small area subjected to measurements and the considerable noise detected near the surface.

Magnetic anomalies mapped at test area "A" (*Fig. 6*) those marked a, b, c, and d are indicative of archaeological features. According to these measurements, the location should be suitable for further magnetic research. It would be desirable, however, to verify these anomalies either by boring (soil core samples) or digging test pits. Results from the sherd analysis suggest that predominantly Árpád Period features may be expected. This observation, however, does not exclude the possibility that remains from other, older periods may be encountered as well.[43] The purpose of test excavations would be to determine exactly the age and nature of features shown by magnetic anomalies.

The range of isolines in *Fig. 6.* is 3 nT, the intensity of characteristical anomalies is +10-20 nT.

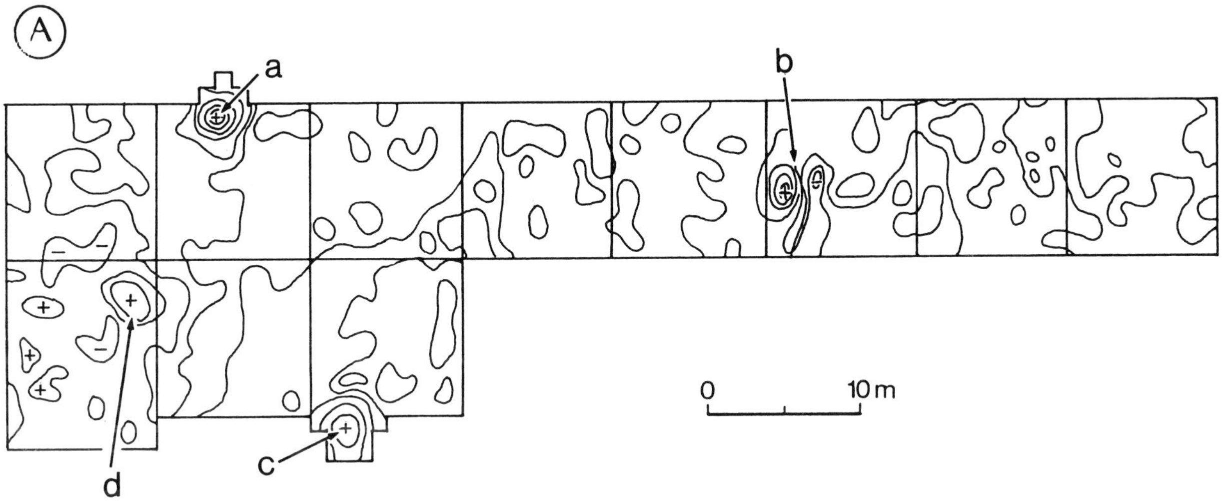

Fig. 6. Magnetic anomaly map of sites "A" at Gyoma 116 (see Fig. 5). a,b,c and d indicate anomalies of archaeological interest.

5.2 Gyoma 264 (Gyoma 121, Egei-mound and environs)

In accordance with the arguments set out in section 4.2, this area was also selected as the scene of magnetic surveys, accompanied by systematic surface collection. Unfortunately, the success of these measurements was questionable from the very beginning due to the proximity of railway tracks. These reservations are further confirmed by the evidence of magnetic maps (*Fig. 7*).

Because the adjacent areas were of small size, the surface distribution of sherds provided no indication about the location of settlements. However, statistical analysis provided valuable preliminary information on the settlement.

All artifacts were gathered in 10 by 10 m squares in each of the three measured areas (*Fig. 8*). Test area "C" was most densely covered by finds. It was in close proximity to the ancient river bed, which is still apparent on the surface. Here only one feature lying parallel with the road could be observed. It probably represents a telephone cable.

43 Sarmatian as well as a number of indeterminate prehistoric sherds were also found among the surface finds.

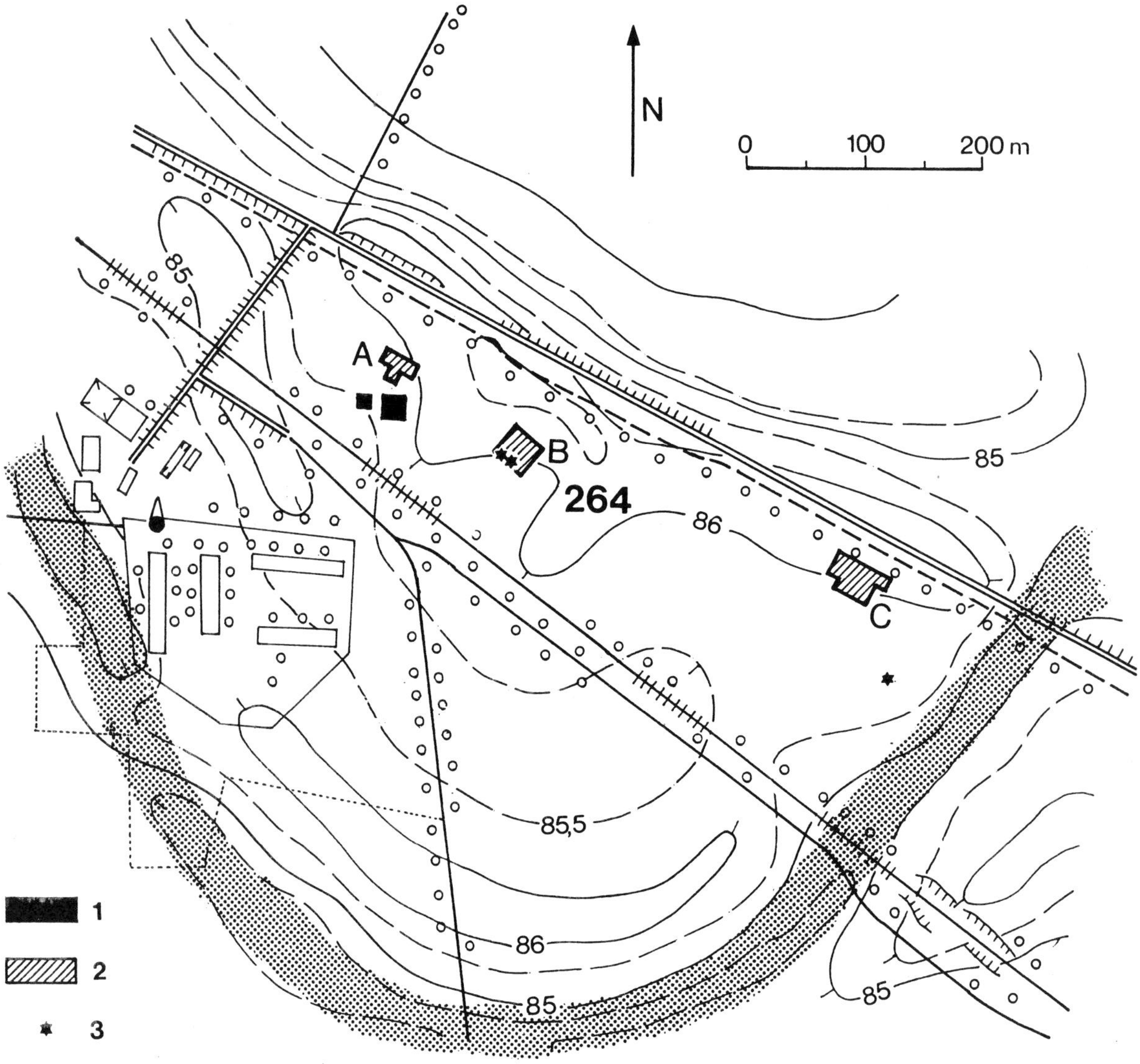

Fig. 7. Location map of archaeological site Gyoma 264 showing the areas of geophysical survey, core sampling and excavation. 1: excavated area 2: A, B, C sites of magnetic survey 3: sites of core sampling

A relatively great number of ceramic fragments were found in test area "A" as well. This spot had been noted previously during field walks, but lay at a greater distance from the railway tracks. This area, however, showed no magnetic anomalies at all.

Area "B" was poorest in terms of finds, although even here the ratio recovered for study was 25 pieces per square metres. Heavy effects caused by the proximity of the electric train line may be detected on this map as well. These make the observation of other features impossible.

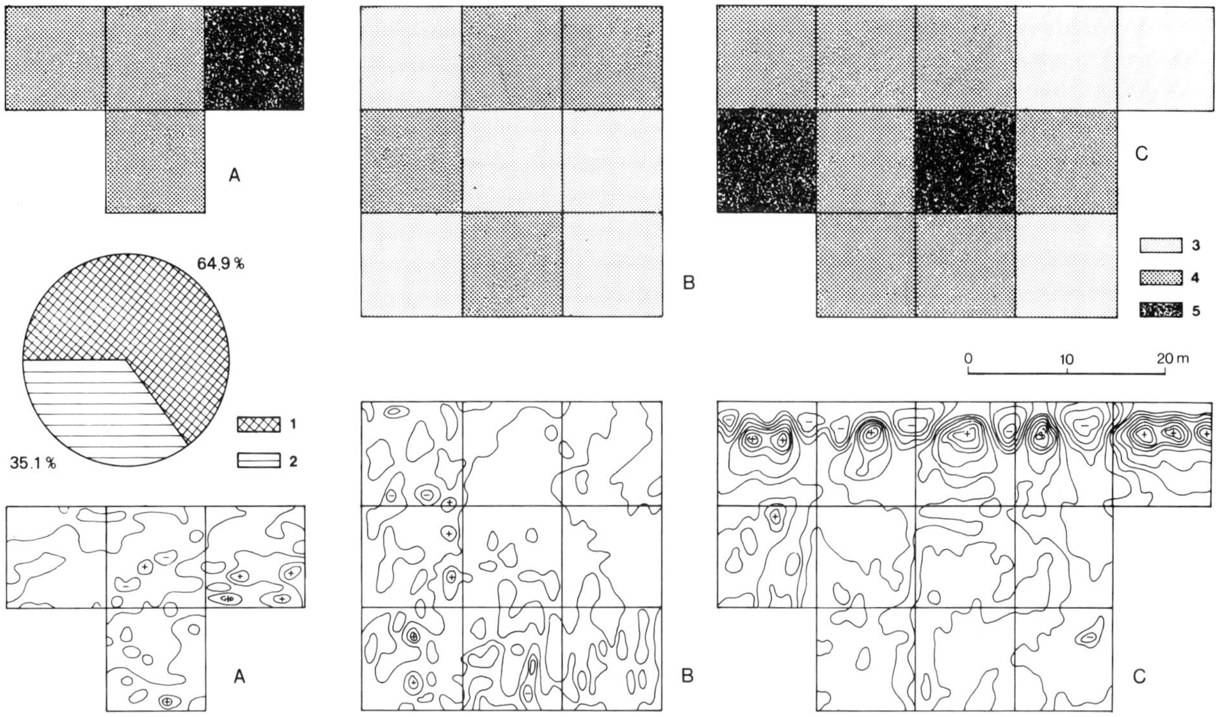

Fig. 8. Density of surface finds per 10 metre square and A, B, C magnetic anomaly maps of site Gyoma 264. (Location see Fig. 7). 1: large and medium size sherds 2: small sherds 3: 1-20 sherds 4: 21-44 sherds 5: 45-55 sherds

Altogether 706 fragments were gathered ("A"= 152, "B"= 173, "C" = 381). Of these pieces, 641 sherds could be evaluated chronologically. The dominance of medium size and large fragments (64.9 %) compared to the contribution of smaller sherds (35.1 %) was noted on all three surfaces during the size related classification of this material[44] (*Fig. 8*).

This phenomenon suggested that the site would be worth excavating in spite of the disturbances caused by ploughing, since the presence of features could be diagnosed unambiguously using the evidence (quantity and relative proportion) of larger sherds.

It shoud be mentioned that a number of Scythian Period (*Fig. 9*) and Celtic sherds (*Fig. 10*) were separated out which confirmed hypotheses concerning the Iron Age occupation of this area.

---

44 Small: 0-3 square cm, medium: 3-5 square cm, large: greater than 5 square cm. These empirical values were used at the site of Endrőd 20 as well.

Fig. 9. Iron Age (Scythian) pottery finds collected by fieldwalking at site Gyoma 264.

Fig. 10. Iron Age (Celtic) pottery finds collected by fieldwalking at site Gyoma 264.

Excavations began during the spring of 1988 in two squares (total surface 300 square metres) near areas "A" and "B" which had been covered by magnetic measurements (Jerem and Kiss *RFüz* 42, 1991, 14).

As had been indicated by preliminary soil core sampling, the humus was as thick as 60-90 cm. This, as well as the density of phenomena at this multi-layer site, required precise observation.[45] At present, some twenty features have been recovered (*Fig. 11*). These varied remarkably both in terms of type and age. They included houses (1 and 5), storage pits (3, 6, 12 and 15), clay extraction pits and pits associated with working areas (2, 11) and ditches (4, 9-10, 14), all representing the previously identified periods. Ceramic material from the classical and younger Gáva culture were found in Pits 13 and 15 respectively. Rubble labelled No. 17, which stretched under the eastern wall of Square I, looked particularly interesting in terms of this period.[46]

Hand formed Scythian sherds, reflecting late Bronze Age traditions were found in the round ditch No. 9 (10). Remains of Celtic (LT C) Period pottery, showing the influence of Scythian traditions, were recovered from the fill of Pit 2. House 5 only appeared as a spot at this stage of recovery. Later it became apparent that it was an Avar period hut.

The early Sarmatian (Dacian) Period was represented by three pits (3, 7 and 21), while an other house and oven surrounded by a smaller ditch (1) also dates to the Avar Period.[47]

5.3 Endrőd 20 (Endrőd 18, Papp-mound and environs)

On the basis of aerial photographs and field surveys the settlement located in the immediate proximity of the former riverbed seemed quite suitable for geophysical research. The measurement grid was marked out in the southern section of the site in the direction of site Endrőd 18 (Papp-mound) so that it extended all the way to the bank of the former riverbed (*Fig. 12*).

Surface finds were also collected in 10 by 10 m squares at the same time that the geophysical survey was conducted (*Fig. 13*).

Results of the magnetic survey are different from those of the previously mentioned examples. The anomalies occur with low intensity but are very well defined. *Fig. 16* shows an interpretation map instead of the isometric plans presented previously. This map shows the intensity, size and location of anomalies which in all probability reflect characteristics of the settlement structure. If this picture is compared to the distribution of surface finds the following observations can be made. The density of these artifacts increases, more or less, in line with the phenomena sensed by the measurements, or more exactly slightly to the north of them (*Fig. 16*). It may be observed that the most intensive occurrence of sherds was 50 to 65 pieces per square. This concentration was recorded in two separate parts of the approximately 30 by 60 m area under study. In areas outside these sections, the density of surface finds initially decreases to 20-40 pieces per square. Finally, the eastern edge of the test area was the poorest in terms of geophysical phenomena and yielded no more than 8-12 sherds per square.

---

[45] This tends to slow down the work. The parent material may be observed at a depth of 140-160 cm.

[46] Our predictions about the importance of this feature were confirmed in a further season of excavation.

[47] This excavation was continued in 1990. This season of work involved opening a new area between two existing squares (Jerem and Kiss 1992).

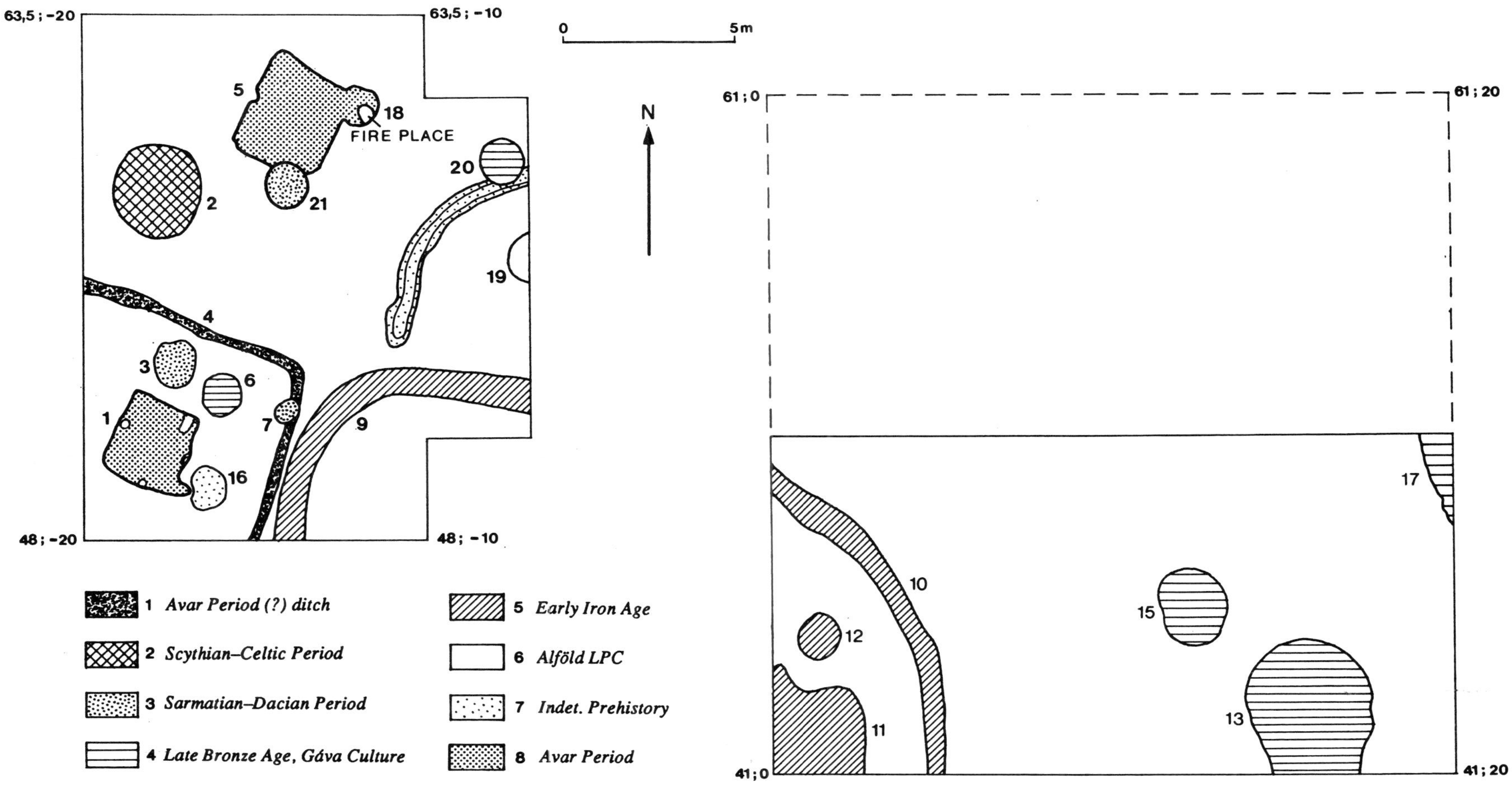

Fig. 11. Map showing features of all periods discovered in the trial trenches at Gyoma 264.

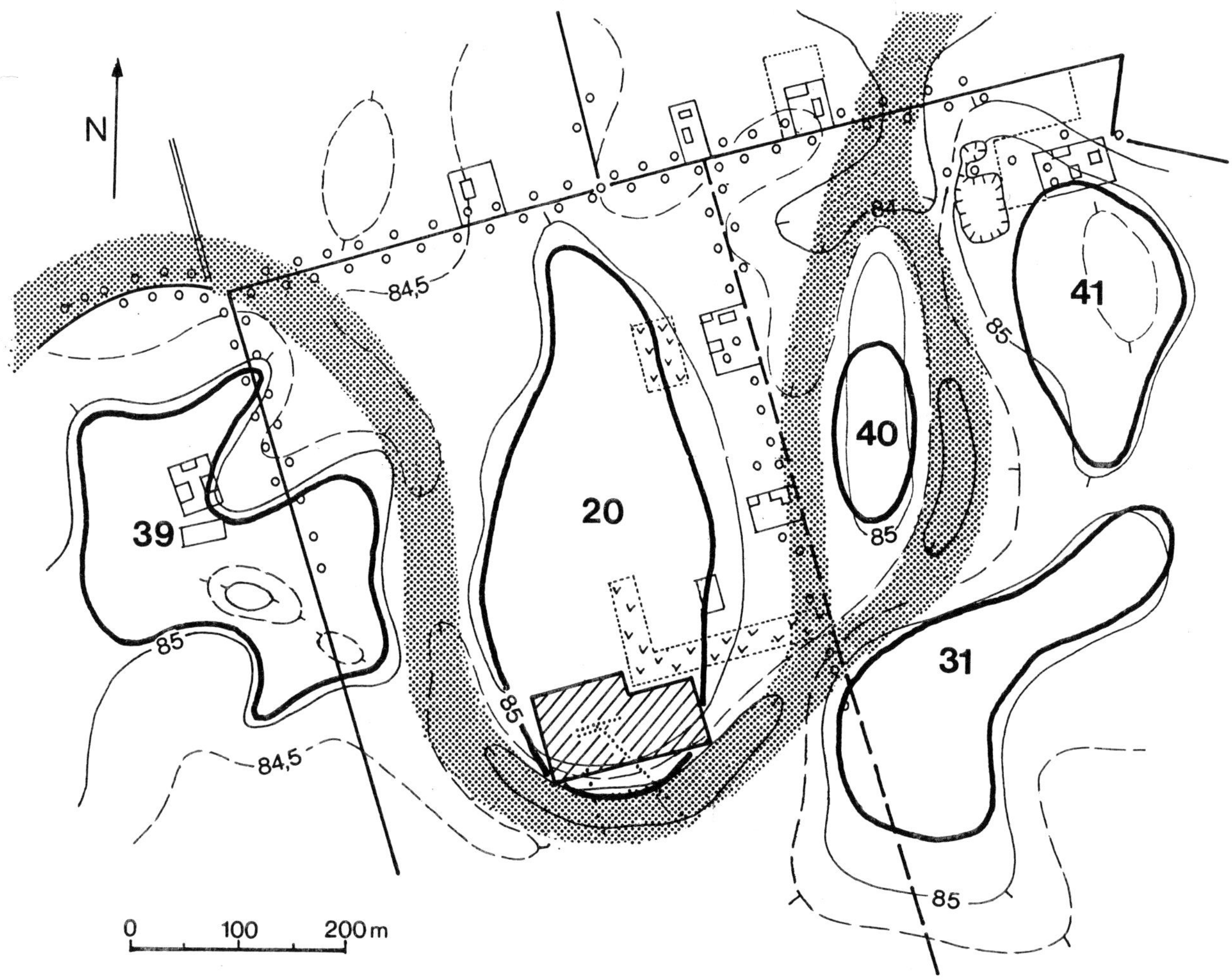

Fig. 12. Location map of site Endrőd 20 showing the area of magnetic survey and core sampling.

Altogether 1834 sherds from this site were examined. In contrast to the artifactual material from Gyoma 264, the majority of ceramics gathered at the Endrőd 20 site consisted of small sherds (1113 pieces, 60.7 %). 518 sherds (28.2 %) were classified as medium size, while 203 pieces (11.1 %) fell into the "large" category (*Fig. 14*). Proportions between the small, medium and large size groups were similar in the material from squares showing the highest density of surface finds.[48] These observations suggest that disturbances at these sites were somewhat greater than in the case of Gyoma 264. To put it more precisely, features at the site under discussion here lay closer to the modern cultivated surface.[49] It is also very likely that settlement structures here were loose and that some of the features were built on the elevated terrace along the river bank.

48 E.g., Squares 36-37; 41-50; 51-55.

49 This assumption was proved by soil core sampling carried out during the autumn of 1989. As a result of erosion, the thickness of the humus layer was only 20-30 cm.

Fig. 13. Pottery finds collected by fieldwalking at site Endrőd 20.

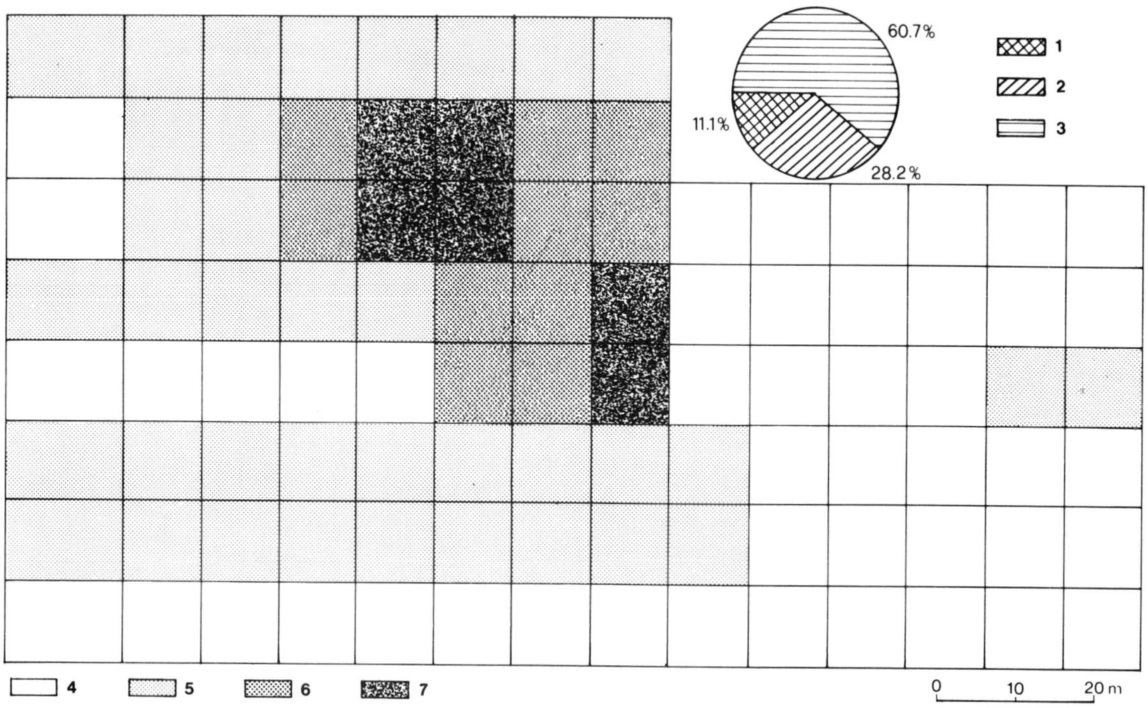

Fig. 14. Diagram showing density of surface pottery finds per 10 metre square at Endrőd 20. 1: large sherds 2: medium size sherds 3: small sherds 4: 1-20 sherds 5: 21-30 sherds 6: 31-50 sherds 7: 51-65 sherds

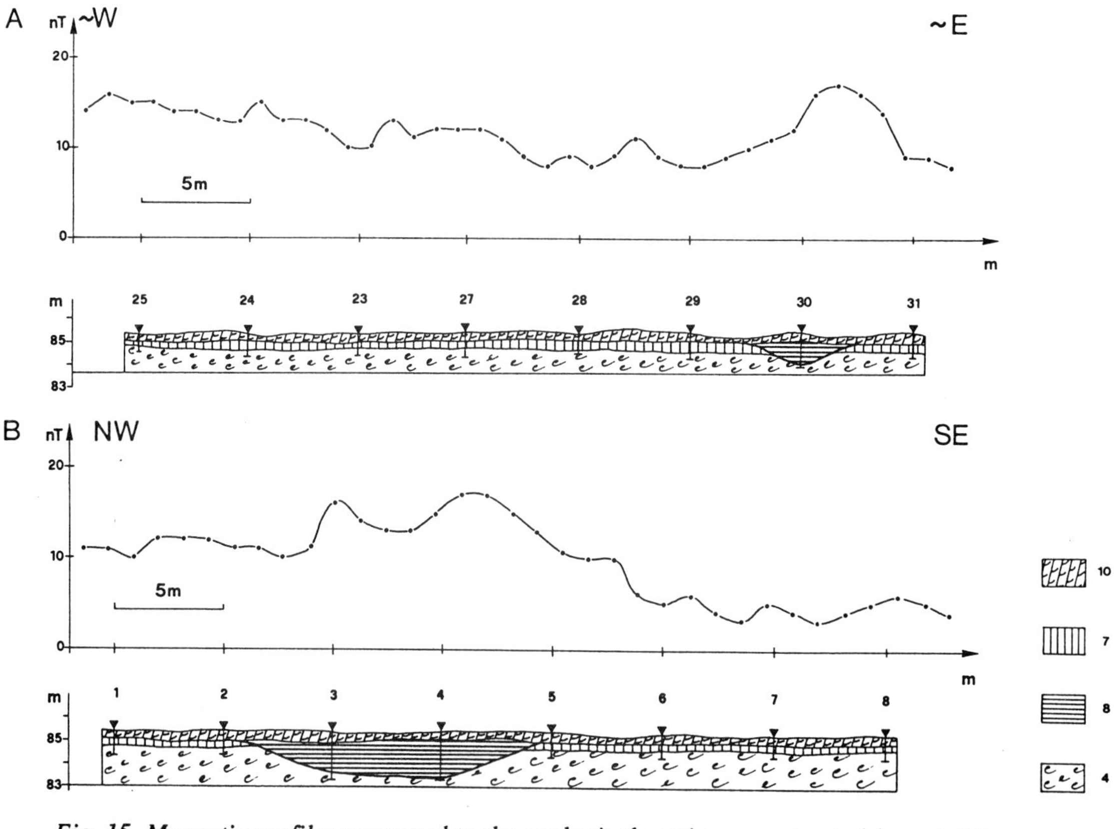

Fig. 15. Magnetic profiles compared to the geological sections constructed from drillings. For identifying profiles "A" and "B" and the numbers of drillings see Fig. 12. and Fig. 16.

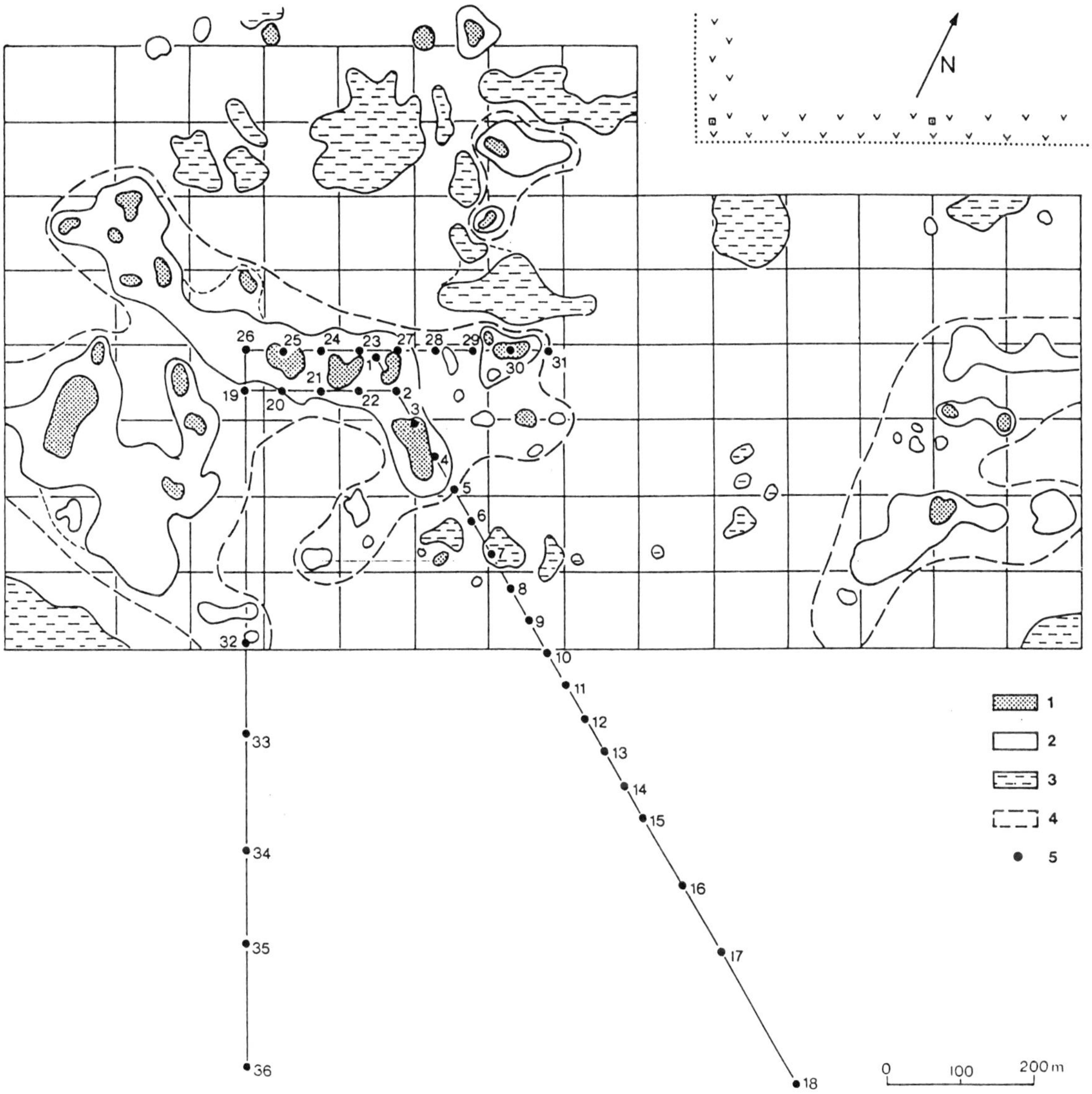

Fig. 16. Simplified magnetic anomaly map with the numbers and positions of core drillings.
1: anomaly is more than 6 nT 2: anomaly is about 3 nT over the average value
3: 3-6 nT negative anomaly 4: suggested outline of the settlement 5: position of core drilling

Additional information on this area was obtained by soil core sampling (*Fig. 15*). It served both to clarify hydrographic relations and to test the results of magnetic measurements.[50]

Fig. 15 A and *B* showing the comparison between the soil cores obtained by boring as well as magnetic anomalies convincingly prove that a portion of the positive anomalies indicate the presence of archaeological features. Thus, one may expect a map showing the extent of the settlement and characteristics of its internal structure to result from a magnetic survey of the entire site. This will be of considerable help in subsequent excavations and provides information about the focal points of the settlement which were not revealed in the thin scatter of sherds.[51]

---

[50] Poster and paper presented at the Heidelberg Archaeometry Symposium in 1990.

[51] The map based on magnetic measurements may show features of Bronze Age and Iron Age dating. This can be tested by a future excavation.

5.4 Endrőd 50 (Endrőd 33, Sós-mound environs)

The extreme density of finds at the foot of Sós-mound (see section 4.4) made it an obvious choice for magnetic survey aimed at identifying the location of features here. However, no anomalies indicative of features could be measured within the 20 by 50 m area shown in *Fig. 17*. The complete destruction of Neolithic layers was shown by soil core samples from a square extending toward the north in the direction of the settlement (also *MRT* 8, 149). Material for the subsequently constructed mound was in fact extracted here. It is for this reason that the undisturbed subsoil may be found directly under the humus of the thin plough soil.[52]

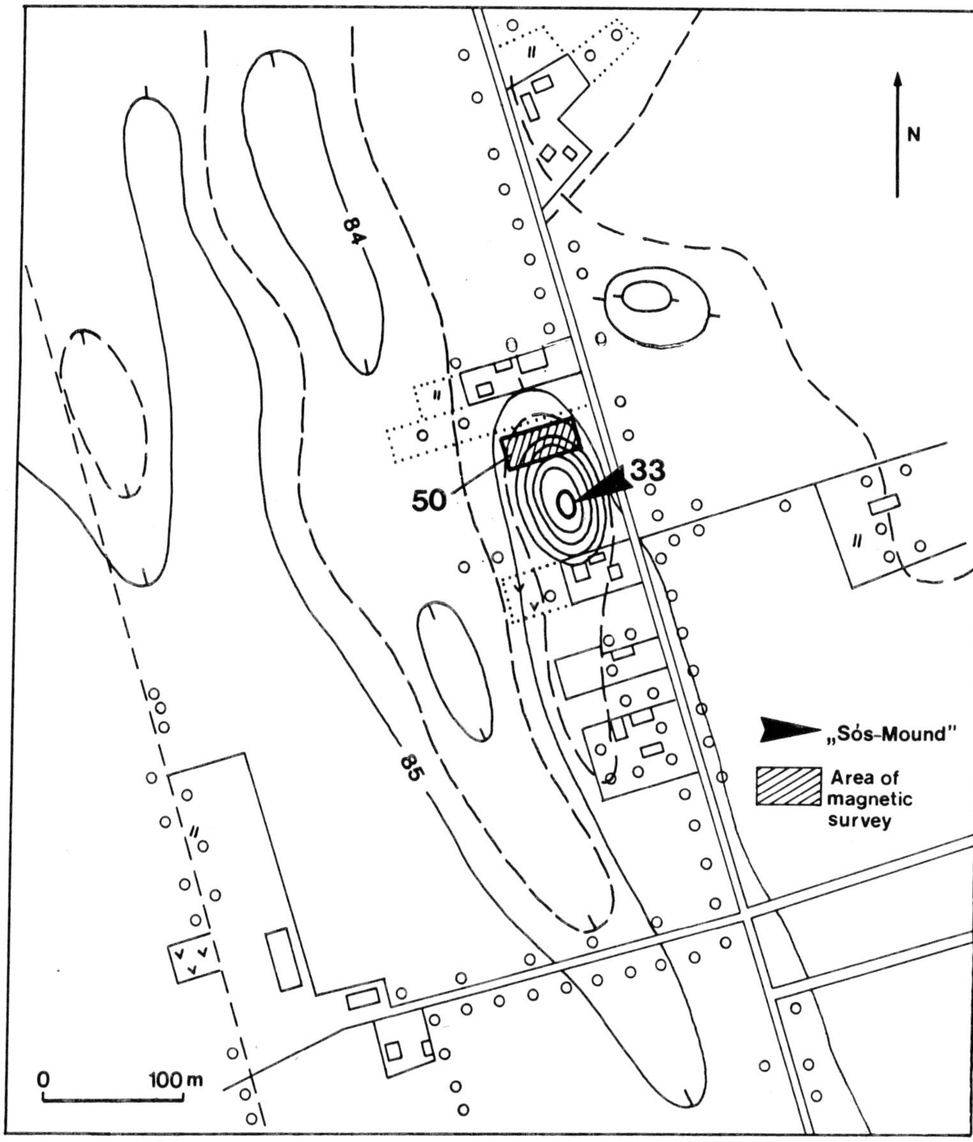

Fig. 17. Contour map of the area surrounding archaeological site Endrőd 33 (Sós-mound) and Endrőd 50 (Neolithic site). The hatched area was examined by magnetic survey.

---

52 On the bases of negative evidence, no additional geophysical measurements are recommended in this area. The great number of surface finds is probably indicative of destroyed archaeological features. Only a rescue excavation would seem to be appropriate here.

6. Preliminary research at the mounds using soil core sampling

6.1 Gyoma 117 (Rigó-mound, earlier called Lippai-mound)

The study of this feature, well known from the literature but previously discussed in contradictory terms (*MRT* 8, 214-219, Pl. 29), was carried out in 1987 and 1988[53] (*Figs. 2 and 18*).

First, a 1:200 scale map was prepared with 10 cm contour line intervals which shows even the smallest elevations and depressions on the surface. This mound has an irregular, elliptical shape with an almost 160 m long northwest-southeast axis and a 100 m long northeast-southwest axis. The southwestern and southeastern sides are steeper while the northern side has an elongated gentle slope. A steep cut, followed by a several metres long horizontal plateau may be observed on the northeastern side. 20 to 30 cm depressions may be found at several points on the surface. These range in length from 10 to 80 m and are of varying widths. Those located near the foot of this mound may have been caused by dirt extraction for the purposes of building the mound (*Fig. 19*). The aim of soil boring in this case was not only the reconstruction of the mound's micro- and macrostructure, but also the identification and separation of areas and layers which were potentially disturbed from an archaeological point of view. This influenced the choice and density of sampling points, especially when the mound's centre was investigated.

The extent and structure of the mound was identified by borings along profiles A and B which lie at 90° to one another. In the next stage, the stratigraphy was refined by additional, densely placed sampling in the central area. The natural soil under Rigó-mound consisted of fine and medium-fine grain sand covered by a 40 to 70 cm thick loess. A transitional layer with an average thickness of 40 cm was found on top of this. It corresponds to the old ground surface at the time preceding the construction of the mound. The horizontal extent of the culture bearing layers deposited over this surface may be estimated as 60 to 80 m. Its greatest thickness exceeds 5 m. As a result of human activity, the redeposited layers contained varying quantities of micro-inclusions like ash, daub and bone remains. Soil characteristics, indicative of periodic flooding observed at the bottom of the mound along the bore lines, must also be distinguished from a structural point of view. This silted clay contained very few micro-traces (*Figs. 20-21*).

With the help of these soil core samples it could be ascertained that the mound had no natural core. There was no significant elevation on the original surface when the burial took place and the mound was built. In its present form, the mound may be considered as having been disturbed several times. Remains of rubble originating from a modern construction were spread over the plateau and could be easily identified. The same holds true for the damage caused by the construction of a datum point on the top of the mound. Disturbances that have reached the core of this mound are more difficult to deal with. Since the traces of this damage could be first observed in borings along the main axes, the existence of a shaft oriented northwest-southeast and extending 5.4 m deep to the layer of yellow loess, was verified by 13 borings, averaging depths of 5 to 7 m in the centre of the mound (*Fig. 21*). A number of the soil core samples thus obtained were subjected to water-sieving: biological, especially malacological and palynological studies supported observations concerning the multiple redeposition and disturbances in the stratigraphy[54](*Figs. 22 A and B*, as compiled by L. Kordos and E. Krolopp).

---

53 Three successive campaigns of boring were organised in order to elucidate the confusing signs of disturbance.

54 Even at 450 cm dept, we still could detect signs of anthropogenic evidence and molluscan fauna of the soil.

Fig. 18. 1-2: Areal view of the burial mound Gyoma 117 and its surroundings. 3-6: Four photographs showing the Rigó-mound from different directions.

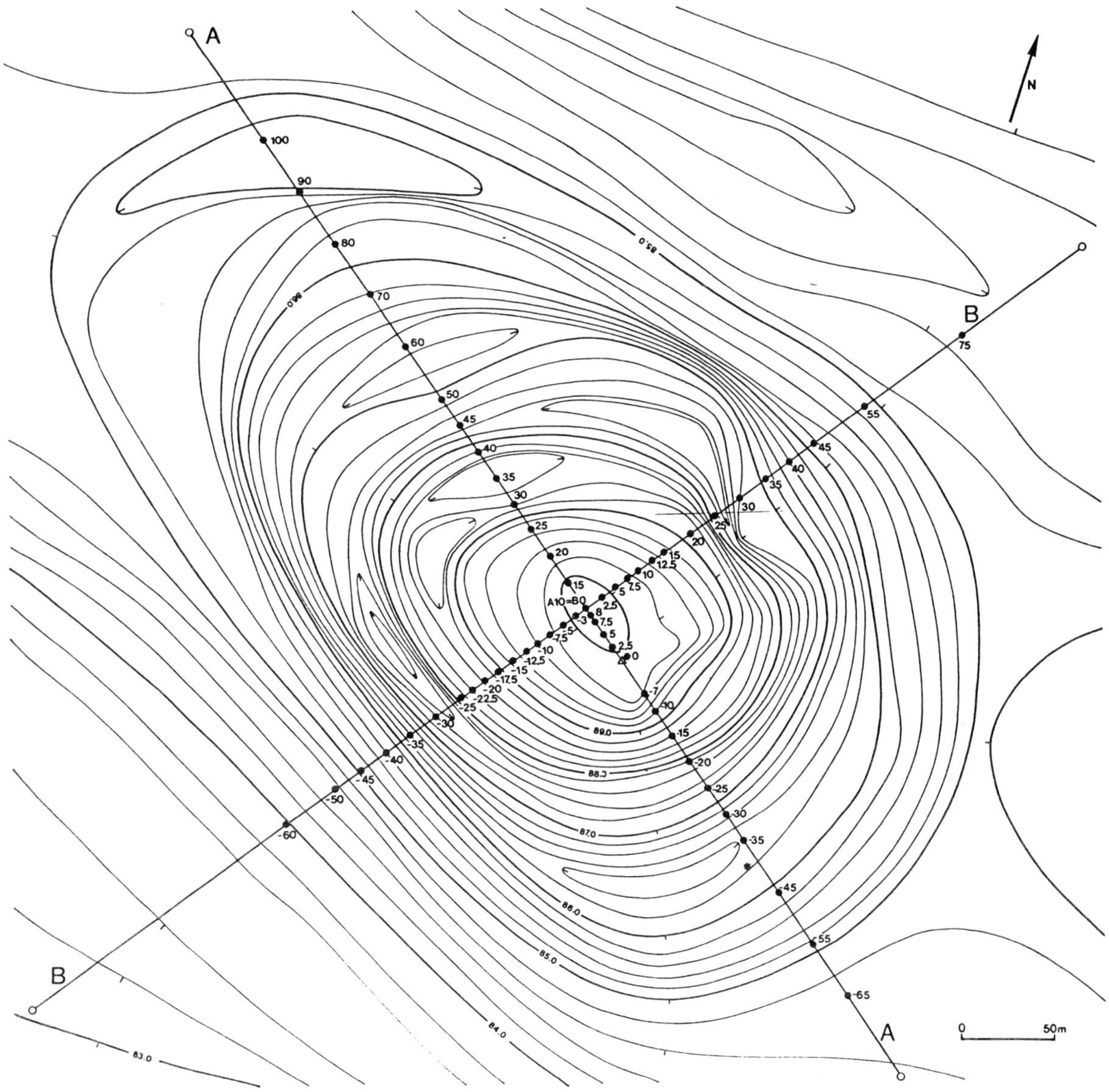

Fig. 19. Contour map of the site Gyoma 117 (Rigó-mound) showing the positions of the soil core sampling.

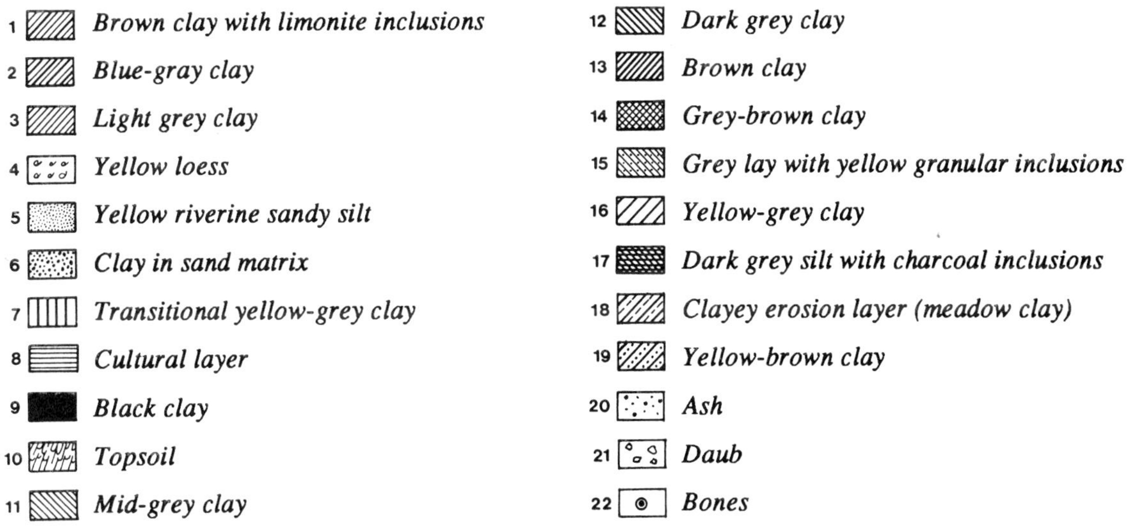

Fig. 20. Key for all geological sections.

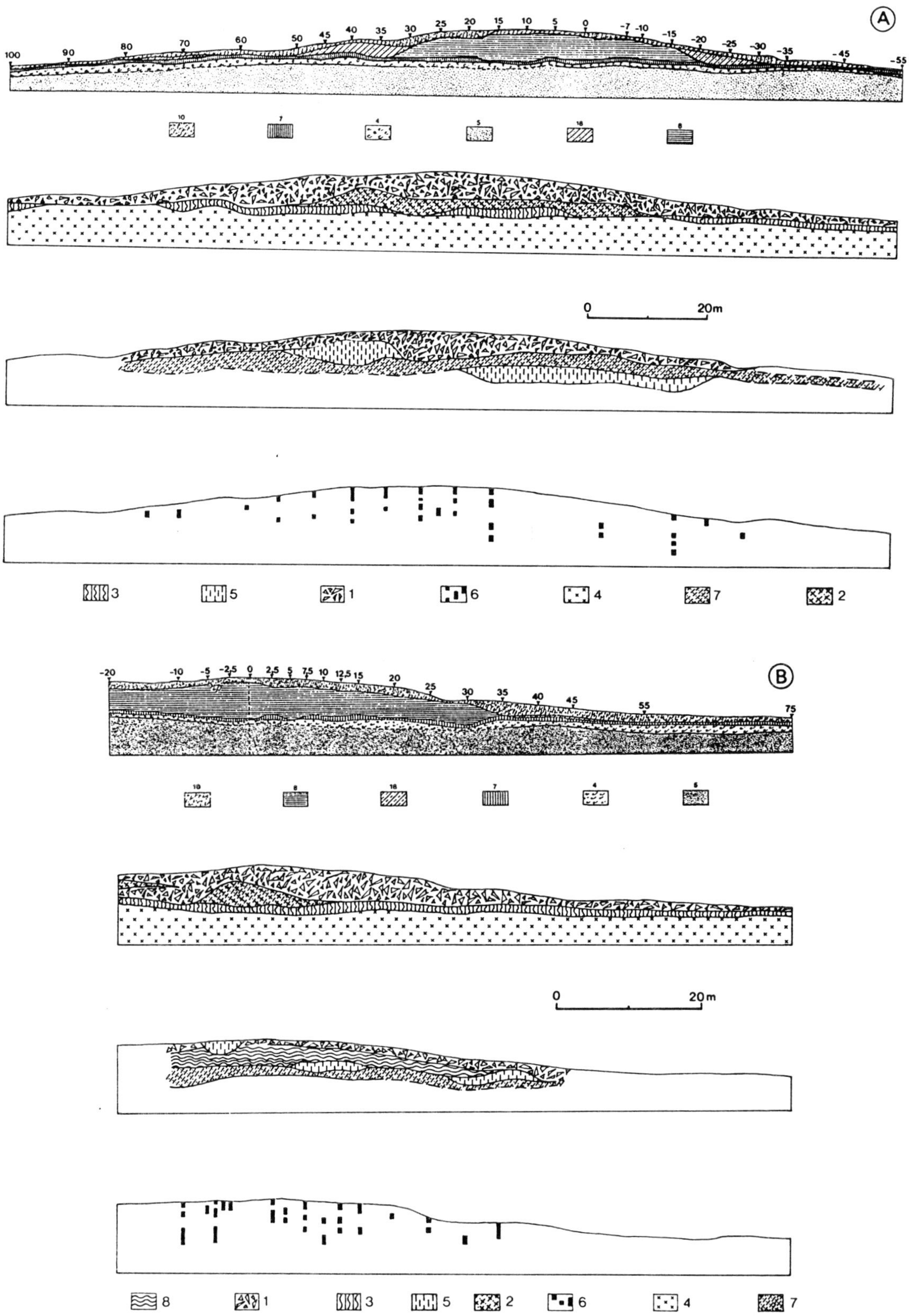

Fig. 22. A and B. Molluscan and micro-vertebrate evidence from site Gyoma 117 (Rigó-mound). 1: molluscan and micro-vertebrate fauna of the plough soil 2: fauna of the plough soil mixed with loess 3: molluscan fauna of the transitional layer 4: molluscan fauna of the pleistocene loess 5: redeposited soil from culture bearing layer 6: position of micro-vertebrate remains in boreholes 7: negative micro-vertebrate faunal evidence 8: micro-vertebrate fauna of the natural soil.

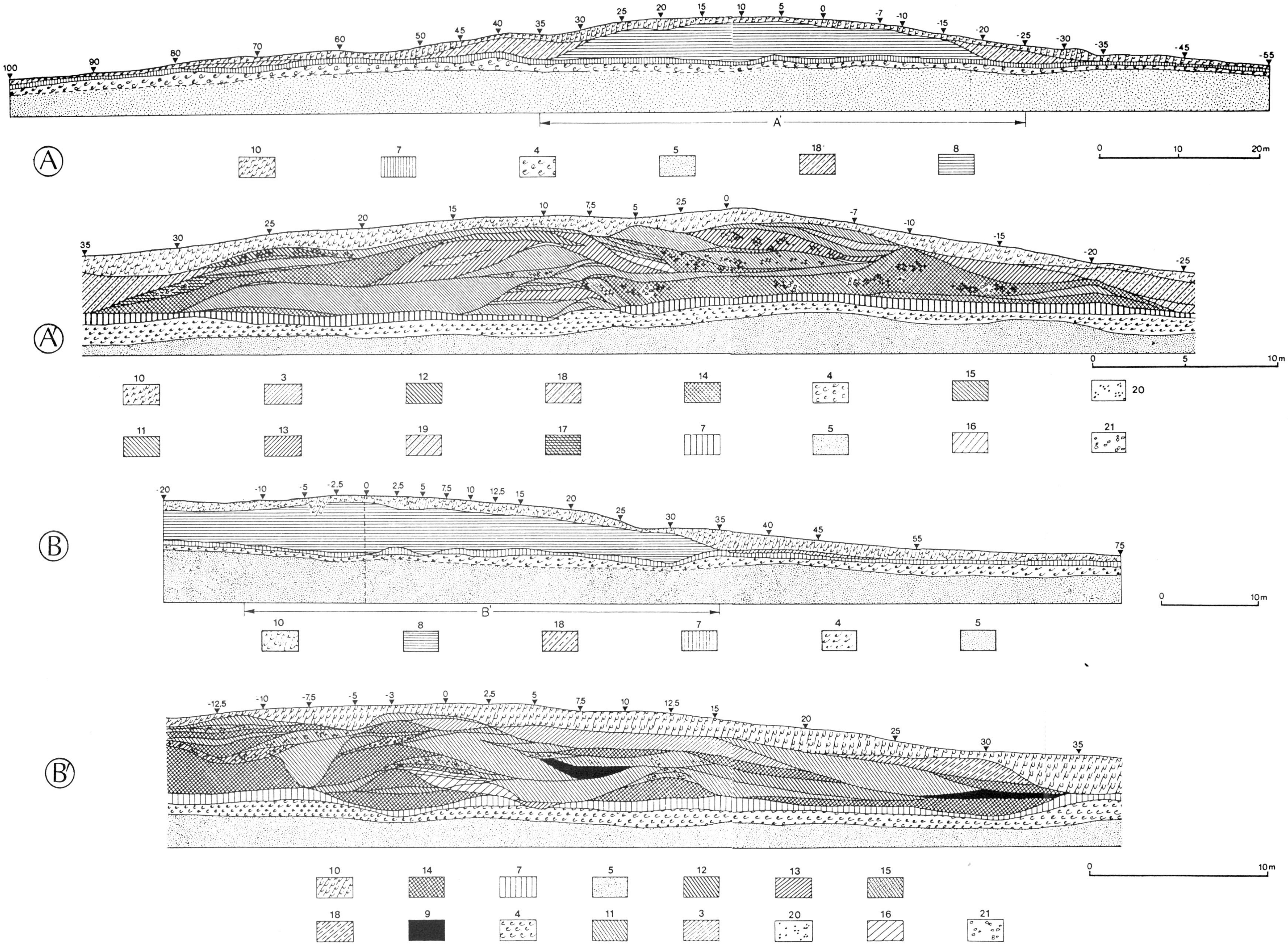

Fig. 21. Geological sections "A" and "B" and figures showing the central part of the sections

(through Rigó-mound) in more detail.

6.2 Endrőd 33 (Sós-mound)

With the exception of its northern slope, the mound is badly damaged. The lower levels have been eroded on all other sides. At present, the site is covered by a black locust forest which inevitably influenced the spatial pattern followed in sampling. The network of roots made the actual boring work difficult as well. Because of the geometrical properties of this feature, the borings were carried out along radially located lines (*Fig. 23*). A total of 70 soil core samples were taken from depths ranging from 1 to 7 m.

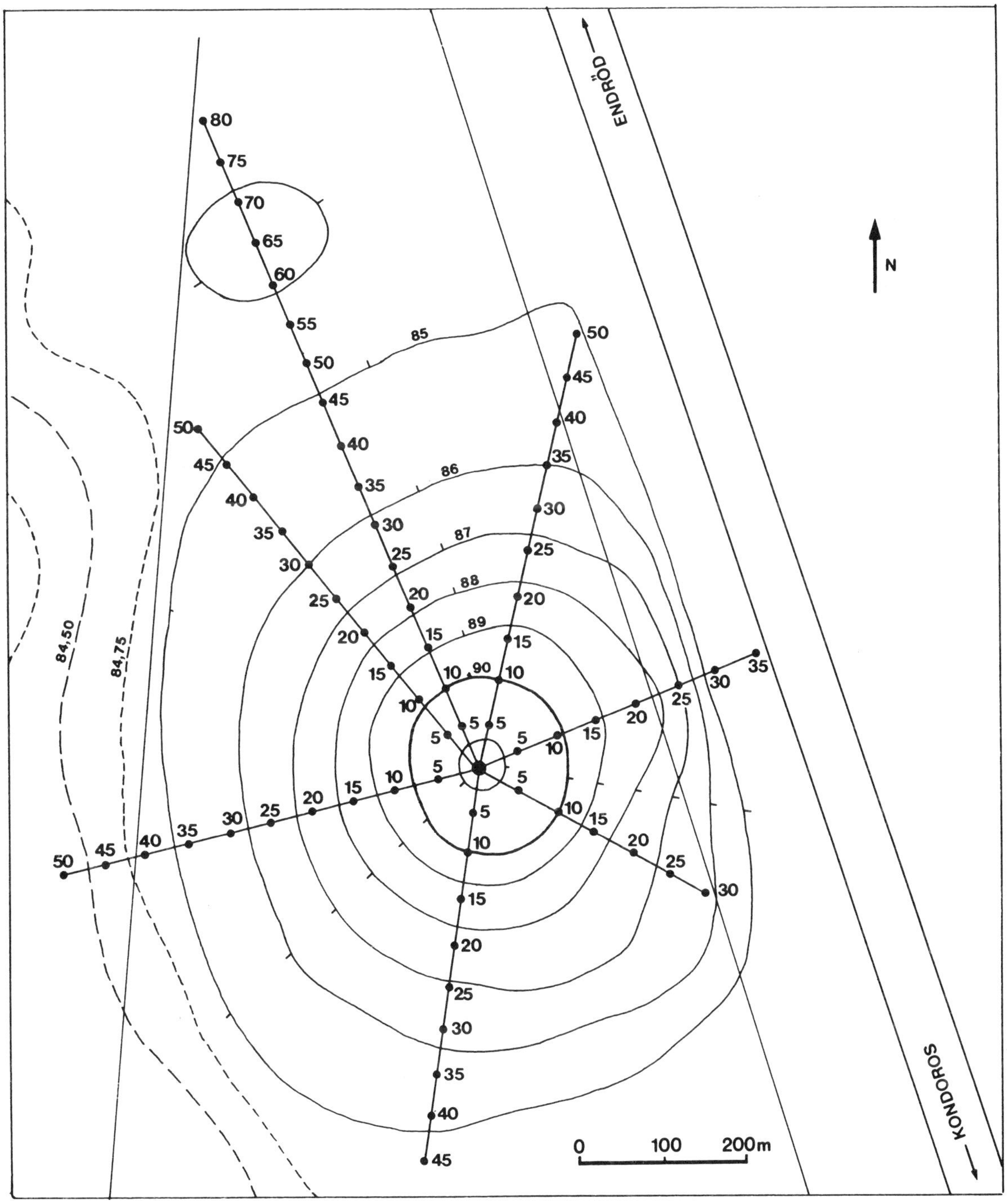

Fig. 23. Contour map of site Endrőd 33 (Sós-mound) with the positions of the soil core sampling.

Borings along line "A" provided an early indication that the structure of Sós mound is different from the internal stratigraphy of the mounds studied previously. A homogeneous layer may be found above the height of 6 m in the centre of the mound. Physical parameters (density, water retention, inclusions) within this layer are fundamentally identical. Differences may only be seen in colour (brown and grey respectively) depending on where the soil was brought from. Profile drawings clearly show that the ground surface in this area is also flat and that the whole mound was created artificially. In contrast with the other mounds under discussion here, it has a monolithic structure.[55] Some deviations were detected only in points D5 and E10 which may be indicative of features subsequently dug into the mound. The layers deposited artificially contained no remains of archaeological significance. Smaller bones, ash and daub fragments were discovered only along line "A". They originate from the disturbed upper layers of the settlement of Endrőd 50. A depression west of the mound may be the remains of a quarry that provided earth for the construction.

6.3 Endrőd 62 (Vaszkó-mound)

This mound has a regular conical shape with a high bank on its southern side. It rises 1.5-2 m above the level of the surrounding area. A silted-up river bed may be found in its immediate environment. The river bed has not yet been examined but hopefully will be subjected to boring. At present, soil core samples on the mound were taken along a longitudinal and two radial profiles at 19 points (*Fig. 24*).

These borings suggest that the mound is composed of a few well defined layers in the profiles which were examined. The old ground surface is even and the loess and sand formations seen in the case of other mounds may be observed here as well (*Fig. 24*).

Of the sampling points, No. 14 yielded human bones at a depth of 135-140 cm[56] while No. 11 was indicative of mixed layers possibly reflecting the fill of a pit at depths of 180 to 350 cm. The sample taken at point No. 18 was water-sieved. Malacological data suggest layers of fill mixed in part with loess to a depth of 150 cm. A potentially older layer of soil was noted between the depths of 150 and 185 cm followed by a transitional layer. The exclusive presence of Pleistocene loess fauna is indicative of the undisturbed natural soil. Further investigation of the mound and its immediate environment are planned for the future. As a preliminary for excavations, low-level aerial photographs were taken in the spring of 1990.

6.4 Endrőd 18 (Papp-mound)

This almost completely regular, conical mound (*Fig. 1*) is located in the inner arch of a former river oxbow which was first cut through and then filled in. Its position could be verified by borings as well (*Figs. 3 and 4/A*). The height of this mound is 5 m above the surrounding area.

The 1:200 scale contour line map of this area was carried out during the autumn of 1987. Two test bores were drilled in the mound's centre and at its foot for the purposes of orientation. Several samples were taken from these for palynological and other palaeobiological studies.[57]

[55] The conical shape and great height relative to the bottom diameter of this mound are different from the usual form of kurgans on the Great Hungarian Plain. This first became evident during fieldwalking and was later confirmed by contour line measurements. The topographic position of this mound was similarly atypical.

[56] Morphological studies on the bone material were conducted by I. Pap. It is also intended to obtain C-14 data from the samples.

[57] It has been found in boreholes "A" and "B" to a dept of 395 cm soil fauna mixed up with shell fragments and the snail species of Pleictocene loess. Under 405 cm, we met only Pleictocene faunal remains, both riverine and continental species, which originated from wet infusion loess.

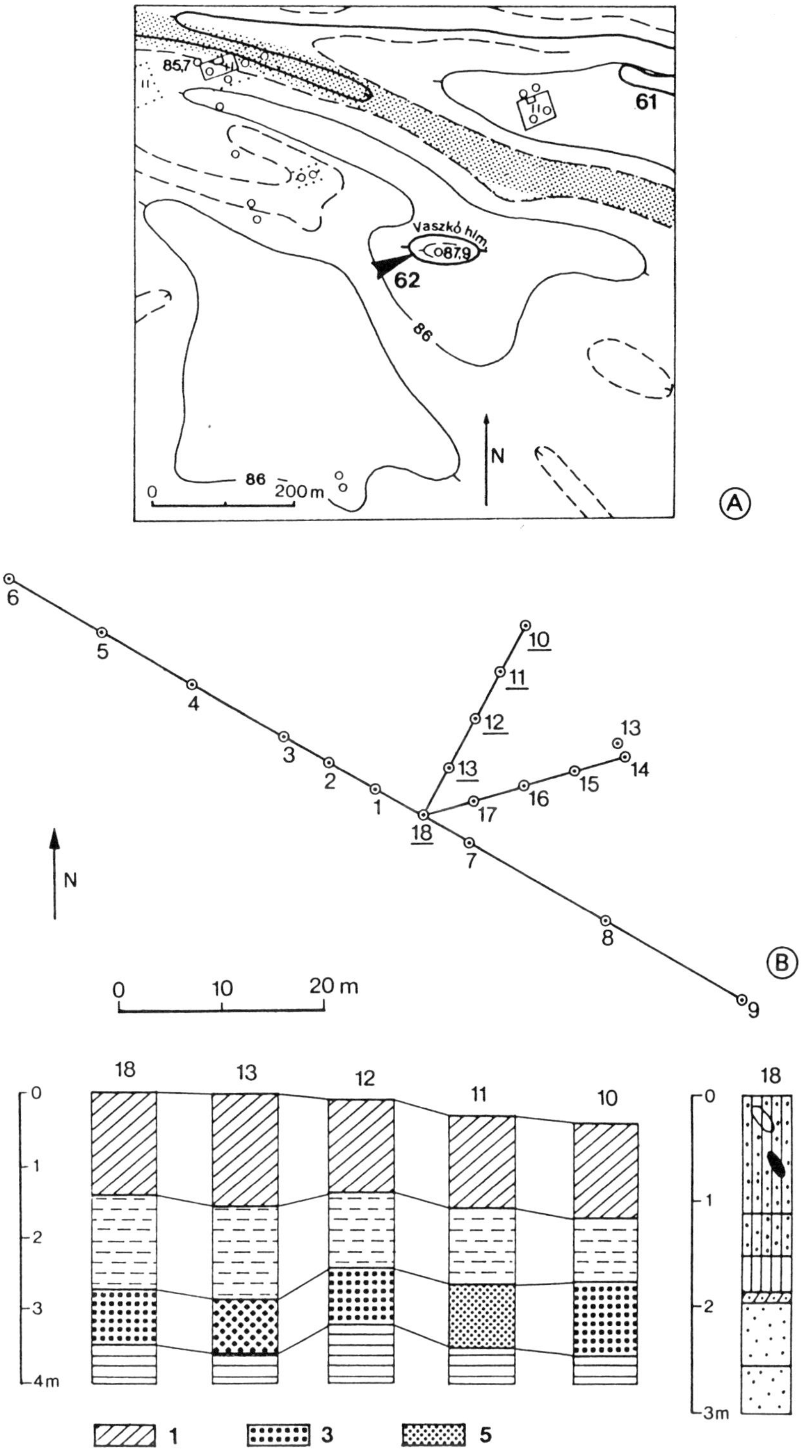

Fig. 24. A: Contour map of site Endrőd 62 (Vaszkó-mound) B: with the positions of the soil core sampling and borehole sections. C: borehole 18 demonstrates the result of biological studies. (Legend: see Fig.28.)

1: dark grey clay in sand matrix — *4: yellow riverine sandy silt*
2: brown clay in sand matrix — *5: disturbed layer*
3: yellow-grey clay in sand matrix — *6: yellow loess*

Detailed research was carried out in the form of soil core sampling in the following year, in line with the requirements of a planned complete excavation. First, a northeast-southwest section was prepared to provide a complete profile of the mound's structure (*Fig. 27*).

It could be established that artificial deposits formed a culture bearing layer extending 4 m in height and that, as in the case of other investigated kurgans, this mound too was built on an almost flat surface. The natural soil is very fine grained sand which was found under the 60-70 cm thick layer of loess.

The profile was extended in the direction of site Endrőd 125 (*Fig. 26*) and shows that the riverbed cut into the loess and underlying sand at this point. Water-sieved samples bear witness to the fact that (apart from a few shell fragments) no faunal remains were found between the depths of 0 to 260 cm. On the other hand, iron conglomerates were found there which had formed in the reductive context of stagnant water deposits. The snail fauna between 260 and 355 cm contained riverine species as well. Thus, it may have formed as part of the river bed's deposit. Samples below this layer contained no faunal remains. The formation of this meander is explained by the stratigraphy revealed by borings carried out southeast of the mound exposing heavy clay soils.

In the second step, a 25 by 25 m grid of 5 by 5 m squares was laid out over the area of a proposed excavation taking account of the mound's shape. The diagonals of this grid fall in line with the two previously studied profiles (*Fig. 25*).

The entire grid was not studied: it was first divided into quarters, then points in these 12.5 by 12.5 m squares coinciding with the original grid were bored. The profiles thus outlined provided a quantity of valuable information. Not only did they permit numerous samples to be taken but they also enhanced the well differentiated reconstruction of the microstructure. This work is illustrated by the individual presentation of profiles where bone and ceramic finds were marked on the basis of their horizontal and vertical positions (*Figs. 27 and 28*).

Samples that yielded morphologically identifiable bone remains[58] are of great significance as well as those in which the relative depths of bone fragments varied within the sample.[59] Additional data were provided by the identification of micro and macro remains recovered by water-sieving.[60] This body of information is best represented by diagrams of results from the malacological study (*Fig. 28*).

Phenomena indicative of disturbances were discovered here as well. In this case, it was present in the form of a 3 m deep ditch-like structure that cut across the entire mound (see sections in *Fig. 27*). Due to the homogeneity of soil strata, however, the exact identification and spatial definition of this disturbance requires more careful, local observation.

---

[58] Samples 4, 16 and 17 provided human bones identified by I. Pap to whom I am very grateful for her help. Also in this case it is hoped to obtain C-14 dates when an appropriate technique has been chosen.

[59] Samples 1, 9, 15 and 20. Unfortunately, the small quantity of bone material does not permit C-14 dating.

[60] Thanks are due to E. Krolopp and L. Kordos who were useful members of the team both during field work and in the evaluation of data.

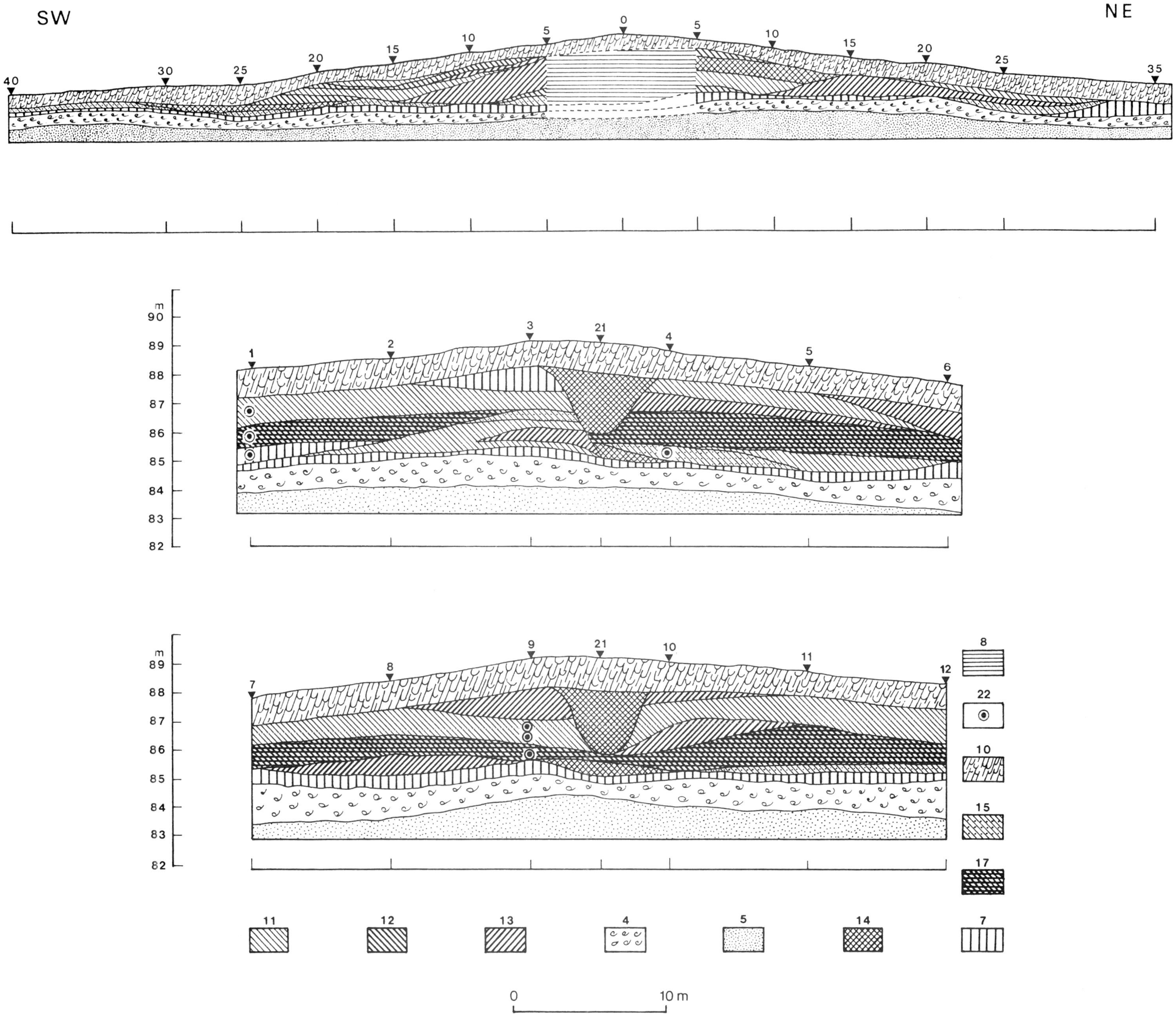

Fig. 27. Geological sections of Papp-mound showing the stratigraphy. Human bones are marked in the section.

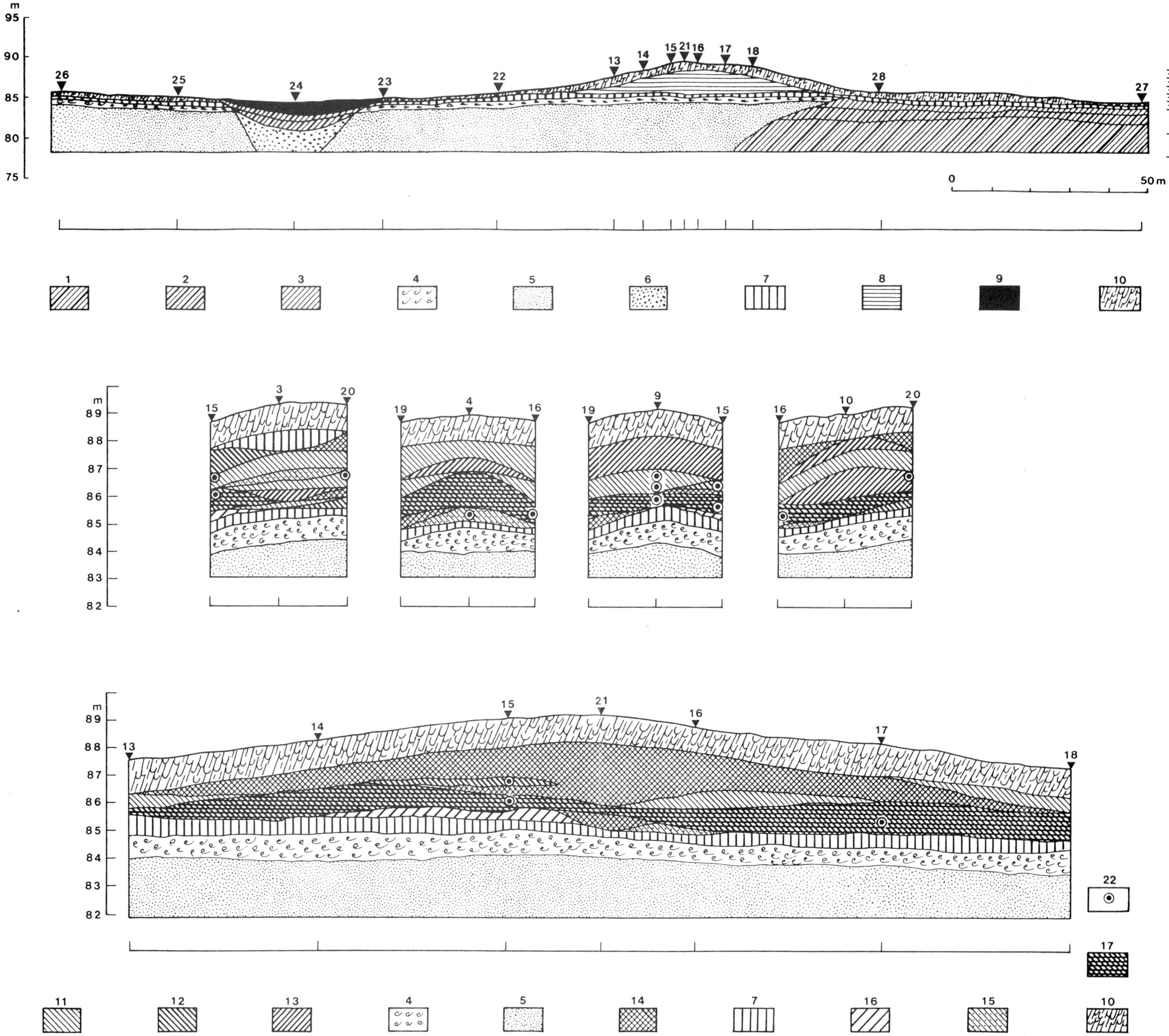

Fig. 26. Geological sections of Papp-mound showing the stratigraphy. Human bones are marked in the section.

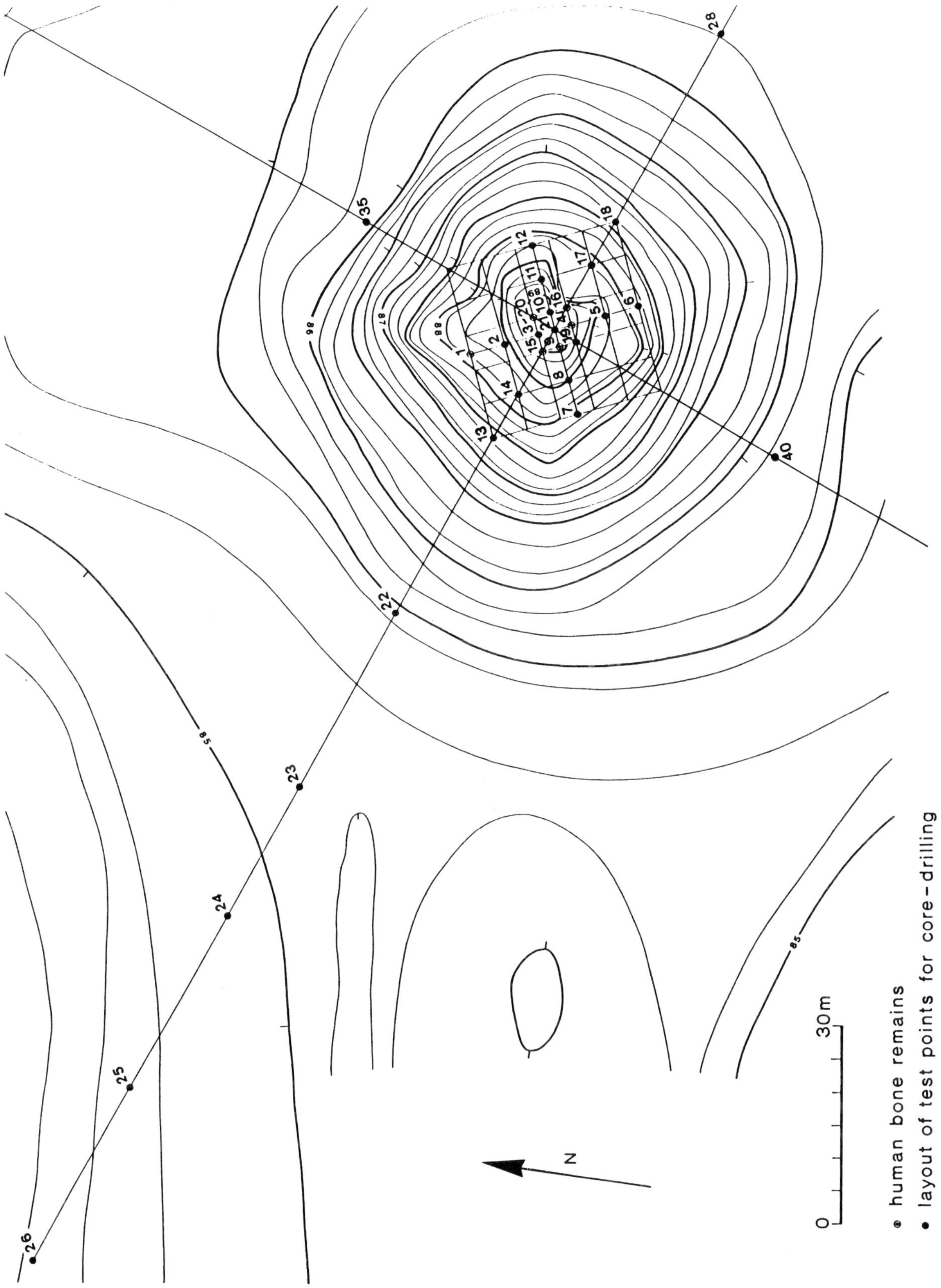

Fig. 25. Contour map of site Endrőd 18 (Papp-mound) with the positions of soil core sampling.

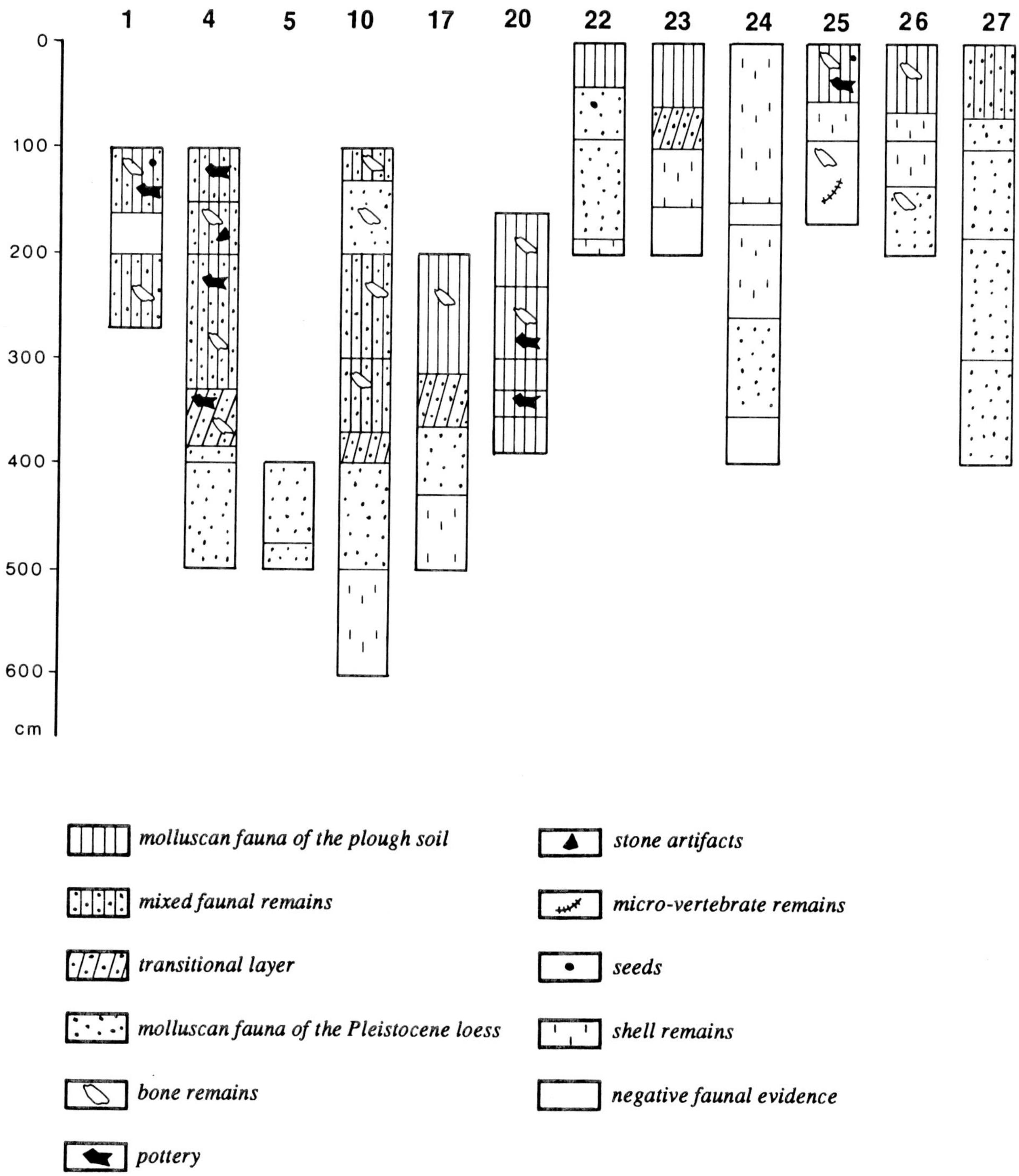

Fig. 28. Core samples showing the results of environmental studies in the Papp-mound area.

7. Conclusions

The combination of research methods employed within the framework of the Microregion Research Project is summarized below in tabulated format (*Fig. 29*).

ACTIVITY	SITE Name	GYOMA															ENDRŐD										SZARVAS
	Id. No.	116	117	121	125	126	127	128	133	238	254	255	256	259	260	264	18	19	20	33	41	50	62	124	125	126	46
1.	Fieldwalking	▲	▲	▲	▲	▲	▲	▲	▲	▲	▲	▲	▲	▲	▲	▲	▲	▲	▲	▲	▲	▲	▲	▲	▲	▲	▲
2.	Statistical evaluation	▲														▲			▲								
3.	Excavation								▲							▲											
4. Prospection	aerial photography	▲	▲	▲				▲	▲	▲				▲	▲	▲	▲	▲	▲	▲	▲	▲	▲	▲	▲	▲	▲
	soil core sampling		▲								▲	▲				▲	▲		▲	▲			▲				▲
	geophysical survey	▲														▲		▲	▲			▲					
5.	River bed boring															▲	▲		▲								
6.	Soil analysis		▲														▲		▲								
7. Palaeoecology	palynology		▲														▲										
	botanical macro remains		▲														▲										
	molluscan fauna		▲													▲	▲						▲				
	micro vertebrate fauna		▲													▲	▲										
	archaeozoology		▲						▲							▲											
	physical anthropology																▲						▲				

Fig. 29. Summary of activities at sites in the Gyoma and Endrőd areas.

Finally, we have attempted to check, in micro-scale, individual sites within specifically defined regional units.

We therefore applied different archaeometrical methods to understand first of all paleotopography as well as transformational and depositional processes (Schiffer 1987).

Prospecting is even more important, in the wider sense, preceding excavations to formulate the appropriate research strategy (Dincauze 1987, 317 ff., Allsop 1990, Taylor et al. 1990). Only multiple data sources can reveal an accurate as possible conceived image of the evolving landscape and changing environmental conditions throughout historical times.

References

Aitken, M. J., 1974. *Physics and Archaeology.*(2nd ed.) Clarendon Press, Oxford.

Allsop, J. M., 1990. Recent Applications of Geoprospection Techniques to Archaeology. In: *Archaeometry* '90, 667-676. Birkhäuser Verlag, Basel.

Barker, Ph., 1982. *The Techniques of Archaeological Excavation.*(2nd ed.) Batsford, London.

Barker, Ph., 1986. *Understanding Archaeological Excavation.* Batsford, London.

Becker, B., Billamboz, A., Dieckmann, B., Kokabi, M., Kromer, B., Liese-Kleiber, H., Rösch, M., Schlichtherle, H. and Strahm, C., 1985. Berichte zu Ufer- und Moorsiedlungen Südwestdeutschlands 2. *Materialhefte zur Vor- und Frühgeschichte in Baden-Württemberg.* Heft 7.

Brown, D., 1975. *Principles and Practice in Modern Archaeology.* Hodder and Stoughton, London.

Butzer, K. W., 1982. *Archaeology as Human Ecology: Method and theory for a contextual approach.* Cambridge University Press, Cambridge.

Clark, A., 1990. *Seeing Beneath the Soil. Prospecting methods in archaeology.* Batsford, London.

Crumley, C. L. and Marquardt, W. H., (eds.) 1989. *Regional Dynamics. Burgundian Landscape in Historical Perspective.* Academic Press, San Diego-New York-London.

Davidson, D. A. and Shackley, M. L., (eds.) 1976. *Geoarchaeology. Earth Science and the Past.* Duckworth, London.

Dincauze, D. F., 1987. Strategies for Paleoenvironmental Reconstruction in Archaeology. *Advances in Archaeological Method and Theory.* Vol.11. 255-336.

Dunnel, C. R. and Dancey, W. S., 1983. The Siteless Survey: A Regional Scale Data Collection Strategy. *Advances in Archaeological Method and Theory.* Vol.6. 267-287.

Ecsedy, I., 1979. *The People of Pit-grave Kurgans in Eastern Hungary.* Fontes Archaeologici Hungariae, Budapest.

Fieller, N. R. J., Gilbertson, D. D. and Ralph, N. G. A., (eds.), 1985. *Palaeoenvironmental Investigations. Research Design, Methods and Data Analysis.* BAR Int.Ser.258.

Gersbach, E., 1988. *Ausgrabungsmethodik und Stratigraphie der Heuneburg.* Heuneburgstudien VI. Römisch-Germanische Forschungen 45.

Gersbach, E., 1989. *Ausgrabung heute: Methoden und Techniken der Feldgrabung.* Darmstadt.

Gladfelter, B.G., 1981. Developments and directions in geoarchaeology. *Advances in Archaeological Method and Theory.* Vol. 6. 343-364.

Goldmann, Gy., 1983. Az alföldi vonaldíszes kerámia fiatal szakaszának leletei Békés megyében. (Funde der Jungen Phasen der Linienbandkeramik des Alföld in Komitat Békés) *ArchÉrt* 109, 24-34.

Harris, E. C., 1989. *Principles of Archaeological Stratigraphy.* (2nd ed.) Academic Press, New York and London.

Haselgrove C., Millett M. and Smith I. (eds.) 1985. *Archaeology from the ploughsoil: studies in the collection and interpretation of field survey data.* University of Sheffield.

Hodder, I. and Orton, C., 1976. Spatial analysis in archaeology. (New studies in archaeology), Cambridge.

Hrouda, B., (ed.) 1978. *Methoden der Archäologie. Eine Einfürung in ihre naturwissenschaftlichen Techniken.* Beck, München.

Huszár, M., 1822. *Hydrographia depressae Regionis fluviatilis Crisiorum...* (Maps in the Institute for Hydrography), Budapest.

Huszár, M., 1985. *Vízrajzi értekezés (Huszár Mátyás leírása a Körösvidékről).* Edited by F. Kósa and translated by P. Lakatos. Gyula.

Jankovich-B., D. and Pattantyús-Á., M., 1990. Archaeological and Geophysical Survey of a Multi-Aged Site Endrőd-170, SE-Hungary. *Prospezioni Arch.* 1/1990, 123-132. Rome.
Jerem, E., Kordos, L., Krolopp, E. and Vörös, I., 1984-1985. The archaeological and environmental investigation of the Iron Age settlement discovered at Sopron-Krautacker. *ArchÉrt* 111, 141-169; 112, 3-24.
Jerem, E., 1986. Bemerkungen zur Siedlungsgeschichte der Späthallstatt- und Frühlatènezeit im Ostalpenraum. (Veränderungen in der Siedlungsstruktur: archäologische und paläoökologische Aspekte). In: Hallstatt Kolloquium Veszprém. *MittArchInst* Beiheft 3, 107-118.
Jerem, E., Pattantyús-Á., M. and Varga, A., 1990. Application of Multiple Archaeometrical Methods in Prospecting Archaeological Sites. In: *Archaeometry '90*, 729-738. Birkhäuser Verlag, Basel.
Jerem, E. and Kiss, Zs., 1992. A short interim report on the 1990 excavation in *ArchÉrt* and *RFüz* (in press)
Jerem, E. and Somogyi, P., 1992. Zur statistischen Auswertung von Keramik aus Siedlungsobjekten. *ActaArchHung* 44, 161-192.
Jerem, E. and Varga A., 1992. Information provided by soil core sampling during preliminary survey of archaeological sites. (*forthcoming*)
Joukowsky, M., 1980. *A Complete Manual of Field Archaeology.* Prentice-Hall, Englewood Cliffs, New Jersey.
Kossack, G., Reichstein, J. and Harck, O., 1980. *Archsum auf Sylt.* Teil 1. Archäologische Geländeforschung 1963-1978. In: Römisch-Germanische Forschungen 39, 173-190.
Kossack, G., Behre, K-E. and Schmid, P.,(Hrsg.) 1984. *Siedlungen im deutschen Küstengebiet vom 5. Jh. v. Chr. bis 11. Jh. n. Chr.* Bd.1. Ländliche Siedlungen. Acta humaniora, Weinheim.
Kulhanek, O., 1976. *Introduction to Digital Filtering in Geophysics.* Elsevier Science Public Co., Amsterdam.
Lászlóffy, W., 1982. *A Tisza.* Budapest.
Macready, S. and Thompson, F.H., (eds.) 1985. *Archaeological Field Survey in Britain and Abroad.* Society of Antiquaries Occasional Papers 6, London.
Maráz, B., 1977. Délkelet-Magyarország La Tène korának kronológiai kérdései. *ArchÉrt* 104, 47-64.
Maráz, B., 1981. A szkítakori őslakosság Latène-kori továbbélése Kelet-Magyarországon. (On the survival of the autochtonous population of the Scythian Age in Eastern Hungary) . *A Janus Pannonius Múzeum Évkönyve,* Pécs. 26, 1981, 97-120.
Mike, K., 1991. *Magyarország ősvízrajza és felszíni vizeinek története.* Budapest.
Mommsen, H., 1986. *Archäometrie. Neuere naturwissenschaftliche Methoden und Erfolge in der Archäologie.* Teubner, Stuttgart.
Nikitsch, R., 1987 a. Photogrammetrische Methoden in der Archäologie. *AAustr* 71, 1-8.
Nikitsch, R., 1987 b. Luftbildarchäologie im Dienste der Bodendenkmalpflege. *AAustr* 71, 9-18.
Nikitsch, R., 1989. Datenerfassung und Datenmanipulation in der Luftbildarchäologie. *AAustr* 73, 13-16.
Papp, M., 1969. A Körösvidék: a felszín kialakulása és mai képe. In: Pécsi, M., (ed.) 1969. *A tiszai Alföld.* 270-274.
Parkes, P. A., 1986. *Current Scientifique Techniques in Archaeology.* London-Sydney.
Pattantyús-Á. M., 1986. Geophysical Results in Archaeology in Hungary. *Geophysics* 51, 561-567.
Pécsi, M., (ed.) 1969. *A tiszai Alföld.* Budapest.
Pécsi, M., and Sárfalvi, B., 1977. *Phisical and Economic Geography of Hungary.* Budapest.

Riley, D. N., 1987. *Air Photography and Archaeology*. Duckworth, London.
Rottländer, R. C. A., 1983. *Einführung in die naturwissenschaftlichen Methoden der Archäologie*. Archaeologica Veneratoria Bd.6, Tübingen.
Schiffer, M., 1987. *Formation processes of the archaeological record*. University of New Mexico Press. Albuquerque.
Schlichterle, H., Bohrungen und Oberflächenbegehungen in der Ufersiedlung Hegne--Galgenacker, Kr. Konstanz. In: Billamboz, A., Czarnowsky, E., Jacomet, S., et alii 1990. *Siedlungsarchäologie im Alpenvorland II*. Forschungen und Berichte zur Vor- und Früchgeschichte in Baden-Württenberg. Stuttgart, 181-198.
Schofield, J., (ed.) 1991. *Interpreting Artefact Scatters: Contribution to Ploughzone Archaeology*. Oxbow Monographs in Archaeology 4. Oxford.
Shennan, S. J., 1985. *Experiments in the collection and analysis of archaeological survey data: the East Hampshire survey*. Sheffield University Press.
Sherratt, A. G., 1982. The development of Neolithic and Copper Age Settlement in the Great Hungarian Plain. Part I. The regional setting. *OxfJA* 1/3, 287-316.
Sherratt, A. G., 1983 a. The Development of Neolithic and Copper Age Settlement in the Great Hungarian Plain. Part II: Site survey and settlement dynamics. *OxfJA* 2/1, 13-41.
Sherratt, A. G., 1983 b. Early agrarian settlement in the Körös Region of the Great Hungarian Plain. *ActaArchHung* 35, 155-169.
Stein, J. K., 1986. Coring Archaeological Sites. *American Antiquity* 51/3, 505-527.
Szénánszky-G., J., 1979. A korai szakálháti csoport települése Battonyán. (The settlement of the Early Szakálhát group at Battonya) *ArchÉrt* 106, 67-77.
Taylor, J. J., Innes, J. B. and Jones, M. D. H., 1990. Integrating Geophysical and Palynological Survey in Wetlands. In: *Archaeometry '90*, 821-830. Birkhäuser Verlag, Basel.
Varga, A., 1989. Talajszondázás egy szarmata lelőhelyen. (Örménykút 52) *A Móra Ferenc Múzeum Évkönyve / Jahrbuch des Móra Ferenc Museums*, Szeged (Ungarn)(in press).
Waldhauser, J., 1976. Topographie der keltischen Besiedlung im Erzgebirgsvorland. *ArchRozh* 29, 294-314.
Waldhauser, J., 1984. Mobilität und Stabilität der keltischen Besiedlung in Böhmen. In: *Studien zu Siedlungsfragen der Latènezeit*. Veröffentl. Vorgesch. Seminars Marburg, Sonderband 3, 167-186.
Waldhauser, J., 1986. Strukture und Ökologie der keltischen Besiedlung während der Stufen Ha D – LT D in Böhmen. *MittArchInst* Beiheft 3, 267-275.
Webster, G., 1974. *Practical Archaeology. An Introduction to Archaeological Field Work and Excavation*. Adam and Black, London.
Wilson, D. R., 1982. *Air Photo Interpretation for Archaeologists*. Batsford, London.

Interdisciplinary survey and trial excavation of a multicomponent site, Endrőd 170

DÉNES JANKOVICH-B. - JUDIT KVASSAY - MIKLÓS PATTANTYÚS-Á.

Abstract

A comparative archaeological and geophysical survey was carried out between 1987-1989 at this site. A detailed surface collection as well as geophysical prospecting was performed at the plot. The geophysical study, using various methods (both magnetic and electronic) was checked later by boring. Essential conclusions concerning the applicability of the individual methods can be drawn from a comparison of the results. Excavation was carried out on three separated parts of the site during 1989. Eleven pits and daub-construction came from the Sarmatian period while 1 pit-house, 7 pits, 2 kilns and an open-air fireplace from the Árpád Period were excavated.

The site

The archaeological site where the investigations were carried out is located on a ca. 1 m high slope bordered on the north by an ancient river-bed, separating it from another archaeological site (see the introductory summary map). To the west is a dirt road, while to the south a canal marks the boundary of the site. Here, along the riverbank is a strip about 400 m long where sherds were found on the surface (*Fig. 1*).

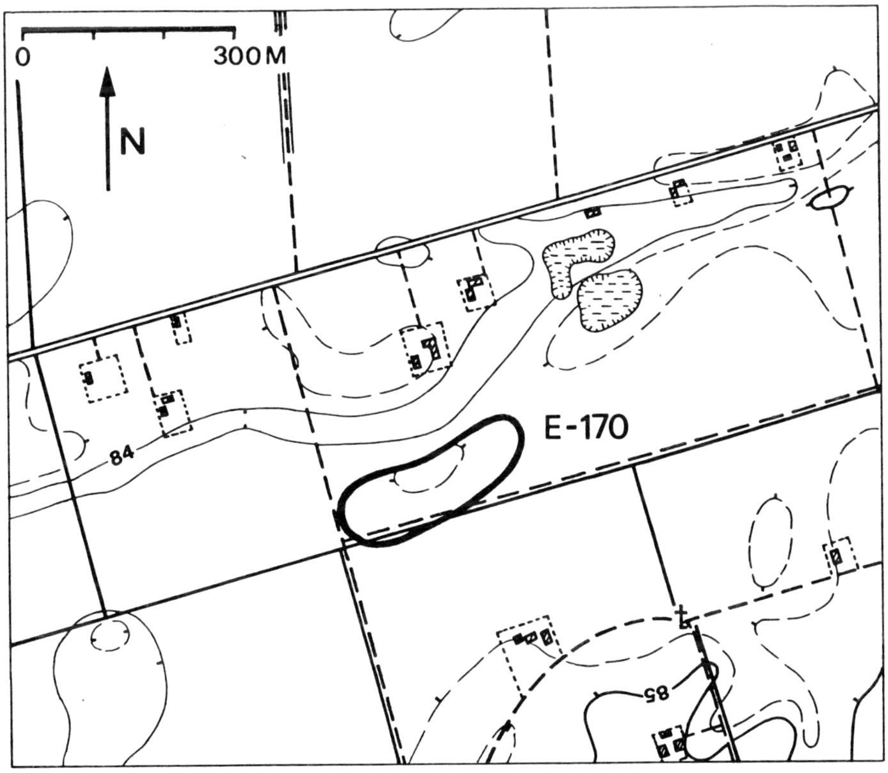

Fig. 1. Schematic map of the archaeological site Endrőd 170.

The site was discovered during the topographical field surveys of 1978. During the surface examinations which were carried out, sporadic Late Neolithic finds and numerous sherds from the Sarmatian (3rd-5th c. A.D.) and Avar (8th-9th c. A.D.) periods as well as ceramic fragments dated to the period of the Árpád dynasty (10th-13th c.) were collected from the surface.[1]

Aims

Our aim was to improve the accuracy of the data obtained during the 1978 surface find collection in order to accurately define the boundaries and central areas of the site based on surface finds and geophysical measurements as well as to justify the results thus obtained by trial excavation.[2]

Applied methods

To accomplish these aims we carried out an intensive, detailed surface collection over two seasons, partial geophysical prospecting of the same sections, some trial boring at the most interesting points, aerial photography and a short trial excavation to check all the data from these examinations.

Surface collection

The work began in 1987. A 5 by 5 m square grid was laid on the area and a 10 m wide strip was chosen from it which followed the direction of ploughing and traversed the whole site in an East-West direction. Every surface find which could be identified (pottery, daub, bone, stone, charcoal, slag, etc.) was picked up and its location plotted on a map. In the short time allotted to us we surveyed an area of 1000 m^2 in this manner (squares XI-XII/20-40, *Fig. 2*). In 1988, a significantly larger area of 9450 m^2 was surveyed within this area, the collection of the previous year was repeated on 200 m over the same area for the purpose of checking. The method remained the same: all finds were mapped and collected. The subareas (strips) chosen for collection represent the central, most intensively occupied parts of the site and clearly indicate the boundaries of the site. These coincide within a few metres with the site boundaries determined during the first comprehensive field survey. These had been located initially from the topographic conditions.

In *Fig. 2*, the area investigated by surface collection (1), and the area of geophysical measurements (2) can be seen. Those areas where both types of survey were carried out are also marked (3) as well as the places of trial excavation (4). Some of the results from these investigations have already been published (5).[3]

---

1 Jankovich-B., D., Makkay J. and Szőke B. M.,: *Magyarország Régészeti Topográfiája 8*. kötet: Békés megye régészeti topográfiája IV/2. A szarvasi járás. Szerk. Makkay, J. Bp. 1989, 176. (The Archaeological Topography of Hungary. Vol. IV/2. Békés County, Szarvas district. Edited by J. Makkay.)

2 The results of these investigations are published in a preliminary report: Jankovich-B., D. and Pattantyús-Á., M.; Archaeological and Geophysical Survey of a Multi-Aged Site Endrőd-170, SE-Hungary. *Prospezioni Archeologiche*, 1/1990, 123-132, Rome. See also Jankovich-B., D., Pattantyús-Á., M. and Varga, A.; Geoarchaeology - An Integrated Survey before Excavation at Archaeological Sites. *Papers of the Dallas-Conference 1989*, 273-275.

3 See footnote 2.

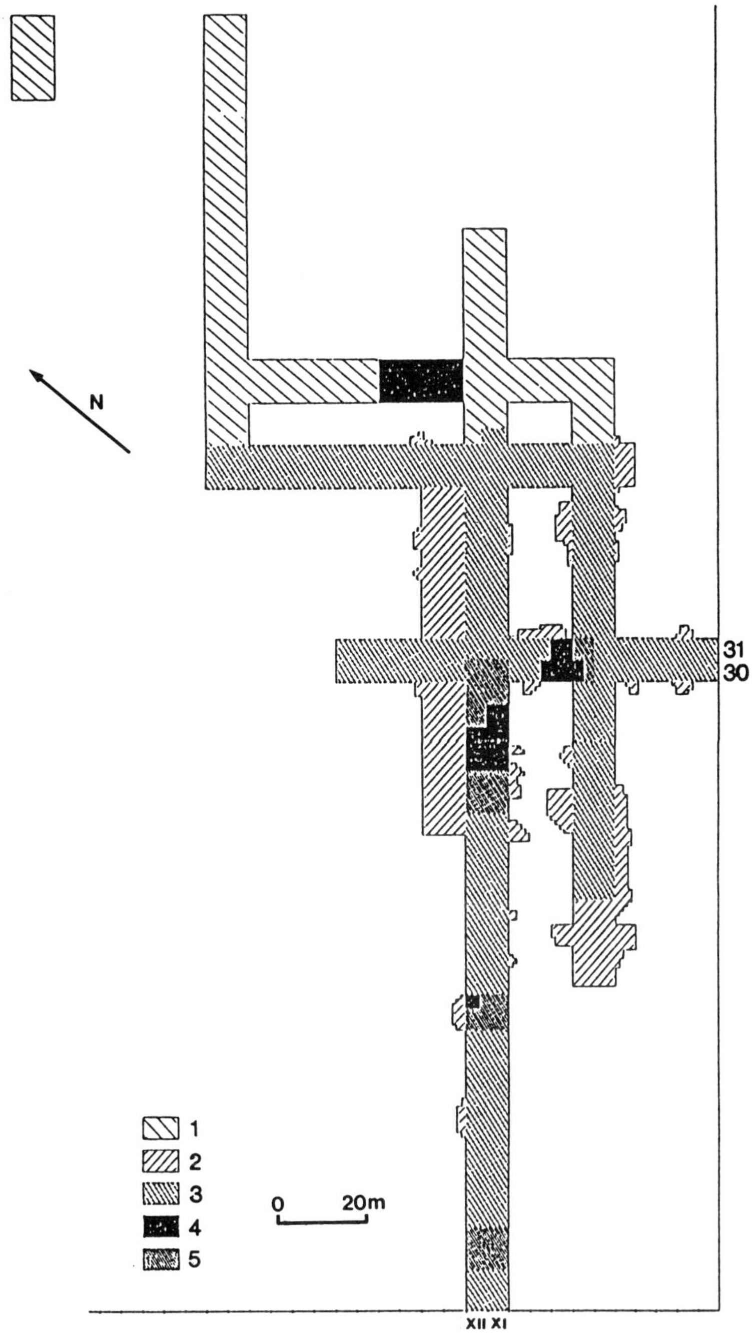

Fig. 2. Map of the surveyed area Endrőd 170.
1: area surveyed by surface collection of finds
2: area of magnetic measurements
3: area investigated by both methods
4: excavated area
5: areas published previously (see footnote 2.)

The concentration of the objects delineates the central portion of the site over two large spots. These can also be seen on the aerial photograph. Here the total number of finds per square (5 by 5 m) exceeds 50 (50-100 pieces/square). The majority of these objects are pottery fragments from the Sarmatian period which suggests that human activity was most intensive during this period. The average find density is about 37 pieces/square, while approaching the boundaries of the site the number of finds falls off to below 10 (*Fig. 3*).

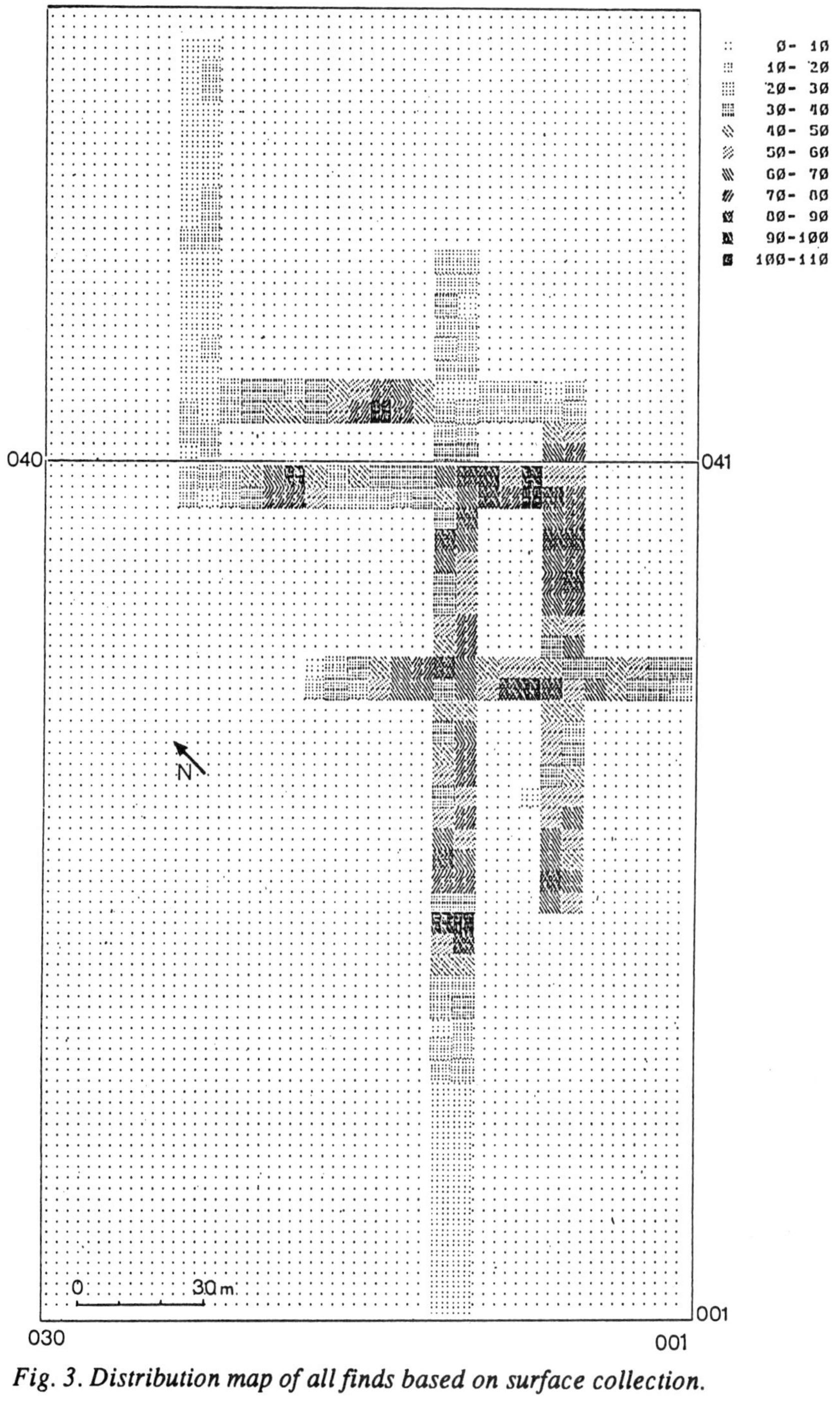

Fig. 3. Distribution map of all finds based on surface collection.

Some significant discrepancies, not only in the amount of all the finds but also in the distribution of the pottery from different periods (*Fig. 4*), or in the number of daub fragments (average: 9/square, see *Fig. 5*). can be observed. It is also clear that the weight of the finds marks the central places of the site (average: 210 gs/square, see *Fig. 6*).

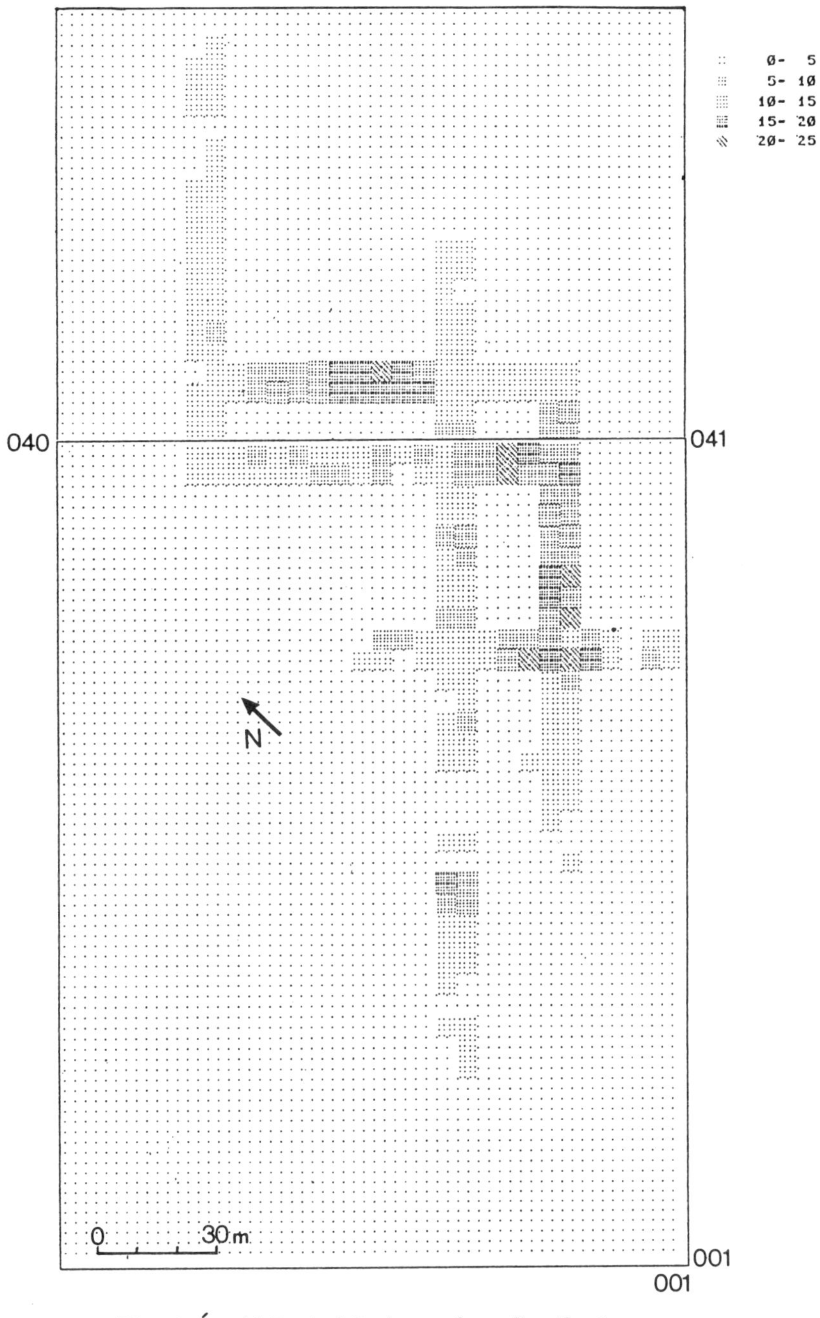

Fig. 4. Árpád Period finds: surface distribution.

There was a hope, even without any proportional calculations and before the geophysical prospecting, that the quantity and the body of surface finds would mark the centre of the site and that the changes in density would allow location of the main structures. This was rendered possible by a grid of relatively small squares (5 by 5 m) since with larger squares these differences would be obscured. Moreover, the presence of groups of finds belonging to the Árpád Period offered an opportunity to unearth constructions of this period. The finds from other periods were not defined separately during this work but most belonged to the Sarmatian period.

At the same time, naturally, there is no hope of locating the structures within the 5x5 m squares based on surface finds. However, it is possible to do so from the sherd maps from a -30 cm depth.

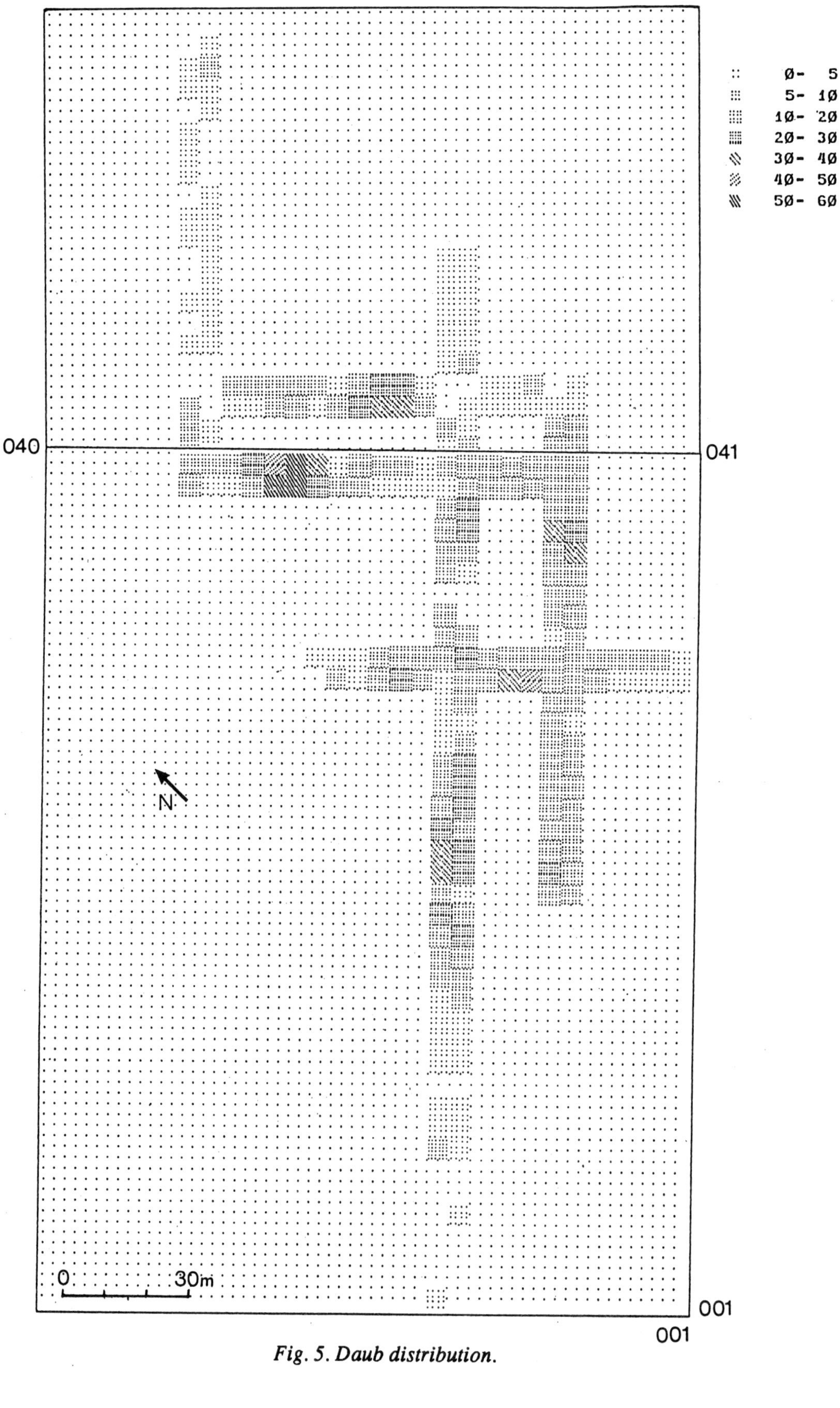

Fig. 5. Daub distribution.

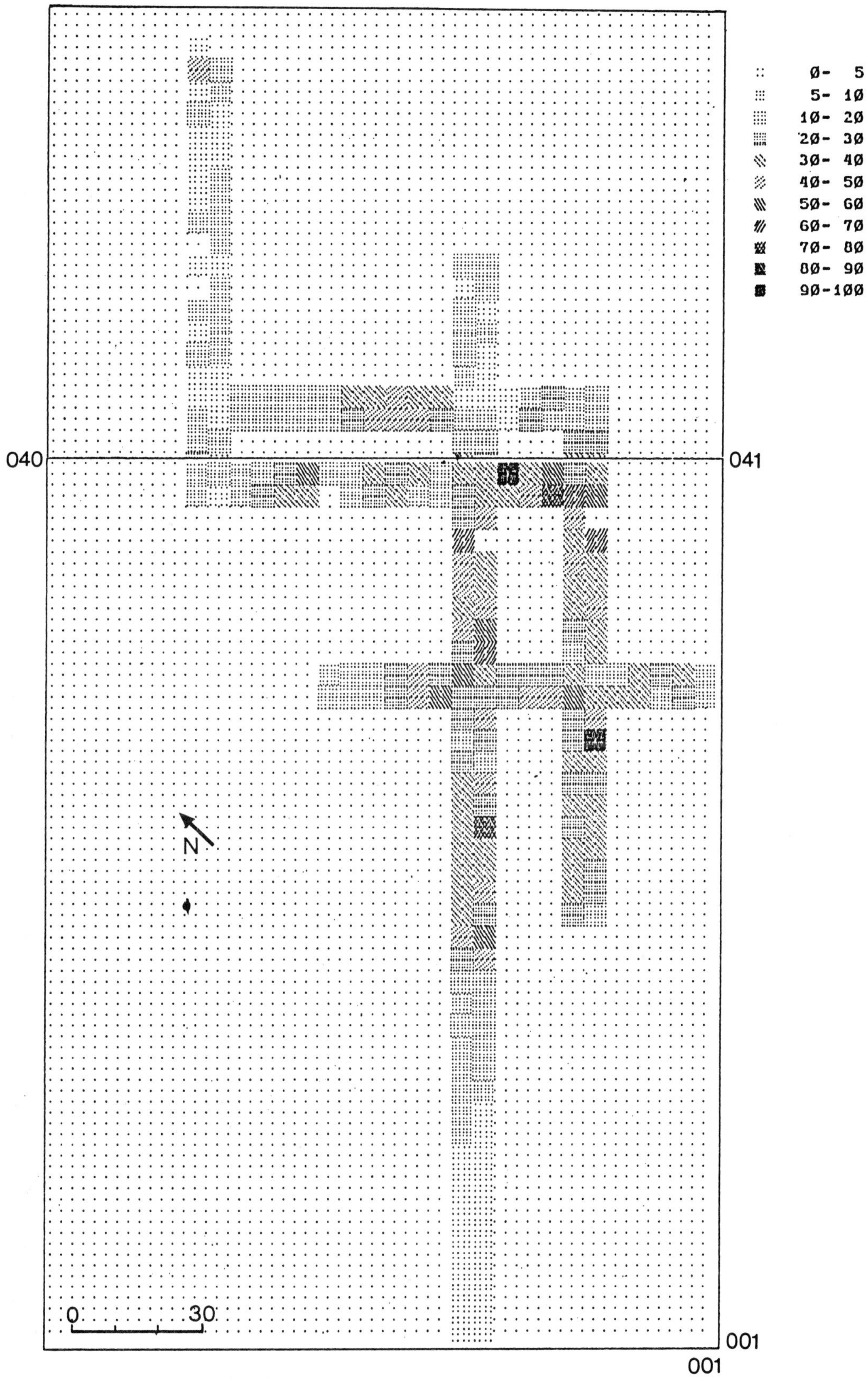

Fig. 6. The weight distribution of all finds.

Geophysical survey

In this paper we would like to compare results from the collection and excavation with the results of the geophysical survey. Thus, distribution maps from some representative areas of the surface collection are presented together with the corresponding magnetic anomaly maps (*Fig. 2:5*). Those areas have been chosen for presentation where high values in both the archaeological material and magnetic measurements indicate a probable feature (find number over 30, magnetic anomaly stronger than 20 nT *Figs. 7, 8*); where only the magnetic measurement provided anomalous values (*Fig. 9*); and finally where both investigations brought negative results (*Fig. 11*).

Table 1. Find distribution in the examined squares

Square	Sherd	Daub	Stone	Slag	Bone	Total
XI/3	7	-	-	-	-	7
XII/3	2	-	-	-	-	2
XI/4	1	-	-	-	-	1
XII/4	2	-	-	-	-	2
XI/14	12	5	-	-	-	17
XII/14	5	2	1	1	-	9
XI/15	23	6	-	-	3	32
XII/15	25	2	1	-	-	28
XI/24	49	21	-	-	6	76
XII/24	37	20	-	-	3	60
XI/25	23	24	-	-	5	52
XII/25	26	6	-	-	6	38
XI/26	45	21	-	-	6	72
XII/26	41	6	-	-	2	49
XI/27	54	7	-	-	7	68
XII/27	33	3	-	1	4	41
XI/28	46	5	-	1	5	57
XII/28	31	4	-	-	5	40
XI/2931	31	11	-	-	2	44
XII/2	31	3	1	2	2	39
XI/30	43	12	-	-	3	58
XII/30	44	6	-	-	4	54
VII/30	47	15	-	-	8	70
VII/31	34	16	-	-	3	53
VIII/30	50	43	-	-	3	96
VIII/31	34	20	-	1	3	58

It is also evident from the Table that an extremely high find density (more than 60 pieces) can be observed in those squares which cover a portion of the high magnetic values. At the same time, on other sections such a coincidence does not occur (e.g. square XII/14. in *Fig. 9*, or squares XII/26-27 in *Fig. 8*).

Magnetic measurements

Magnetic measurements were carried out using a Geometrics G-816 proton-resonance magnetometer at a sensor height of 0,6 m, in a 1 by 1 m grid. The surveyed area is about 7000 m^2 (*Fig. 2*), but only characteristic parts of it are shown. Experimental resistivity measurements were also carried out over a smaller area to investigate the detectability of magnetic bodies by electronic methods. In the same area we attempted to identify these magnetic bodies using an auger.

Fig. 12 shows the schematically simplified magnetic anomaly map of the surveyed area. Higher density of magnetic bodies coincides with both the above-mentioned areas of high find density and with the promising area indicated by topographic conditions. A significant number of bodies (features) can be found, however, in the area which was described as located outside the settlement during the original field survey (*Figs. 9 and 10*). The intensity of the anomalies lies between 5 and 40 nT in this area and in some places a negative anomaly is associated with the positive peak.[4] In *Fig. 12*, only the isolines of the most significant anomalies are plotted. Based on the position of these anomalies, the borders of the promising area and extent of the settlement can be estimated.

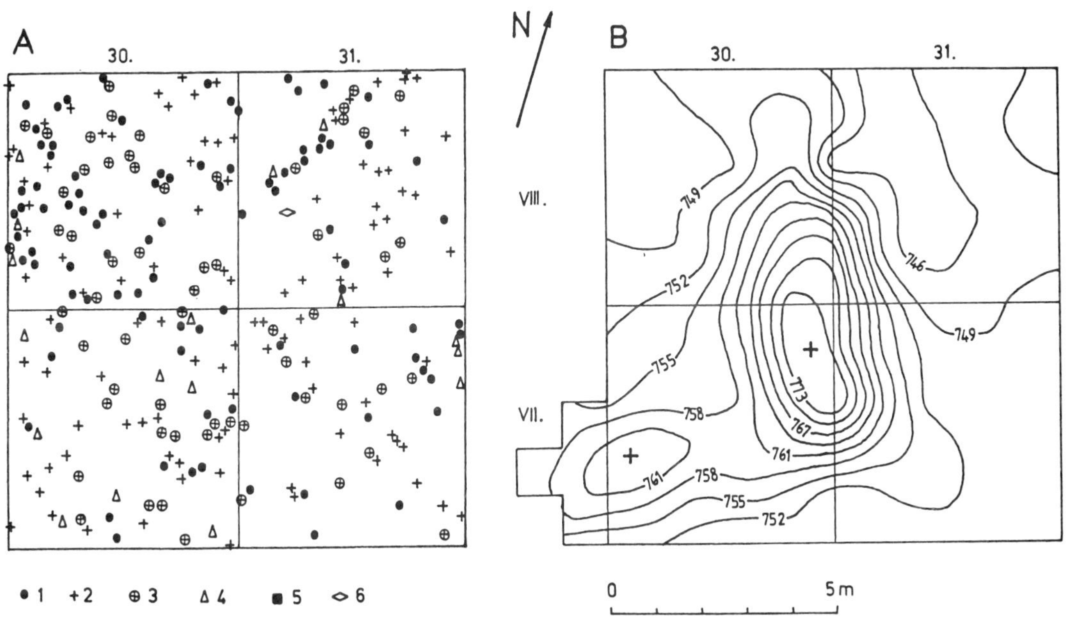

Fig. 7. Scattering of the surface finds (A) and magnetic anomaly map (B) of squares VII-VIII/30-31.
1: daub — *4: animal bone*
2: sherd — *5: stone*
3: pottery from the Árpád Period — *6: slag*
The high density of surface finds and the magnetic anomaly agree well with each other.

The positive anomaly in *Fig. 7* which consists of two peaks with maximum intensities of 15 nT and 25 nT respectively, is supported by a very high density and number of finds, 60-90 pieces/square. Thus, the approximate location of the features can be determined here without magnetic measurements but from the surface finds alone.

---

[4] Pattantyús-Á., M., Geophysical results in archaeology in Hungary, *Geophysics*, 51:3, 1986, 561-567.

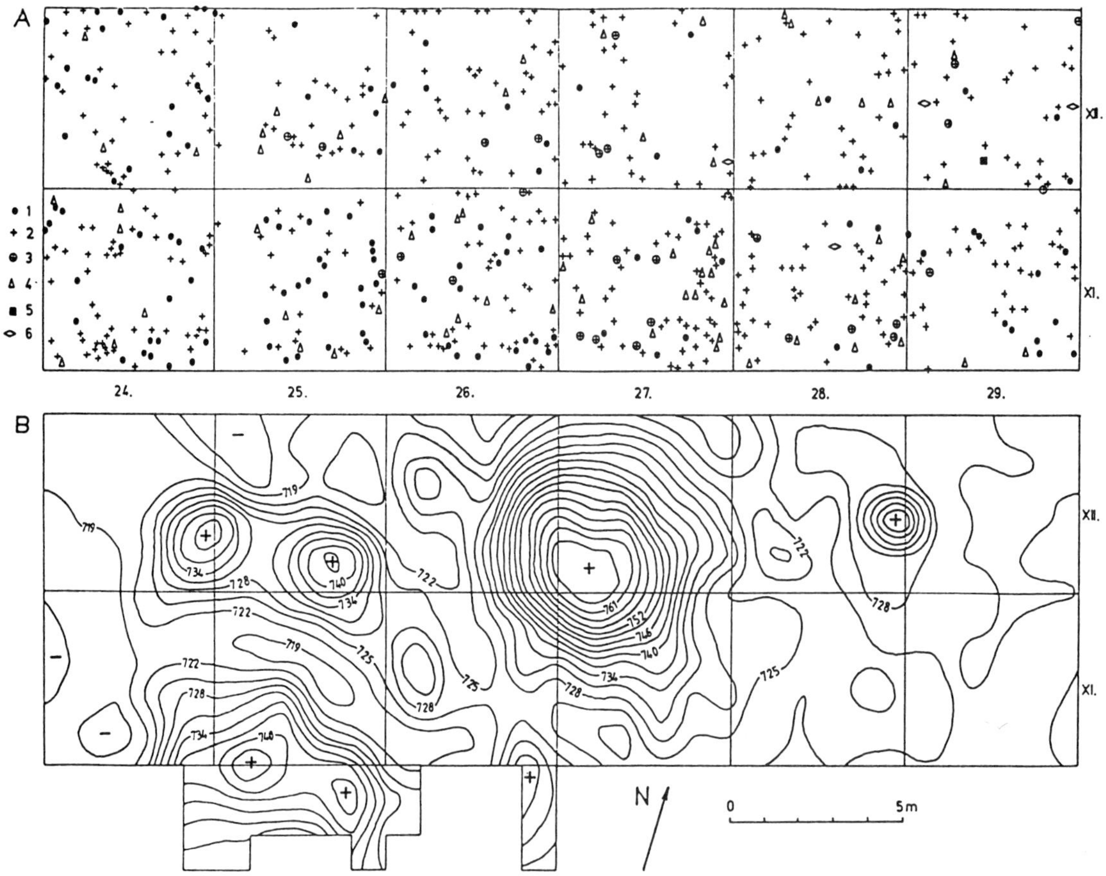

Fig. 8. Scattering of finds on the surface (A) and magnetic anomaly map (B) of squares XI-XII/24-29. (Legend: see Fig. 7.)
Magnetic anomalies indicate the location of the features more exactly than surface find density.

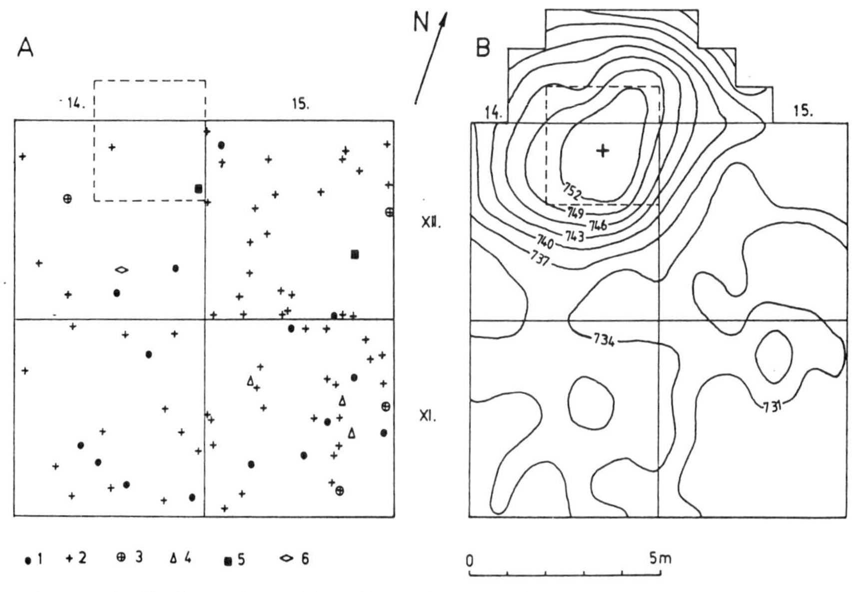

Fig. 9. Scattering of the finds on the surface (A) and magnetic anomaly map (B) of squares XI-XII/14-15. (Legend: see Fig. 7.)
The excavated area is marked by a dashed line. Only the magnetic anomaly provides information on the feature. The surface find density does not have a high value at the place of the anomaly.

It is not so simple to draw conclusions from the maps shown in *Fig. 8*. A high density of finds is shifted away from the magnetic bodies in some places. The intensity of the anomalies is 20-40 nT while find density is 40-70 pieces/square in the western two-thirds of the area and 40-50 pieces/square in the eastern part. From the magnetic results it can be inferred that the objects are located in the western part of the subarea. The density of surface finds, however, does not always support this.

The maps of *Fig. 9* demonstrate that neither the topography nor the surface finds indicate the features everywhere. Control excavation carried out on the positive anomaly of 20 nT uncovered a fireplace within a probable house (*Fig. 10*). This kind of fireplace was found at another archaeological site by means of a resistivity survey later supported by electromagnetic measurements.[5] At the same time, the number of finds is only 10-30 per square (see *Table 1*), and this area lies about 40 m from the topographically promising area.

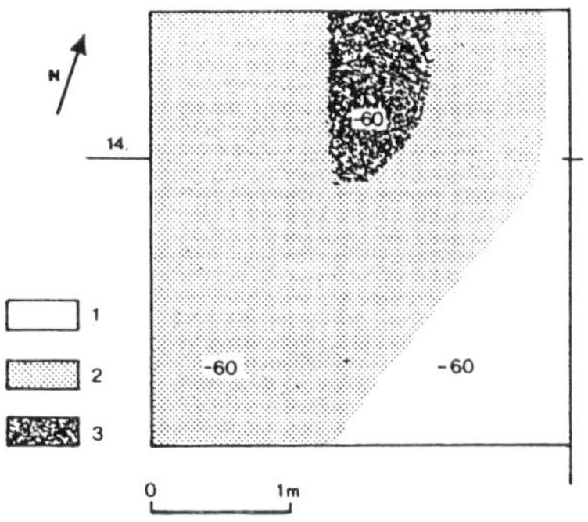

Fig. 10. Picture of the feature excavated in square XII/14.
1: yellow subsoil
2: feature (house?) dug into the subsoil
3: remains of a heavily burnt fireplace.

Fig. 11 demonstrates that the negative results produced by the two methods can provide support for each other as well. In this subarea there is no significant magnetic anomaly present while the number of surface finds is insignificant. Thus, this subarea does really lie outside the promising part.

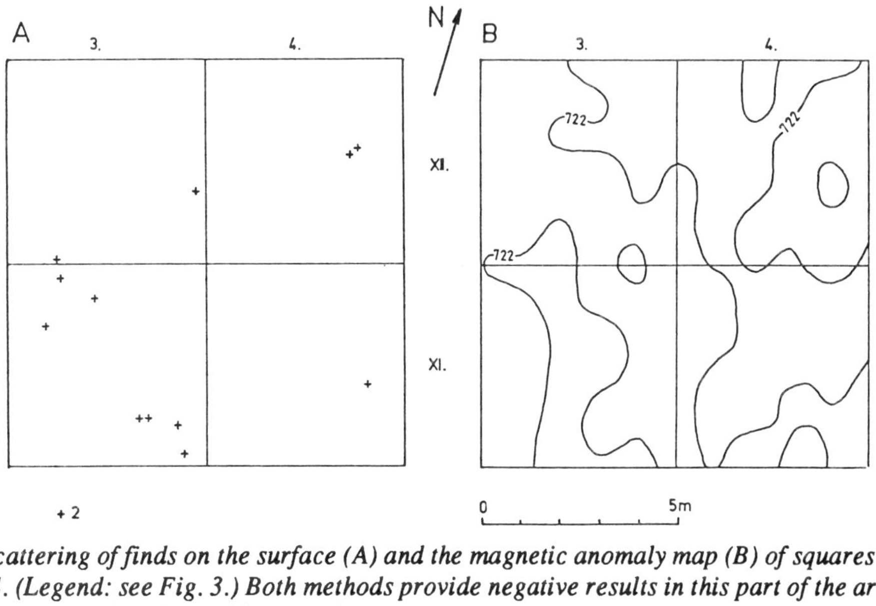

Fig. 11. Scattering of finds on the surface (A) and the magnetic anomaly map (B) of squares XI-XII/3-4. (Legend: see Fig. 3.) Both methods provide negative results in this part of the area. There are probably no archaeological features here.

---

5 Pattantyús-Á., M. in: *Annual Report of the Eötvös Lóránd Geophysical Institute of Hungary for 1986*, 215-216.

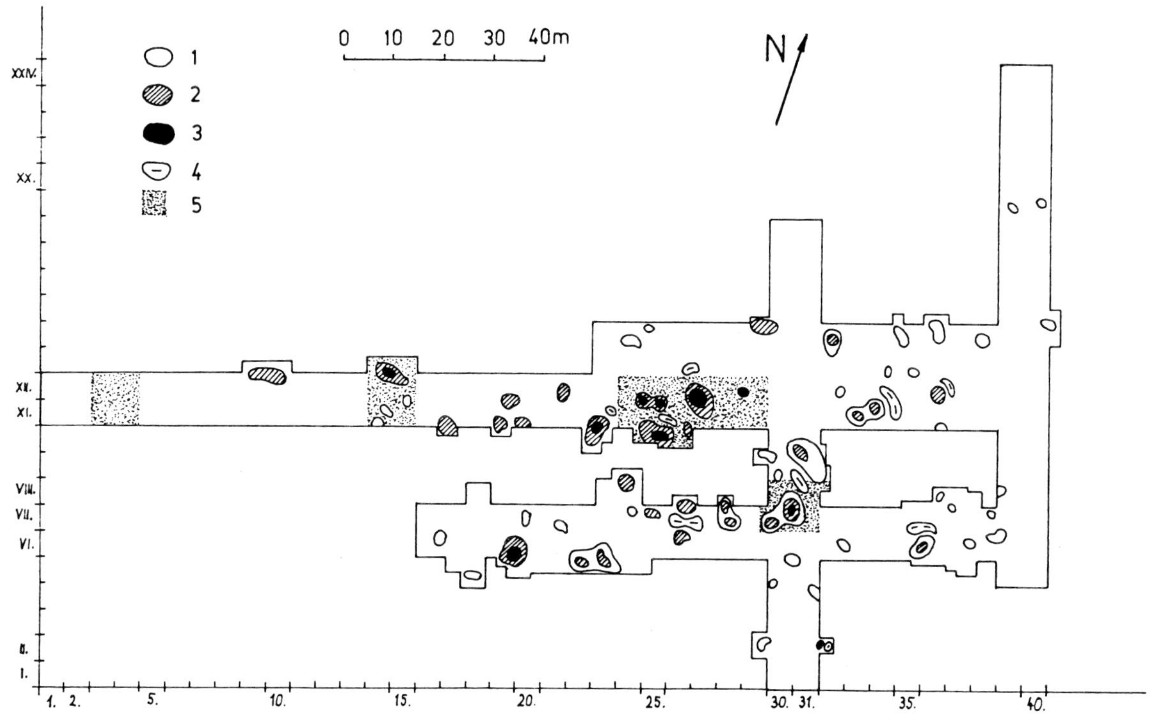

Fig. 12. Simplified magnetic anomaly map of Endrőd 170.
1: anomaly weaker than 10 nT
2: anomaly between 10 and 20 nT
3: anomaly stronger than 20 nT
4: negative anomaly (weaker than 10 nT)
5: areas discussed in detail and shown in other figures.

Comparative measurements

Resistivity measurements were carried out in the subarea shown in *Fig. 8* in order to detect objects using this method too. Measurements were taken with the dipole-dipole array, using an electrode separation of 1 m and several dipole separations.[6] The instrument was a Hungarian-made dc resistivity metre DIAPIR-10R. The results are presented in *Fig. 13*. Here the magnetic map and the apparent resistivity maps, corresponding to different depths in the same area, can be seen one below the other. Locations of auger borings along the profile are plotted as well.

Conclusions that can be drawn from the maps on *Fig. 13* are:

– Magnetic bodies can also primarily be traced by conparing resistivity and magnetometry measurements (bodies A, B and C). Body D could be a near-surface iron object because it does not cause a resistivity peak, in fact the resistivity decreases at a greater depth;

– The resistivity anomalies (i.e. the archaeological features) appear shifted northwards from the magnetic anomaly. The shift is directly proportional to their depth (the shift of A and C is greater while that of B is smaller);

---

[6] Csathó, B. and Pattantyús-Á., M., Possibilities and experiences of geophysical prospecting at archaeological sites in Hungary, in: *Archaeometrical Research in Hungary*, National Centres of Museums, Budapest 1988, 13-27.

– The depth of the bodies can be estimated from the maps. C lies deepest, A appears on the resistivity map at a depth corresponding to 2.5 m, while body B disappears at that depth, i.e. it is shallower. The same conclusion may be drawn from the approximate depth calculations of the magnetic anomalies.

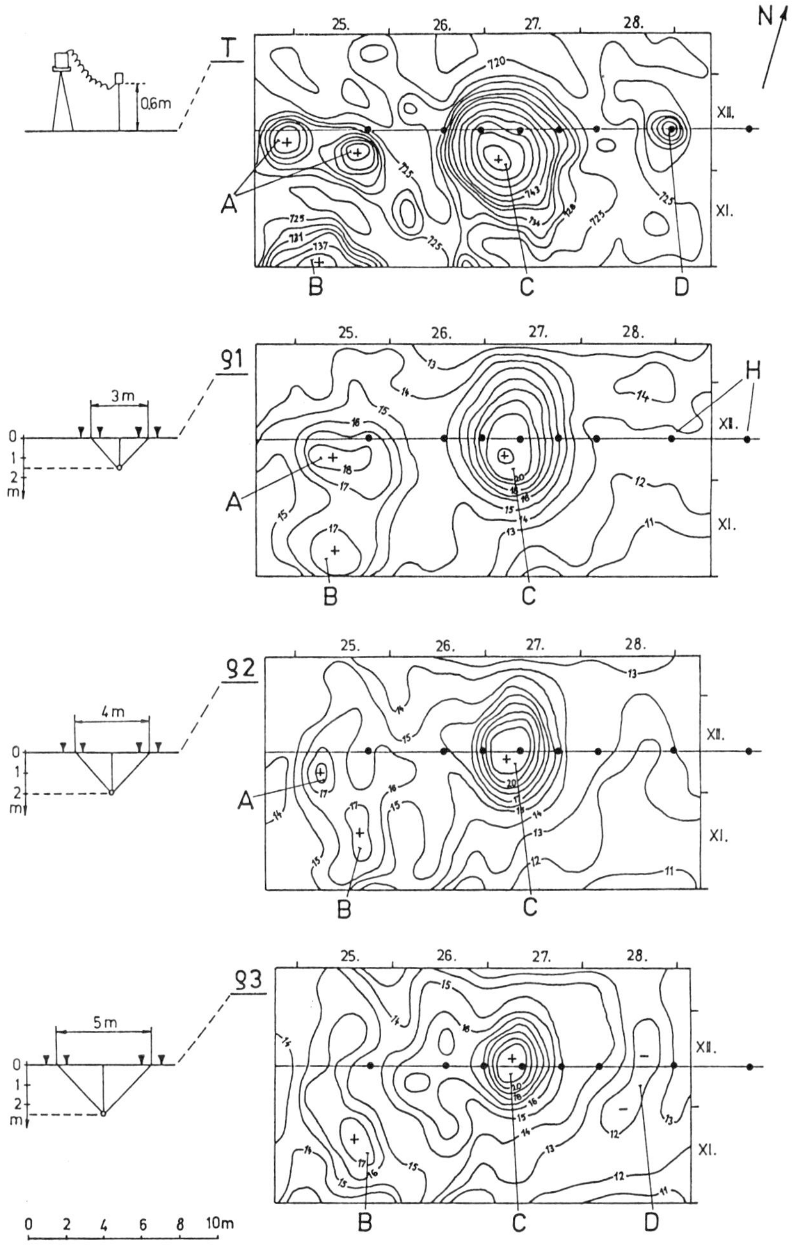

Fig. 13. Anomaly maps of a selected area of Endrőd 170.
T: anomaly map of the total magnetic field
1: resistivity map at a depth of 1,5 m
2: resistivity map at a depth of 2,0
3: resistivity map at a depth of 2,5 m
A, B, C, D: characteristic magnetic anomalies also recognizable on the resistivity maps.
H: boreholes
Magnetic measurements were carried out with a sensor height of 0,6 m. Resistivity measurements were performed in a dipole-dipole array with different dipole separations. Auger borings were made along the profile marked on the maps (see Fig. 14).

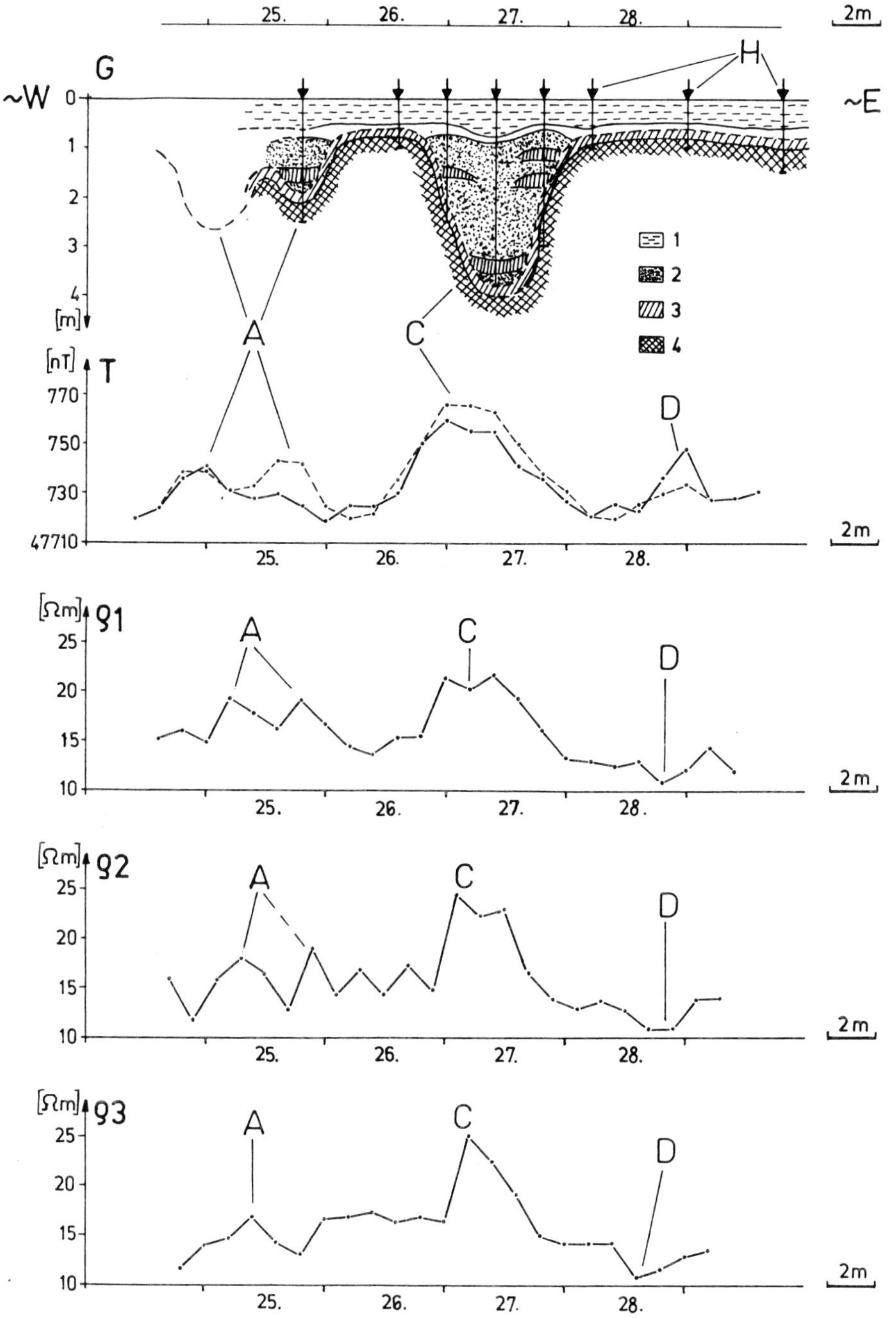

Fig. 14. Geological, magnetic and resistivity data along the profile marked in Fig. 13.
G: geological depth section constructed from auger borings
1: upper soil (humus)
2: burnt porous soil
3: mixed (transitional) soil
4: yellow subsoil
T: magnetic profile
1: resistivity profile at a depth of 1,5 m
2: resistivity profile at a depth of 2,0 m
3: resistivity profile at a depth of 2,5 m
A, C, D: magnetic and resistivity anomalies (the same as in Fig. 13).
H: boreholes

From the comparison of the individual methods their characteristic features can be quite clearly established: resistivity measurement failed to reveal the double nature of body B. In fact, at greater depths, even bodies A and B are hardly separable. On the magnetic map, on the other hand, the bodies can be clearly discriminated. Conversely, the magnetic map indicates a body D, which very likely has no archaeological explanation. It could not be exposed by drilling (A drilling located about 1 m north of the profile also yielded no results).

Fig. 14 shows the geological depth section constructed from the auger borings placed 2 and 4 m apart along the profile marked on the maps in *Fig. 13*, the magnetic profile measured along here (or rather 1 m to the south) and resistivity profiles at three different depths. Conclusions can be drawn which are similar to those from *Fig. 13*. Here, it can be seen that object C lies at the greatest depth (around 4 m) which agrees with the 150-180 cm of the central-part depth obtained from the magnetic depth calculation. The central-part depth of the eastern portion of body B is 80-120 cm from the magnetic depth calculation with its deepest part found at about 2 m according to the boring. At the place of magnetic anomaly D a negative anomaly appears in the resistivity profiles.

Excavation

In 1987, simultaneously with the magnetic measurements, a half-day trial excavation was carried out at the place where the magnetic anomalous value was especially high (*Fig. 9*). Here at a depth of 60 cm, the spot of a feature (*Fig. 10:2*) appeared dug into the yellow subsoil (1). Within it remains of a heavily burnt fireplace (3) were found. Measurements exactly indicated both the feature (pit, house?) and location of the fireplace. This feature was not excavated.

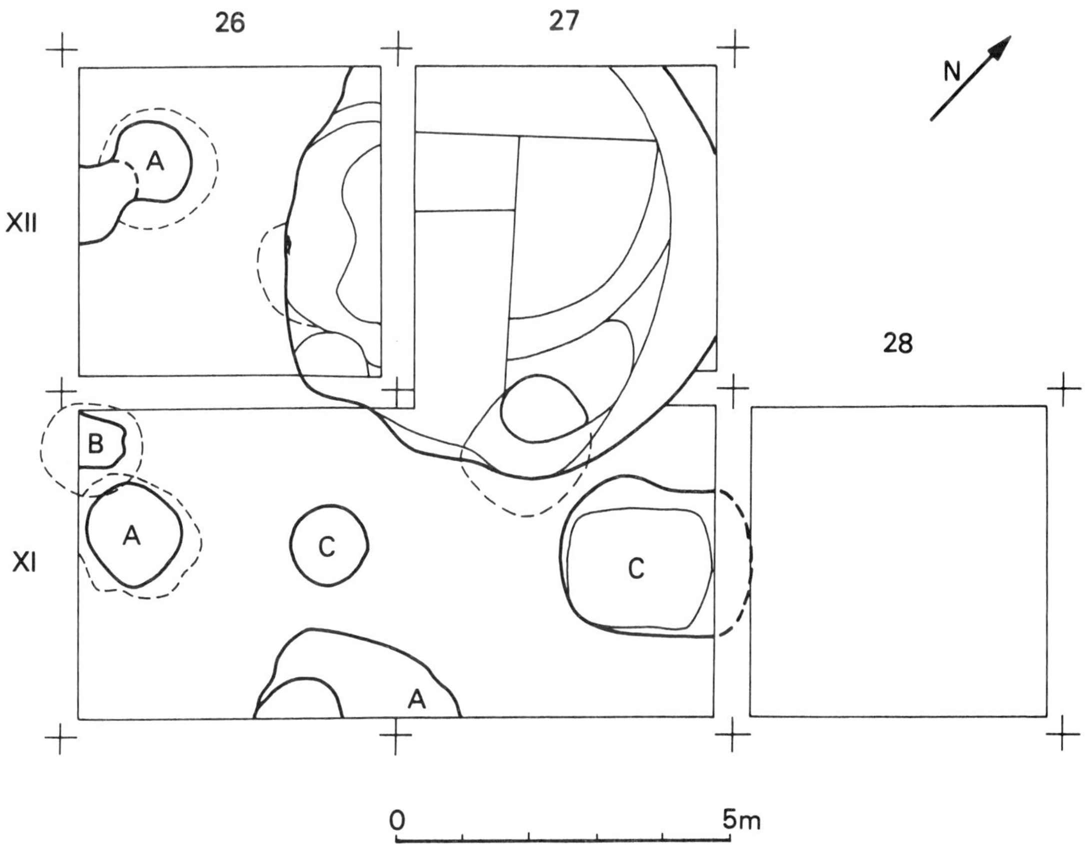

Fig. 15. Compiled drawing of the archaeological features uncovered in square XI-XII/26-28.

Excavations were started at three spots in August of 1989. (*Fig. 2*)[7] That area was first chosen where the surface collections, the geophysical survey, the boring and the aerial photography all promised significant - surely Sarmatian - finds (sections XI/26-XII/27). There were also many finds on the surface (especially daub and sherds from various periods) while the magnetic anomaly displayed high values at the second spot (sections VII/30-31). No geophysical prospecting was carried out at the third spot, although, the density of the surface finds (Árpád Period), was greater than average (*Fig. 4*).

The following structures were found in the rows of sections XI-XII (*Fig. 8: 15-16*):

Section XI/26: Pit "A" a Sarmatian beehive pit in the uppermost layer in which there were, nevertheless, sherds from the Árpád Period. It contained a large amount of pottery.

Pit "B", also a Sarmatian beehive pit had its swelling section cut by pit "A". Its fill consisted of alternating yellow and dark brown, loose layers, and contained fewer finds than that of the previous pit. There was a few mm thick layer of soot on the bottom while the sides were red and slightly burnt up to a few centimetres from the bottom.

Pit "C" was a shallow, barely 30 cm deep pit from the Árpád Period containing a halved sheep and a halved cattle skull.

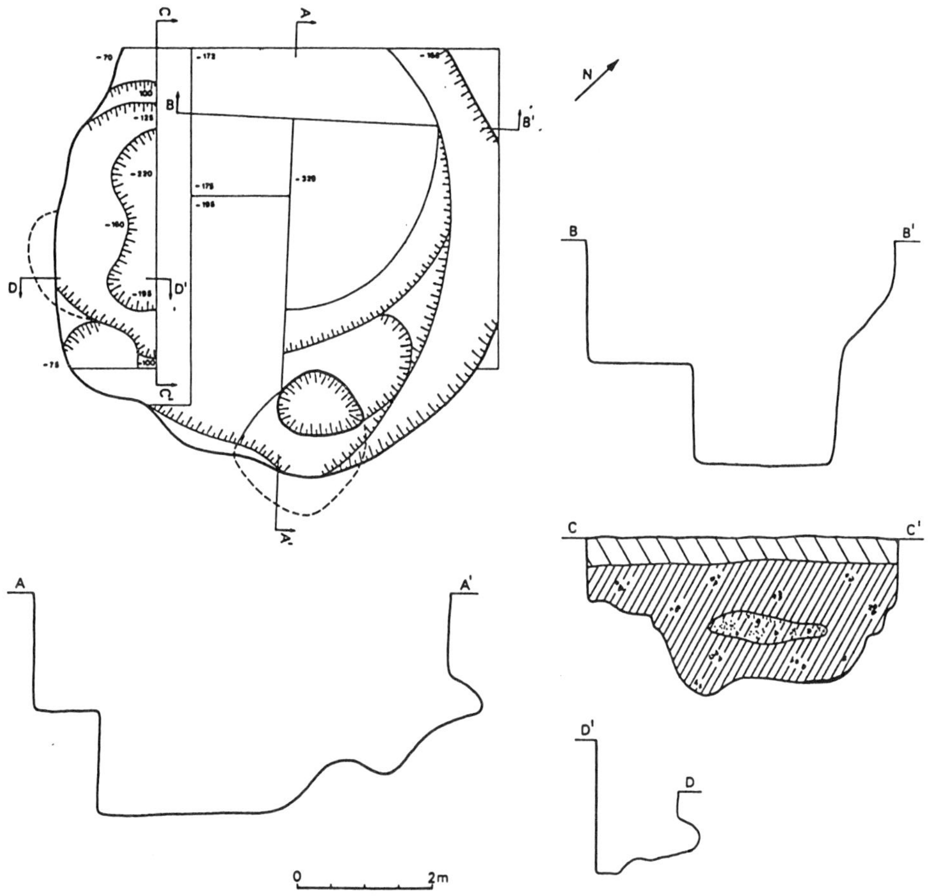

Fig. 16. The plan and profiles of the Sarmatian pit excavated in square XII/27. (See Fig. 20 for key.)

[7] This paper concentrates on comparison of results from the various methods employed. The publication of the archaeological material will be the task of another study. The Sarmatian finds will be published by A. Vaday.

Section XI/27: Pit "A" was just 25 cm deep. The plain bottom displays the traces of slight compacting. Its western edge falls in section XI/26 where a small round pit filled with ash and earth was sunk in the yellow sterile soil. Its greater southern portion has not yet been excavated. It contained few Sarmatian sherds.

Pit "C" was a shallow Sarmatian pit rich in finds, a small part of which fell in section XI/28, where the concentration of the finds only signified its boundary.

Section XII/26: Pit "A" was a Sarmatian beehive pit with a thin layer of soot at the bottom. It cut off a small part of the much smaller and shallower pit "B" at the edge of the section.

Section XII/27: This huge Sarmatian pit occupied not only the whole section but spread over to neighbouring areas in all directions. We succeeded in unearthing its edges on two sides. The fill is of an evenly mixed yellowish-brown, loose earth, sometimes with light grey ash spots or grains of daub. It was deepened until 330 cm, when the appearance of ground water hindered further work. No discernable layer series could be observed in this pit (*Fig. 16*).

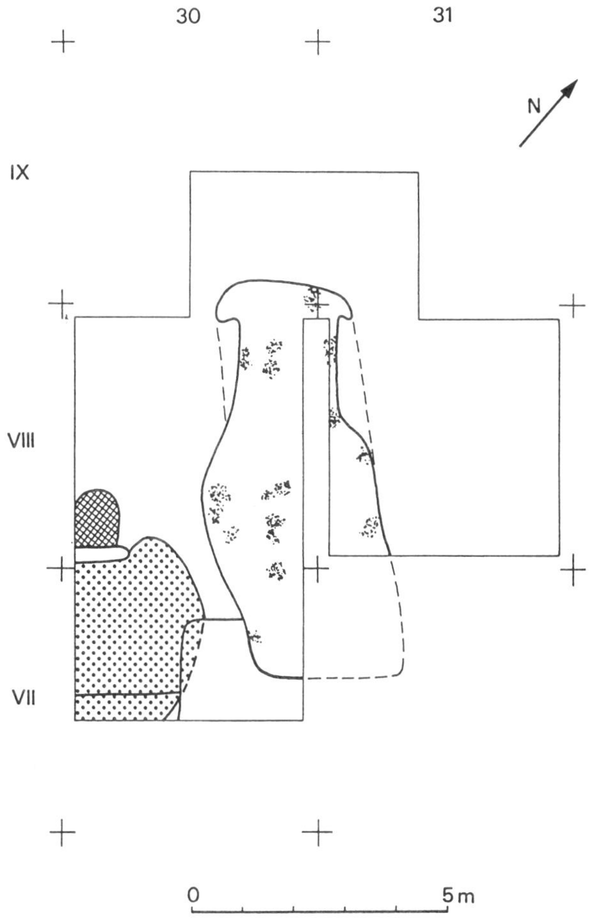

Fig. 17. Archaeological features excavated in square VII-VIII/30-31.

Comparing the results of excavation with surface collection and geophysical prospecting (*Fig. 8*) it can be seen that the magnetic anomalies located the position of the larger pits which contain a great number of sherds, daub fragments or ash, to within nearly a metre. At the same time, they did not indicate smaller pits with fewer finds (e.g. section XI/26. pit "C"), the effect of which was overshadowed by a neighbouring structure with very high values (XI/27. pit "C"). Neither the even scattered surface finds nor the average values indicated the presence of these structures (*Figs. 3, 5-6*).

The following structures were found in the second spot during the excavation (*Fig. 17*):

The discoloration of a pit from the Avar period was noted in the SW corner of section VII/30., but it could not be taken out (*Fig. 17:8*).

VIII/30. Above the previous pit, a portion of an open-air oven from the Árpád Period, falling into the section, was excavated.

At the same place, a 7,5 m long, 2,5-3 m broad and 15 cm thick dauby surface was observed. Based on many of the sherds it contained, this dauby surface was estimated to be the ruins of a Sarmatian house. The same is attested by the cinders found among the daub fragments. However, there was nothing – neither postholes nor floor level – under the daub layer where even the finds disappeared. The construction extends over the neighbouring sections, as well. Only part of it could be unearthed and just the northern edge was taken up. Both the surface finds (*Figs. 3-6*) and the magnetic anomalies (*Fig. 7*) indicated the constructions in this area of the excavation. The magnetic anomalies again achieved the exactitude of a metre in the case of the dauby surface and the oven.

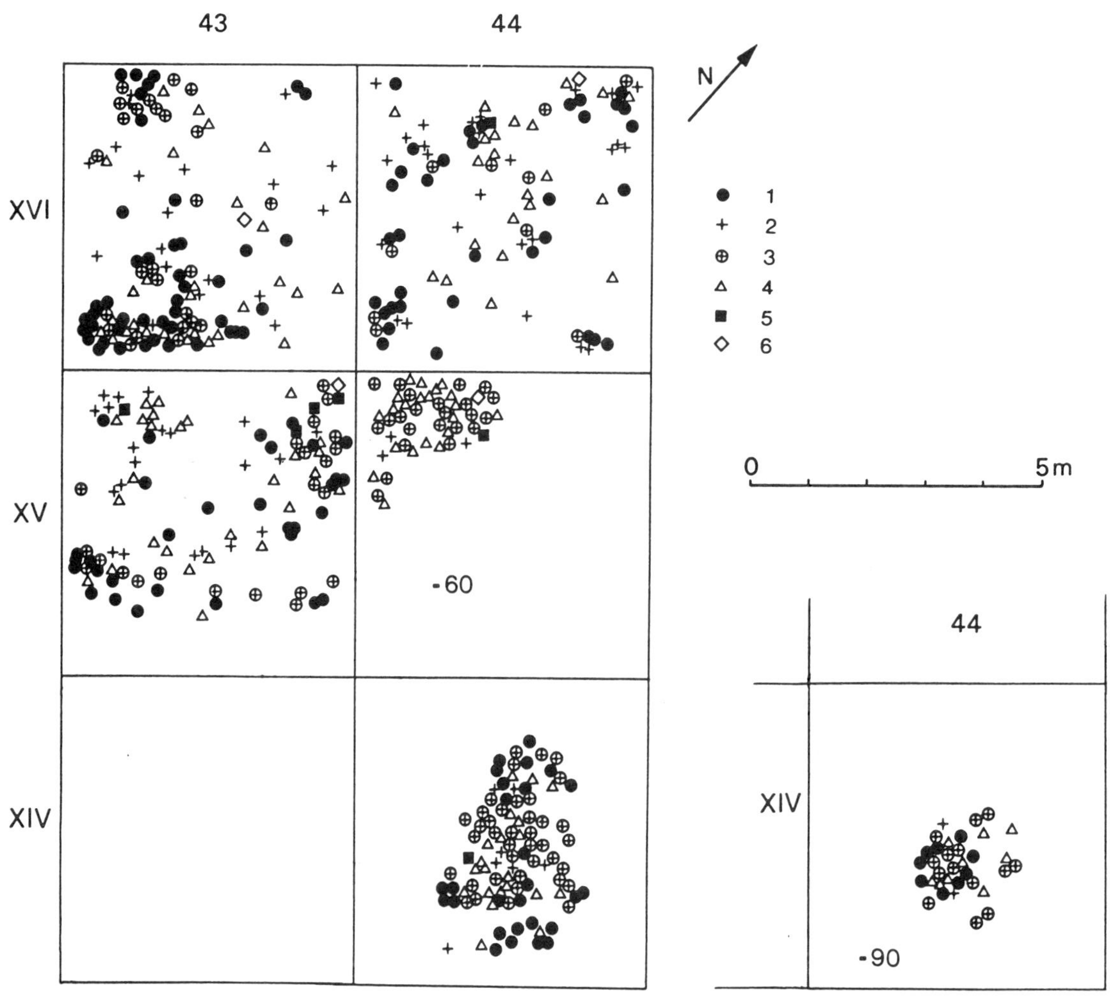

Fig. 18. Find distribution in square XIV-XVI/43-44 at a depth of 30-60 cm. The clustering of finds in square XIV/44 at a depth of 60-90 cm (that is the ovens pit) is shown in the quadrant located in the lower left corner. (See Fig. 7. for key.)

At the third excavation spot where no geophysical prospecting was done, the great number of finds from the Árpád Period suggested the existence of constructions from this period (*Figs. 3-6*). At the depth of 35 cm the ground level from the Árpád Period was found directly under the plough zone. It contained pieces of daub, animal bones and sherds. The pits and other constructions were dug down from this level.

As the soil here was of the dense black meadow solonets type, the discoloration of the features can only be detected with difficulty. The borders of the constructions above -60 cm could be defined only with the help of careful mapping of the finds. The finds were evenly scattered between the surface and -30 cm while they were definitely concentrated inside the constructions between -30 and -60 cm (*Fig. 18*).

The following constructions were found in this area (*Fig. 19*):

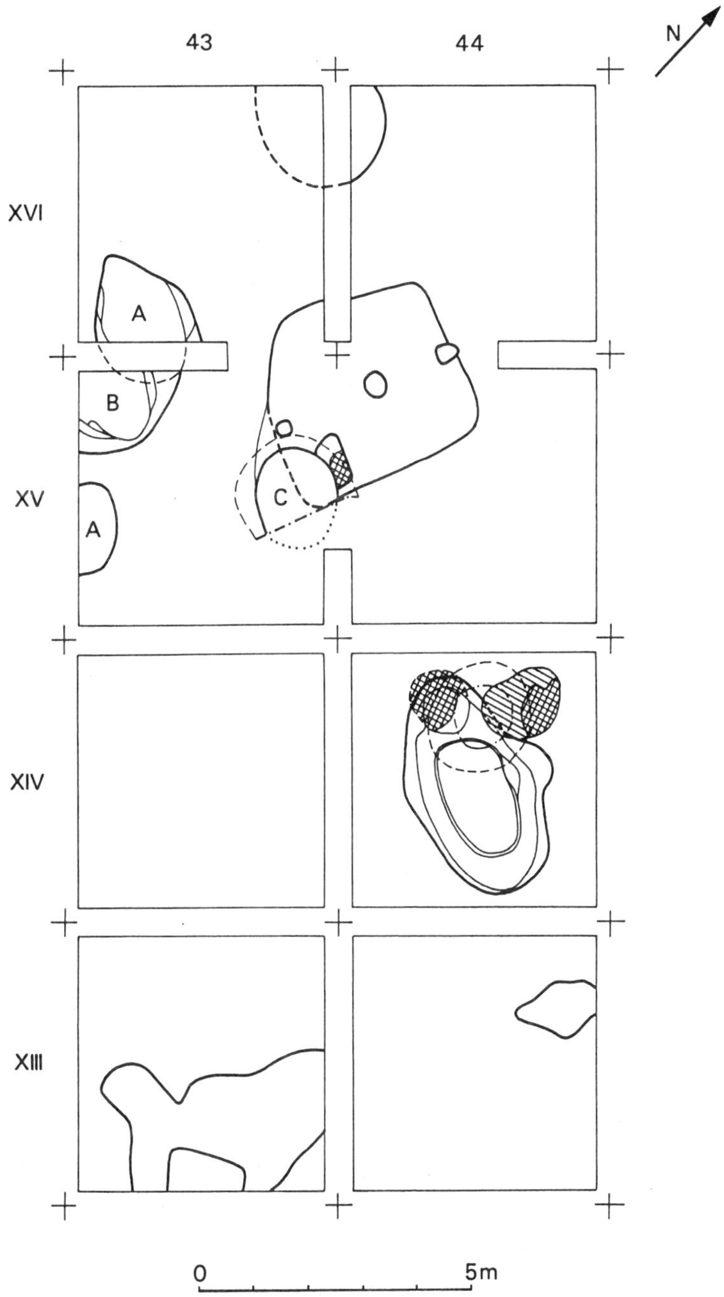

Fig. 19. Archaeological features excavated in square XIII-XIV/43-44.

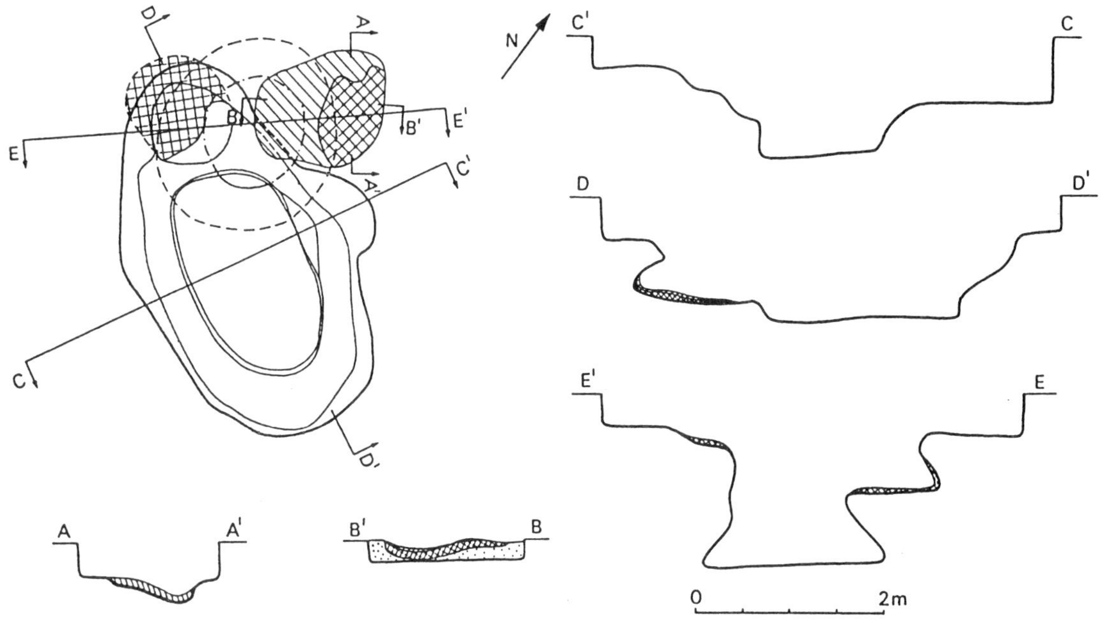

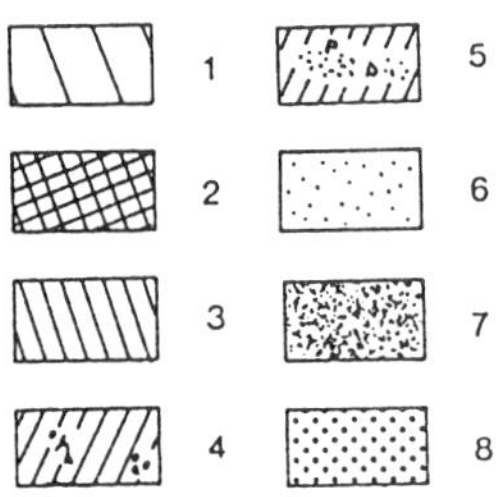

Fig. 20. The Árpád Period baking surface, oven as well as the Sarmatian pit excavated in square XIV/44.
Key:
1: Plough zone
2: Burnt baking surface
3: Cracked up baking surface
4: Light brown fill mixed with many sherds
5: Light brown fill mixed with many daub fragments
7: Carbonized wood imprints
8: Spot of an unexcavated pit

Section XIII/43: A shallow pit with many animal bones. The edges could be delineated only at the bottom (Árpadian Period).

Section XIV/44: a 140-110 cm long trough-shaped fireplace was found at a depth of 35 cm, just at the bottom of the topsoil. Based on the finds found around and chiefly under it, the construction belonged to the Árpád Period.

Another open-air oven has a common shape and structure. It contained a few sherds, some of which were hand-made, and also animal bones. It is earlier than the above mentioned fireplace, since the latter spreads over the ash pit by a few centimeters. Nevertheless, the sherds found in the plastering and in the pit associate it with the Árpád Period.

This Sarmatian pit was found under the oven. It is a regular, circular, hive-shaped pit containing a dozen of sherds. A few mm thick layer of soot was observed on its bottom (*Fig. 20*).

Section XV/43: Pit "A". Only half of this pit fell into the section and was uncovered. It lay at a depth of 35 cm while the few finds found in it connect it to the Árpád Period.

Pit. "B". Only a quarter of this pit could be excavated, as it continued under the west wall of the section and it had been partly demolished (on the North) by a pit from the Árpád Period. The construction is from the Sarmatian Period.

Pit. "C". This circular, beehive pit cuts the corner of the house to its north and the western part of the house's oven (See section XV/44.). The fill consists of alternating layers of black mud and yellow sand. The bottom is plain, covered by a slightly burnt, sometimes double plastered surface. The lower surface was covered by a few mm of soot. It has been dated to the Árpád Period and partly excavated.

Section XV/44: This 350x280 cm pit-house dates to the Árpád Period. The roof was supported by three posts in a Northeast-Southwest direction. The two lateral posts were sunk 40 cm under the floor level, while the central post had a rounded base going down only 10 cm. The floor was so lightly compacted that it was barely noticable. Except for the three posts supporting the roof no other post-holes were found.

The oven was a 50x40 cm small rectangular firebox sunk below the floor level in the SW corner. It must have faced the entrance (*Fig. 21*).

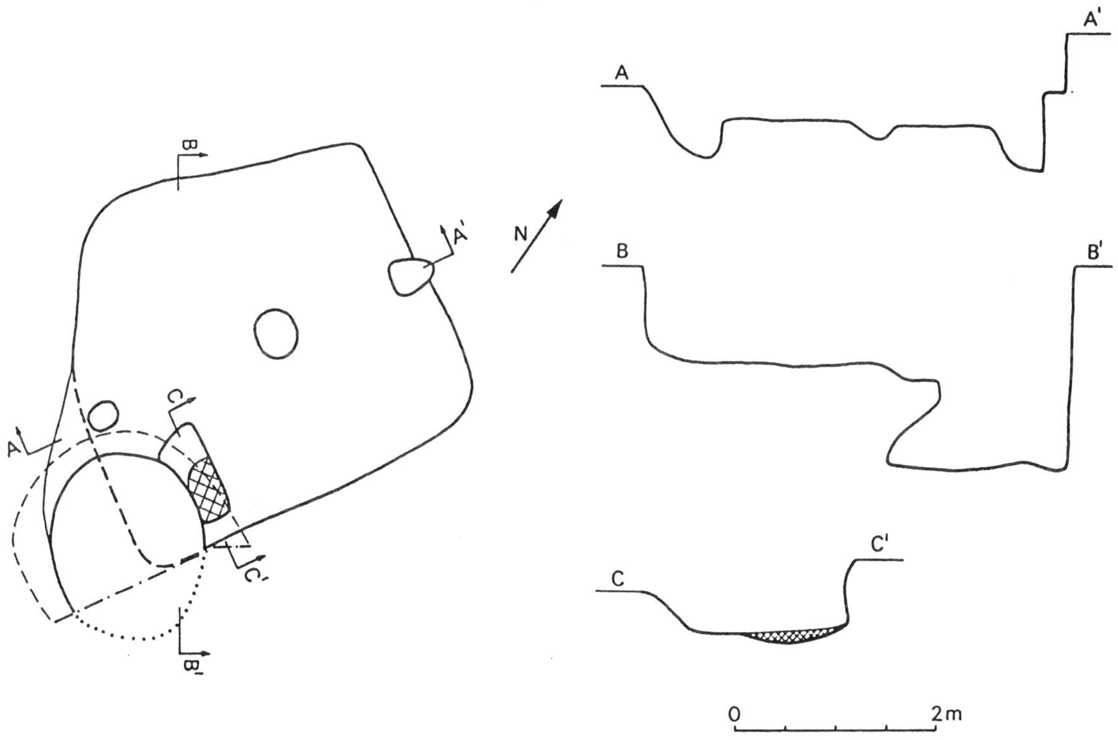

Fig. 21. Semi-subterranean Árpád Period house and its profile excavated in square XV/44.

Section XVI/43: Pit "A" was a 110 cm deep refuse pit with many daub fragments and a 5-10 cm thick black soil which had been brought in and compacted on the bottom. The pit extended so slightly under the southern side of the trench that it was no longer observable in section XV/43.

Section XVI/44: There was a beehive pit in the NW corner containing no finds although the fill (alternating black mud and yellow sand) suggests the Sarmatian Period. It was only partially excavated.

The above data show that the unearthed structures are mostly from the Árpád Period, that is, the excavation results are in harmony with those from the surface collections.

Conclusions

The results of excavation seem to confirm that:

1) A carefully executed traditional topographical field survey may delineate the borders of a site with great exactitude.

2) Site survey, that is intensive collection based on well chosen samples, may define not only the borders of a site but its central area and the location of more important constructions as well. It is also suitable for revealing the horizontal stratigraphy within the site as well. It was observed that the surface finds are not found at any considerable distance from the constructions they came from.

3) Geophysical prospecting (magnetic, electronic) defines all archaeological phenomena containing a greater quantity of burnt material (daub, ash) to within a metre both vertically and horizontally. However, smaller constructions cannot be detected.

4) Aerial photography, made at the right time, serves to illustrate the resolution of the above problems and with luck may also improve it.

It can be ascertained from the foregoing that the results of the surface find collection emphatically agree with the conclusions that can be drawn from the magnetic anomaly map. In future investigations of archaeological sites it should be decided which method is quicker, more economic and more effective. Where density of the surface finds suggests a feature, sooner or later a detailed excavation will be carried out. Thus, strictly speaking, no magnetic survey is necessary in such places. A magnetic survey is, however, very important towards the boundaries of the area where the density of objects decreases. Features located at the boundaries of a promising area can be found using magnetic measurements alone, because a surface find density will not indicate them unambiguously. A magnetic survey might thus be of great importance in delineating a promising area.

The dc resistivity measurement – in conformity with its theoretical limitations – provides results very similar to the magnetic survey method. Because of its clumsiness, however, it is suitable rather for follow-up, confirmation surveys. In many cases it is very useful to verify results of various geophysical surveys by shallow borings because archaeological excavations can often be carried out only later. Borings allow a more accurate interpretation to be made immediately and the conclusions can then be extrapolated to a larger area.

Acknowledgement

The laborious work of surface collection was performed with the help of B. Darázsy (archaeologist), B. Dukay (draftsperson; both Archaeological Institute of the HAS) and G. Gábor (conservator, Munkácsy Museum, Békéscsaba). They are all gratefully acknowledged for their conscientious and good humoured collaboration.

Excavations at the Körös culture settlement of Endrőd - Öregszőlők 119 in 1986-1989

A preliminary report

JÁNOS MAKKAY

This excavation was carried out over four seasons between 1986 and 1989 within the framework of the so-called Microregion Research Project of the Archaeological Institute of the Hungarian Academy of Sciences. The exact name for this location is Endrőd-Páskum, Iványi köz and it was given the number 3/119 in volume IV/2 of the Hungarian Archaeological Topographic Survey.[1] The work lasted for twenty weeks altogether, between July 14 and August 19 in 1986, between July 13 and August 19 in 1987, between July 4 and July 29 in 1988 and finally between July 17 and August 29 in 1989. The number of workmen ranged between eight and twelve, thus averaging ten. This means some 1000 man days during the hundred days period.

A surface of 2500 square metres was investigated over 55 squares which equals a 2.5 square metres per capita daily recovery rate. The depth of the site ranged between 60 and 200 cm, so when a rough mean thickness of 1 metre is estimated one may reckon with a mean performance of 2.5 cubic metres, which may easily be the upper limit of the output by which the excavated material may still be appropriately retrieved. For this achievement thanks are due to crew recruited in the village of Endrőd and especially the first class skilled workmen Géza Valuska, György Nagy and Sándor Czikkely.

The small site was discovered in 1976 within the framework of field surveys related to the Hungarian Archaeological Topographic Survey program. The small, oval elevation, hardly one metre in height (relative to its surroundings) lay in a northeast-southwest direction. Its surface and especially the top was covered with Körös culture sherds. At the time of discovery, this phenomenon was connected to the intensity of occupation. Subsequently, however, it turned out to be the unfortunate result of tillage. During the course of this work, a number of steel wedges of an average 80 cm length and 30 to 40 cm from each other are pulled along at a depth of approximately 80 cm in the soil. This method is aimed at loosening the soil below the plow zone condensed under the pressure of heavy machinery such as tractors and combines. Even if this technique does not destroy everything between the upper plow zone and a depth of 80 cm it certainly causes breakage and dislocation. Fortunately, as experience shows (at least in the outskirts of Szarvas, Endrőd and Gyoma) none of the tractor drivers have ever worked conscientiously enough to reach the 80 cm depth required for this kind of tillage. The affected depth was only around 50-60 cm. In this way it was possible that at the site under discussion here only a depth between 30 to 60 cm was disturbed. In addition, the yellow virgin soil may often be found at a depth of 60 cm at this site and it would not have been sensible to mix this layer with the cultivated strata. Consequently, at the site under discussion the 30 to 60 cm depth corresponds to the layer in which the burnt remains of two houses (House 1 and 2) built on the original prehistoric surface (i.e. the buried soil) were found. Due to tillage, structural data on the two houses could practically not be gathered. In addition, artifacts associated with the houses which must have been quite well preserved just a few years ago, were badly damaged as well.

---

[1] *MRT* 8: *Békés megye régészeti topográfiája. A szarvasi járás* IV/2 (The Archaeological Topography of Békés county, the Szarvas district IV/2). Edited by János Makkay, Budapest, 1989, 166.

As is known, the whole area of Endrőd-Öregszőlők is located on the border of two regions defined by soils of different origins. It spreads over the northern edge of the deposition fan of the Maros river (more exactly the earlier "Eastern Great River") which are interwoven by smaller stretches and even the floodplains of the Körös river valley. If scant artifactual material of the Körös culture site of Endrőd 121 (which is only located approximately 200-300 m from the site examined here)[2] is not taken into consideration, then site 119 is the southernmost settlement of the Körös culture in the Gyomaendrőd area of the Körös river valley. At this point, the closest Körös culture site in a southern direction is Csorvás-Orosházi útfél.[3] The peripheral location of our site made it suitable for excavation and emphasized the necessity of research here. Another reason why this site was selected for study is that it is in the proximity of Endrőd 39, which was investigated by five rescue and test excavations during a time period between 1975 and 1978. Thus, the site under discussion here may contribute data to solving the chronological questions that emerged there at that time. Finally, an important argument was that this site spreads over a small area and, thus, meets the desirable requirement of complete recovery as stated in the Microregion Research Project.

Since the site is 70-75 metres long in a northeast-southwest direction and 40-50 metres broad in a northwest-southeast, the approximately 2500 square metres excavation area essentially meant full recovery. Only a small, that is, a shallow and marginal portion of some features (refuse pits) was left unexcavated (for example Pits 4, 10, 13 and 18). Namely, it could be observed during the whole excavation that no greyish-ashy layer was present under the plow zone and above the bottom soil of the settlement in the peripheral parts of these marginal squares. Instead the bottom soil was covered only by a thick 50-60 cm deep hard, black layer. Although this deposit once undoubtedly contained vessel fragments and animal bones, their presence could not be documented anymore. (From this point of view Squares 1, 5, 9 and 13 are especially empty). Although the, at most, five metres wide band surrounding the excavated squares may still contain sporadic finds or even small features, their excavation would have been disproportionately expensive.

Due to the damage caused by tillage but a few years ago, data on some details of the Körös culture houses were not available anymore. It is certain that two such houses stood within the site. Both of them were built on the prehistoric surface and had an oblong shape plan. Their long axis was oriented north-south. The length of these houses was 10-12 metres and they were 4-6 metres wide. According to observations neither ovens nor fireplaces were located within these buildings. Both houses stood above a refuse pit: House 1 stood above the relatively small Pit 5, while House 2 overlapped with the greater part of the huge Pit 12. This position should provide an excellent opportunity for relative dating in itself. However, it turned out that in the case of House 2 and Pit 12 several sherds came from the pit while associated fragments were recovered from the house located above it. A difference discovered between the two houses was that while no post holes could be identified at House 1, several occurred in House 2. Some of these fell within the area of the house, while others were discovered on the outside. These post holes, however, did not follow a definite outline. Plasters and/or burned floors were not apparent either, thus, the floor level of these prehistoric buildings could only be defined on the basis of the artifacts' position. From this point of view, House 2 may be considered especially rich. Here fragments from at least thirteen vessels were scattered on the floor surface and to some extent among the ruins of the house. At the same place a number of net weights occurred in several clusters (among these the examples shown in *Plate 1,2, Plate 17,1, Plate 22,3,*

---

[2] Ibid., p. 167.

[3] *RFüz* 32, 179, 9. Compare: J. Makkay: Kontakte zwischen der Körös-Starčevo-Kultur und der Linienband-keramik. *ComArchHung* 1987, Map 1, No. 10.

Plate 22,4, and net weights in *Plate 41,4*). Similarly to previous observations, the large vessel put by one of the northeastern post holes in House 2 must have had a ritual function. János Banner noticed at a site near Hódmezővásárhely (Kotacpart-Vata tanya) that single large vessels were found in a standing position in a particularly intact position.[4] A similar custom may be observed in relation to the face decorated, large vessels of the Alföld Linear Pottery (ALP) culture as well.[5] The fact that this Endrőd specimen was found near a post hole may, of course, be a coincidence. One should not, however, rule out the possibility that this object was of ritual significance within the framework of a so-called "post-sacrifice".[6] On the other hand, it is possible that the fragments of large vessels were placed in the post holes in order to fix the post itself (as may be seen with the fragments of a bowl shown in *Plate 36,1*).

Among the other settlement features, perhaps only Oven 8 is worth mentioning since at this point no similar construction is known from the Early Neolithic. It must be emphasized, however, that the Körös culture dating of this undoubtedly complicated oven is certain due to a complete pedestalled bowl found in its peripheral section.

A remarkably great number of features at this site may be considered sacrificial pits or places. They occur in a variety of completely differing forms, the simplest of which is Sacrificial Pit 8, where a large vessel was put in a round pit of about 80 cm diameter only slightly dug into the sterile soil. This large vessel was accompanied by a four-legged cup found in an upside-down position and a couple of animal bones (*Plate 40,5*). Of the seven other sacrificial pits, one differs from the other six: this is Sacrificial Pit 1 located in the middle of Pit 5. It is a deep and round narrow pit filled with successive repeated layers of sacrificial deposits which also contained numerous snails. Each of these layers was covered with a thin layer of yellow clayey soil. At the bottom of this pit the cranial half of a dog's skeleton was found in an *in situ* position. On the very top, a clay lamp was recovered (*Plate 32,1*) with a fragmented cup among a mass of fish bone and burnt remains. The lamp was found in an upside-down position as well.[7] This sacrificial pit bears some resemblance to the sacrificial pit uncovered under the layer of a Körös culture house at the site of Endrőd 39.[8] The difference seems to be that in that case a single sacrificial act was observed (which was presently related to the construction of the building), while a repeated rite may be identified in the case under discussion here. Consequently, it should not be considered a building sacrifice. Another difference is that a bloodless rite was performed at Endrőd 39 while the bloody sacrifice here was accompanied by a deposition of snails. It is this aspect, that is, the repetition of the sacrifice that points to the non-construction sacrifice character of this offering. It belongs within the framework of another

---

[4] J. Banner: Ásatás a hódmezővásárhelyi Kotacparton (Excavations at Hódmezővásárhely-Kotacpart). *DolgSzeged* 11 (1935), 105, 115, Plate XIV,7, Plate XV,6-7, Plate XVI,1-2 and Id.: Das Tisza-, Maros-, Kőrös-Gebiet (sic!) biz zur Entwicklung der Bronzezeit. Szeged, 1942, 17.

[5] N. Kalicz and J. Makkay: Gefässe mit Gesichtsdarstellungen in der Linienbandkeramik aus Ungarn. In: *Idole, Prähistorische Keramiken aus Ungarn.* Wien, 1972, p. 15.

[6] J. Makkay: Opferpfosten in Gebäuden und Opfergaben in oder neben Pfostenlöchern. *ActaArchHung* 40, (1988), 3-21.

[7] In detail see J. Makkay: Zwei neuere Opfergruben der Körös–Starčevo-Kultur. *Bylany Seminar 1987.* Collected Papers. Prag 1989, 243-248. Cf. D. Gronenborn: Neue Überlegungen zur Funktion von Schlitzgruben. *AKorr* 19, 1989, 339-342.

[8] J. Makkay: Foundation sacrifices in Neolithic houses of the Carpathian Basin. In: *Valcamonica Symposium III - 1979. Proceedings. The Intellectual Expressions of Prehistoric Art and Religion.* Valcamonica, 1979. Capo di Ponte - Brescia, 1983, 157-160; Id., Bauopfer in der Lengyel-Kultur und seine Beziehungen zu den Bauopferformen der Körös Kultur und der Linienbandkeramik. In: *Internationales Symposium über die Lengyel-Kultur, 1984.* Nitra-Wien, 1986, 170, Fig. 1.

sort of ritual and as such its dating is independent from the period of House 1. As mentioned previously, however, this chronological relationship could not be clarified.[9]

The other sacrificial features (that is Sacrificial Pits 2 to 7) are practically similar to each other, even if there are important differences between them as well. This type, which could be best termed a "ritual shaft", has not yet been described from the period of the Körös–Starčevo culture although such features were probably found by János Banner during his previously mentioned excavation at Kotacpart Vata tanya. It may not be excluded that feature E4 is such a narrow shaft as well and it is almost certain that he discovered such a sacrificial shaft in the case of Pit 5, which had diameters of 2 and 1.4 metres and a depth of 3 metres. He noted that "at a depth of 1 metre this pit broadened to 2.4 metres at which breadth it was 2.5 metres deep. Then it ended in a hardly 0.5 metre broad but 2.8 metres long ditch at a depth of 3.16 metres".[10] Some of the sacrificial shafts contained no artifacts aside from the usually small ceramic fragments of the Körös culture. One may assume, however, that these sherds came into the feature independently of possible rituals. It is characteristic of these pits that they narrow abruptly at around the half of their depth. Consequently, they are only a few centimeters broad at their deepest point. In this deeper, narrow part a stratigraphy of alternating dark, organic and virgin clay deposits may be observed. These latter are usually yellow and deepen inward in an arched contour from the sides (and short ends) of the pit. (The curved profile line of such a yellow layer intersected by a straight plane is shown on the photos of *Plate 40,1-2*). The question was raised of whether the alternating black and yellow fill of these shafts (of otherwise unknown function) could be natural in origin, that is, deposited by erosion. This hypothesis, however, is contradicted by the fact that the top fill of Sacrificial Pit 6 is of almost purely yellow clay. Such material could not have been directly available on the surface of the buried, Neolithic soil in the proximity of the shaft, i. e. it must have been brought there artificially (*Plate 40,1*). The shafts may be classified into two groups on the basis of their contents. Some of them contain artifacts, among other things sacrificial remains (Sacrificial Pit 4: a large complete vessel with a net weight in it, *Plate 25,1*; Sacrificial Pit 3: bones from the cranial half of a sheep skeleton). The other group of shafts contains no finds at all. As has been elsewhere mentioned, it was not possible to establish if the small idol (*Plate 28,1*) and the rare, relief decorated fragment (*Plate 31,2*) as well as a small vessel with a human face (not discussed at length here) found in Pit 10, could originally have been placed in Sacrificial Pit 5.

The orientation of sacrificial shafts is ambiguous although they are undoubtedly oriented in a north-northwest – south-southeast and southwest-northeast direction respectively. One of their ends is always narrower than the other, the narrow end, however, may point in any of the main directions. It is remarkable that (including Sacrificial Pit 1 with its peculiar shape) such sacrificial pits or shafts almost always were dug into refuse pits or at least in their proximity (Pit 5 + Sacrificial Pit 1, Pit 4 + Sacrificial Pit 2, Pit 12 + Sacrificial Pit 3, Pit 13 + Sacrificial Pit 4, Pit 9 + Sacrificial Pit 7). This, however, is not a systematic rule since there are refuse pits without associated sacrificial shafts (Pit 14 and Pit 10; unless this latter is accompanied by a sacrificial shaft east of the southern section of Square 24 and remained unexcavated). On the other hand, it is likely that sacrificial shafts which are in part or completely located in refuse pits (Sacrificial Pit 2 and Sacrificial Pit 5) are independent of these refuse pits and were dug into them secondarily following their partial or complete filling. There is no doubt that the recovery of features belonging to this pit type at the site contributed significantly to the

---

9 Banner, op. cit. 1935, 105, although this feature was very long and Copper Age pottery is mentioned.

10 Id. 104.

understanding of religious rituals in the Körös culture although at this point it is impossible to explain why such narrow shafts are known predominantly from the area of the Central European Linear Pottery culture.[11]

The majority of find material was brought to light from ten large Körös period refuse pits: Pit 4, Pit 5, Pit 9, Pit 10, Pit 12, Pit 13, Pit 14, Pit 17, Pit 18 and Pit 20. The actual size of these pits did not display a linear relationship with the abundance of their contents. The artifactual material from Pit 4, for example, was more modest than the inventory from Oven 8, at least during its second phase of deposition. It is, in general, characteristic of the uppermost fill of these refuse pits that they are composed of ashy layers which form a correspondingly grey and loose deposit. Subsequent lower strata are made up of a brownish but still loose material. The stratigraphy remained brownish but became denser toward the sterile bottom soil. The number of sherds otherwise forming the majority of finds, also became smaller in a downward direction. Smaller or larger burnt patches were common within the stratigraphy. In a few cases they may have been fireplaces themselves, however, most of these burnt patches originated from cinders which kept on burning after having been dumped into the pit. Typically, layers of fish scales may be observed as well. Easily restorable or even almost intact vessels were often found in these pits. In several cases, asymmetric wear on one side of such vessels was indicative of the fact that although these pieces were eventually thrown in the refuse pit they had remained on the surface for a considerable time in their original position. Pit 12 was extraordinary both in terms of the richness of its ceramic material and its dimensions. At other settlements in the Endrőd - Szarvas region (especially at the sites of Szarvas 23 and Endrőd 39) 220 and even 300 cm deep refuse pits have been observed, although their depth most commonly fell around 2 metres. At this site, only one feature, Pit 13, was deeper than 200 cm and this is obviously connected with the height of the water table at that time. A regular difference between ALP and Körös culture features may be observed as well. As opposed to the majority of ALP refuse pits and pit dwellings, the fill of Körös pits is loose and ashy with a high organic content. This is what made it necessary to subject twenty kg soil samples from the 1987-1989 seasons to water sieving. The residues from this project have not yet been selected and identified.

Both in the case of the two houses and refuse pits a detailed analysis of artifacts (especially that of the ceramic material) remains to be carried out. This naturally would be beyond the focus of a preliminary analysis. In addition, conservation and restoration of these ceramics are still in progress. It is for this reason that it would be premature to take a stand in questions which require a detailed analysis, especially those concerned with chronology. Namely, the two houses and refuse pits obviously represent more than a single period. One may hypothesize, for example, that only one of the houses stood at the site at any given time so that the question is which of the two buildings was the earlier. It is also likely that some of the refuse pits were originally opened to obtain clay for the construction of houses. However, in the case of House 1, Pit 5 was probably not this kind of a quarry while in relation to House 2 this function for Pit 12 may be ruled out as well since the houses themselves are partly or completely located above these pits. This fact in itself may serve as a starting point for chronological conclusions. The function of the remaining refuse pits, on the other hand, poses a persistent problem. These must have been used in the construction and renovation of houses as well as other activities. It is, however, impossible to tell which of the refuse pits may have played such a role. The largest pits providing the richest find material are undoubtedly located at the southern edge of the settlement, although this was possibly the area located in front of the south facing entrances of the houses and as such may have been one of the important scenes of daily activities.

---

[11] See footnote 7.

As far as the relative chronological position of some of the pits and houses is concerned the following data are worth considering. Some of these are provided by completed vessels whose sherds came to light from different depths in the same pit. It is well known that fragments of a vessel thrown out into a refuse pit may end up at considerable distances from one another. They may even be mixed with the lower layers of a pit in the process of filling. However, one should assume such movements when the pit is still "alive", that is, its undoubtedly high organic content has not completely settled yet due to organic decay. This period of fill and deposition should not take longer than 20 to 25 years, in other words, a human generation. On the basis of restored vessels (even more evidence is provided by sherds not yielding a complete vessel) in at least four pits, differential vertical deposition may be observed. Fragments from the same vessel were often found one metre above each other and this difference in depth was sometimes even greater (*Plate 12,3:* 120 cm; *Plate 13:* 140 cm). If one hypothesizes that the Körös Period occupation of this small settlement was continuous and in order to put each of the more significant pits into a time sequence (assuming that only one of the pits was used at a time and functioned for twenty years), a maximum 200 years life span may be estimated for this settlement.

Another source of data is the inventory of completely restored vessels as well as other artifacts the fragments of which were brought to light from two different features. This tabulation shows that a particularly great number of such associations may be identified between House 2 and Pit 12 (at least ten vessels and their fragments as well as further sherd which cannot be fully restored). There was a very characteristic find, a bowl fragment decorated with a stroke burnished pattern, whose matching parts were brought to light from Pit 12 and Pit 13 respectively (*Plate 34,1*). It must be emphasized that in certain cases the possibility that this occurred coincidentally can be ruled out. A prime example would be the large bowl shown in *Plate 36,1*, large fragments of which (as mentioned previously) were put against the wall in the deepest part of Pit 12, while other parts were placed in the post hole which probably belonged to House 2. Thus, one may see that the construction period of this house (which could not have been longer than a season but was probably even shorter) overlapped with the deposition times of Pit 12 and Pit 13. This points to the fact that more than one pit were in simultaneous use at the settlement. At the same time, House 2 covered the majority of Pit 12. Thus, the chronological sequence may be the most simply reconstructed in the following manner:

1/ within the ensemble of three features, Pit 12 is the earliest and was opened before the construction of House 2. In fact, its clay served as building material. At the time of construction, however, Pit 12 was not completely filled. It is for this reason that parts of the pottery placed into the post hole belonging to Pit 12 matched with pieces from the depths of Pit 12.

2/ During the lifetime of House 2, the parts of Pit 12 not covered by the building gradually filled up. Since this concerns the majority of the Pit deposit, one may assume that its chief content was construction debris so that only the Pit section outside the house remained empty. In fact, the sherds shown in *Plate 36,1* were found in Square 33, that is, beyond the hypothetical outline of the house. This logic of the deposition process, however, is not supported by the evidence of complete vessels. Parts of these were found among the house rubble, while other fragments came to light, sometimes from considerable depths, in the pit located below.

3/ Finally, Pit 13 follows the period just described. This feature was not necessarily opened during the lifetime of House 2 (no associated sherds could be identified). On the other hand, there is one find linking Pit 12 and Pit 13: fragments of the same vessel (*Plate 34,1*) were found in both features showing their parallel use, even if only for a short time.

There was only one artifact, a small vessel, whose related fragments were found in House 2 and Pit 10 respectively (*Plate 26,17*). This find suggests that Pit 10 (which was in the immediate proximity of the entrance of House 1 probably facing south) is partly or completely coeval with House 2.

It is remarkable that features in the eastern part of the settlement (that is House 1 and Pits 4, 5, 9) show no evidence of such interrelatedness.

As a preliminary conclusion, it may be assumed that the life of this small settlement may be divided into two periods. House 2 and the two or three pits associated with it (Pits 12, 13 and perhaps Pit 10) may be dated to the earlier phase. This dating is supported by two fragments with fine white painting on a burnished brownish slip. Although these two pieces cannot be fitted directly to each other they certainly belong to the same vessel. One of them came to light at a depth between 60 to 90 cm in Pit 12 (Square 32), while the other was discovered in the layer below 140 cm (Square 29). A third stray fragment of the same vessel was found in 1989 on the border between Squares 41 and 43 at a depth of 30 cm and so cannot be evaluated. According to the current view, this kind of painted ware belongs to the early, so-called "white-painted" period of the Körös–Starčevo culture.[12] The only problem with this dating is that at the same time both House 2 and Pit 12 yielded a number of finds (e.g. forms of the so-called "metallic ware" *Plate 1,2* and other, similar sherds) which usually are not considered types characteristic of the early period. One should also note, that ceramic types characteristic of the Körös–Starčevo cultureŁs late period (and also known as Protovinča types), occur infrequently (not to mention the types of incised-impressed-stroke burnished ware already known) also from early Körös–Starčevo assemblages. Among the restored vessels only two may be so classified: one from Pit 4 and the other from Pit 13 (*Plate 1,1, 5*).[13] Thus, life at this small settlement had ceased shortly before the latest phase of the Körös–Starčevo period, that is, before the Protovinča phase. On the other hand, life went on at the nearby settlements of Endrőd 39 and Endrőd 6. This scenario is somewhat contradicted by the fact that a typical ALP sherd was found among the characteristic sherds of the Körös culture in the undisturbed fill of the southern part of Square 18.[14] This area was not disturbed by the oven built nearby during the Period of the Árpád Dynasty. It also yielded the small idol shown in *Plate 28,7* and *Plate 29,9* respectively. Lacking further evidence, this assemblage of a few fragments may be regarded as signs of a short, chronologically isolated latest Körös occupation which followed the settlement's actual lifetime, or, in turn, the first appearance of Alföld linear sherds of the Szatmár phase can be dated immediately before the Protovinča phase.

Alternating dark (sacrificial deposit) and yellow clay (covering) layers in the sacrificial pits are usually interpreted as remains of periodically repeated (seasonal, yearly etc.) sacrifices. Thus, a time sequence of sacrificial shafts, assuming that they followed directly, without interruptions, and that the layers represent actual seasons or years, would provide the site's total life span. This analysis, which would obviously provide very unreliable results, has not yet been carried out. It seems, however, that the layers probably would not point to a time interval exceeding four generations. Burials suggest an even shorter occupation numbering at most nine (Graves 1-4, Graves 9-11, Grave 13 and remains of a newborn infant in Pit 12) four of which were the graves of adult individuals. As we know, all existing data show that no separate cemeteries were opened during the Körös Period. The deceased were buried at the settlement either in refuse pits or (as is the case with the site under discussion) close by. Consequently, we either have to accept that the one or two Körös culture burials brought to light sometimes

---

[12] J. Makkay: The Protovinča problem – as seen from the northernmost frontier. Paper read at the 1988 International Conference titled "*Vinča and its world*". *Int. Symposium*, ed. by D. Srejović-N. Tasić. Beograd, 1990, 113-122, Plate 4, 2a-b.

[13] Ibid. Plate 1, 2 and 4.

[14] See J. Makkay in footnote 3, Fig. 4, 3-8.

at large-scale excavations represent real population size or we must rely on completely non-testable hypotheses. One of these is the so-called "platform burial" (This theory suggests that the corpses were placed on a platform and thus consciously exposed to scavenging animals). It must be emphasized, however, that no evidence whatsoever is available to support such hypotheses. Thus, it is still less speculative to conclude that the number of graves uncovered is realistic in the sense that it excludes a several centuries long period or even a number of periods of existence for a settlement which had only two houses, eight sacrificial pits and ten major refuse pits. At this point, the continuity of occupation (i. e. the sequential existence of houses) cannot be proven. Even if we suppose that these two houses are not coeval but followed each other diachronically (continuously or separated by a certain time interval), the occupation at this settlement may not be estimated as longer than twice two generations, that is, twice fifty years. It is likely that House 2 and the pits associated with it belong to the earlier period, while House 1 and its group forms the next stage. A small spot indicative of settlement in Square 18 marks the subsequent period. It is possible, of course, that detailed analysis of the find material will slightly modify this temporary chronology.

Summarizing the settlement history of Endrőd 119, it should be mentioned that following the Körös period only sporadic finds of the Bodrogkeresztúr culture are known from various squares at the site. These are decorated sherds from individual vessels. Refuse pits and a burial from the late Bronze Age were also recovered. The next period is from a Sarmatian occupation represented by only a few poor features. One or two houses may have stood at the site during the late Avar Period as well as during the Period of the Árpád Dynasty. Once these were abandoned, a small cemetery was opened during the 10th-11th centuries. Subsequently, the top of the small elevation lay abandoned with neither a settlement nor a cemetery in evidence.

A list of more important features mentioned in the text, plate captions and a short description of the find material

Pit 4

Represents the group of artifacts found below a depth of 60 cm in the refuse Pit located in Squares 3/1986, 4/1986 and 17/1986. It is probable that finds from a depth of between 30 and 60 cm belong to this assemblage as well, however, they were treated separately in the artifactual material. Late Bronze Age finds occurred in the humus above its eastern end at a depth of between 30 and 60 cm. It covers Sacrificial Pit 2 which is probably contemporaneous with Pit 4. Fireplace 1, located at the southern end of this Pit is probably coeval with this feature as well.

Pit 5

A Körös Period refuse Pit found in Squares 6/1986, 7/1986, 10/1986 and 11/1986. Körös type finds occurring in a secondary position in the late Bronze Age Pit 6, Pit 7 of uncertain dating and in Grave 7 dated to the 10-11th centuries were separated. The 50 to 70 cm stratification of House 1 located above Pit 5 could also be clearly distinguished. Sacrificial Pit 1 was located in Pit 5.

Pit 9

Refuse Pit recovered from Squares 12/1986, 16/1986 and 19/1986. Its find material occurred at 40 cm below the surface. The relationship between Pit 9 and Sacrificial Pit 7 could not be clarified. Grave 9 was located at the northern edge of this feature although it was not dug into the strata of the pit.

Pit 10

This refuse Pit was recovered from Squares 20/1987, 21/1987, 22/1987, 23/1987 and 24/1987 below the depth of 40 cm. Only a small portion of its eastern end remained unexcavated. This, however, represents only a shallow layer. South of this pit, Sherd Concentration 1 and Oven 4 were discovered, both probably older than Pit 10.

Pit 12

Enormous refuse Pit recovered from Squares 26/1987, 27/1987, 28/1987, 29/1987, 30/1987, 31/1987, 32/1987, 33/1988 and 36/1988. Its find assemblage could be clearly separated from the stratigraphy and material of House 2 located partly above it. The distinction was based on the colour and character (such as burnt rubble, wattle-and-daub, well defined clusters of sherds in the house layer, grey ashy fill in the pit) of strata located around a depth of 60 cm. (The majority of house remains were found at a depth of between 30 to 40 cm in a burnt layer of wattle-and-daub). The section wall between Squares 32/1987 and 33/1988 was destroyed by land cultivation at the end of 1987. Thus, parts of Pit 12 recovered from Square 33 during the 1988 season could not be exactly connected to the sections excavated previously. However, there is no evidence pointing toward a separation of these two parts. In spite of this, artifacts from Square 33 and Square 36 were kept separately. Sacrificial Pit 5 was dug probably secondarily into this southern section.

Pit 13

This is a refuse Pit recovered in Squares 35/1988, 37/1988, 38/1988 and 39/1989. Its western section was completely excavated while a small and shallow part in the south remained intact. It contained Oven 7 near its southwestern edge. Large lumps of wattle-and-daub fired to a glassy state were found south of this oven in the pit's stratigraphy. Oven 6, dating either to the late Avar Period or to the Period of the Árpád Dynasty alternatively, was found at the southeastern edge of this pit.

In summary, Körös Period pits include Pit 1 (in Squares 1/1986 and 2/1986), Pit 4, Pit 5, Pit 9, Pit 10, probably Pit 11 (in Squares 25/1987, 26/1987 and 28/1987), Pit 12, Pit 13, Pit 14, Pit 17 (in Squares 50/1989 and 53/1989), Pit 18 (55/1989), Pit 19 (52/1989 and 54/1989) and Pit 20 (46/1989).

The late Bronze Age is represented by Pit 6, Pit 15 and Pit 16.

Pit 2, probably Pit 3 and Pit 21 belong to the Sarmatian Period. (This latter was dug into Square 55/1989).

House 1

This house is represented by a characteristic layer of partly burnt rubble in Squares 3/1986, 4/1986, 6/1986, 7/1986, 8/1986, 10/1986, 11/1986, 12/1986, 14/1986 and 15/1986 at a depth of between 30 to 60 cm. This falls between the plow zone and the sterile soil at 60 cm as does the uppermost greyish level of Pit 5. This layer of rubble was sunk in an oblique plane deepening toward the east above Pit 5 and reached at times a depth of between 70 to 75 cm. This shows that the deposit in the Pit had not yet completely settled at the time of the construction of House 1. Layers of the house were heavily disturbed by tillage down to a depth of between 45 to 50 cm. Grave 1, Grave 2, Grave 3, and Grave 4 were found at the eastern end of this layer of rubble. They were badly damaged by tillage. In addition, their bone material was in part burnt and, in part, slightly dislocated. These latter two phenomena suggest that they belong to House 1 destroyed by fire, that is they date to the Körös Period. No post holes could be observed in association with this house.

House 2

This feature is represented by characteristic house rubble in Squares 25/1987, 26/1987, 27/1987, 28/1987, 29/1987, 30/1987, 31/1987, 32/1987 and 33/1988 at a depth of between 30 to 60 cm. It was badly disturbed by tillage. One may assume that post holes found in these squares as well as in Squares 34/1989, 41/1989 and 42/1989 belong, at least in part, to this house. House 2 was destroyed by fire and inside installations were at their usual place at the time of destruction. This is shown by a number of vessels (a minimum of 13 at least partially restored specimens), and the numerous net weights which lay in several clusters among the ruins. It is likely that the large vessel which was found in the northwestern corner of Square 25 and which could be largely restored (*Plate 25,2*) was a "post-sacrifice" placed by one of the post holes in this house (in the northeastern corner of Square 26). The upper part of this vessel was destroyed by tillage, however, its side is decorated by a twisting snake formed by pinched motifs below an otherwise peculiar pattern. It is very likely that Pit 2 in Square 34 was another post hole associated with House 2. It is important to mention that some fragments of a large bowl (*Plate 36,1*) were placed in the upper part of this post hole, while other sherds of the same vessel came to light in Square 33 at the bottom of Pit 12. This location is bordered by or coincides with the western end of Sacrificial Pit 5. The sherds stood vertically by the western wall of Pit 12 (or possibly in the western end of Sacrificial Pit 5) where its edge merged into the sterile soil.

House 3

This house with an oven recovered in Square 2/1986 belongs either to the late Avar Period or to the early Period of the Árpád Dynasty. Its dating is uncertain due to the lack of ceramics necessary to place it within a chronological context.

House 4
House 4 was found with an oven and floor in Square 55/1989. It extended in part over the Körös Period Pit 18. On the basis of its ceramic material, it may be considered a standard late Avar Period house.

House 5
This feature was recovered from the northern half of Squares 29/1987 and 31/1987. It penetrated the late Bronze Age Pit 15 and partially cut through the layer containing House 2. Its northern wall remained (probably with remains of its oven) under the profile wall.

Sacrificial Pit 1
This feature was located in the southern end of refuse Pit 5 dated to the Körös culture. The Sacrificial Pit 1 had not cut through the stratigraphy of House 1, however, its relation to the fill deposited in Pit 5 was impossible to appraise. If Sacrificial Pit 1 belongs to House 1 (as building sacrifice), then the following chronological sequence may be suggested: Pit 5 - Sacrificial Pit 1 - House 1. Sacrificial Pit 1 differs from the rest of the sacrificial pits typologically due to its round, rather than narrow, elongated form. Another difference is that as opposed to other sacrificial pits at this site (with the exception of Sacrificial Pit 4 dug in the proximity of Pit 13), Sacrificial Pit 1 contained an artifact. This was a somewhat damaged lamp (the upper part of its small bowl was missing) found in an upside-down position in the uppermost layer of the sacrificial Pit (*Plate 32,1*).

Sacrificial Pit 2
This feature is located within the greater part of Pit 4. Several layers with small, scattered Körös Period sherds were found in this sacrificial pit which make its dating reliable. However, its chronological relation to Pit 4 could not be clarified. The two features, in fact, may be contemporaneous. Sacrificial Pit 2 is certainly not earlier than Pit 4 and was probably dug into it. This may not, however, reflect a great difference in time: it is possible that Pit 4 was not yet completely filled when Sacrificial Pit 2 was opened.

Sacrificial Pit 3
Sacrificial Pit 3 is located in Square 31/1987. It contained no vessels only the cranial half of a sheep skeleton together with its skull. It dates unambiguously to the Körös culture.

Sacrificial Pit 4
It was located on the border of Squares 31/1987 and 35/1988 under the profile wall, thus, the western part of Sacrificial Pit 4 was destroyed during the winter of 1987. This Pit contained a large vessel with a completely intact net weight in it (*Plate 25,1*). It stood on a thin yellow clay layer covering the bottom of this feature. The rim of this vessel was already missing at the time of excavation – it is possible that it was removed by tillage. This feature is undoubtedly a Körös Period sacrificial pit.

Sacrificial Pit 5
This sacrificial Pit was dug into Pit 12 in Square 33/1988. Its section, carefully recovered outside Pit 12, contained no artifacts. On the other hand, in sections of this sacrificial Pit cut into the refuse pit (or in the proximity of the hypothesized western end) Pit 12 contained remarkably well preserved finds, including the only intact clay figurine found in the entire excavation (*Plate 28,1*). A tiny vessel with a pointed rim and a human face on it was also recovered from here. Even if these two pieces do not belong to Sacrificial Pit 5, their deposition in this refuse Pit may be related to it. The same holds true for the fragments of the previously

mentioned large vessel which were deposited either in the deepest part of Pit 12 in the end of Sacrificial Pit 5 or in its proximity (*Plate 36,1*; see also the description of House 2!).

Sacrificial Pit 6
It was located in Squares 41/1989 and 44/1989. This Pit had a stratigraphy but contained no artifacts. It undoubtedly belongs to the Körös Period.

Sacrificial Pit 7
This sacrificial Pit was found in Squares 19/1986 and 52/1989. Its eastern end falls under the narrow profile wall between Squares 19 and 52. The eastern end of Sacrificial Pit 7 could not be located during the excavation of Square 19 in the 1986 season (assuming that this could be expected at a depth of 60 cm). This feature also had a stratigraphy but contained no artifacts. It unquestionably belongs to the Körös Period.

Sacrificial Pit 8
This is a sacrificial Pit different from the previously mentioned features: a large restorable vessel, a footed bowl (located upside-down north of the large vessel) and a few animal bones (toward the south) were placed into a roundish Pit at a depth of 110 cm, dug slightly into the sterile soil. This feature also belongs to the Körös Period.

Grave 1, Grave 2, Grave 3, and Grave 4 were located in Square 10/1986. Their detailed stratigraphic position is given under "House 2". They belong to the Körös Period. One may not rule out the possibility that they are not formal burials but are connected to the destruction of the house.

Grave 1
Orientation: east-west, infant without grave goods.

Grave 2
Orientation: west-east, infant without grave goods.

Grave 3
Orientation: west-east. Skeleton laying on its ventral side. The skull and legs were swept away during tillage. Some of the remaining cranial fragments as well as the ribs show marks of burning. This points to a connection with House 1, implying that the skeleton belongs to the Körös Period.

Grave 4
Infant skull located south of the previously mentioned three burials at the same depth. This ranges around 50 cm below the surface in the case of all four graves.

Grave 5
This burial was recovered from Square 15 during the 1986 season. Orientation: northwest-southeast. No grave Pit could be observed. The body lay on its dorsal side at the very bottom of the humus layer. The skeleton is extended. Its skull, left arm and left side ribs were disturbed by a ditch from the Period of the Árpád Dynasty or the late Avar Period. No grave goods were found. The depth of the burial was between 50 to 55 cm. It dates to the 10th-11th centuries.

Grave 6
The body lay in an extended position on its back part below the pelvic region. The upper part of the skeleton rested on its left side while the skull was similarly turned on its left side. The skull was found some ten cm deeper than the rest of the bones from this skeleton. No grave goods were found in this 50 to 55 cm deep burial. An infant skeleton was identified left of the body in the region of the thigh.

Grave 7
The oblong grave Pit with rounded corners was dug into Pit 5. At a few spots, remains of a wooden coffin could be observed, especially in the area of feet at the level below the bones. The only grave good discovered was an iron buckle in the pelvic region. Orientation: northwest-southeast, depth 139 to 140 cm. This grave dates to the 10th-11th centuries.

Grave 8
This burial is located in Square 13/1986. Orientation: northwest-southeast. No grave Pit could be identified since the skeleton lay in the humus. Both forearms (below the elbow joint) were found under the body. No grave goods were discovered.

Grave 9
This burial was found at the northern edge of Pit 9, however, it was not placed directly into this refuse pit. The skeleton lay directly on the yellow sterile soil in a greyish, 10 to 25 cm thick ashy layer at a depth of between 60 to 85 cm. It is in a strongly contracted position with the face down. The left arm is pulled up and bent at the elbow joint with the phalanges of the left hand found under the face. The right forearm was shifted from its original position. The right femur lay below the left femur and tibia while bones of the right ankle were found under the left ankle joint. Several bones show original fragmentation. These include the right and left tibia, the right femur, and bones of the right forearm. This damage may have been caused by the central part of the skeleton being located over a layer softer than its surroundings, thus, corresponding parts of the vertebral column and the rest of the skeleton sank. No grave goods were found. Orientation: north-northwest – south-south-east, depth: 85 cm.

Grave 10
Grave 10 was found in Squares 37/1988 and 38/1988 at a depth of 120 cm. It consisted of a small child's contracted skeleton laying on its left side with pieces of its skull scattered apart. Southeast of this find, the humerus and some other bones of an adult were found. These latter are heavily calcinated, purple in colour. This "burial", that is to say group of bones, were undoubtedly thrown into a refuse pit: no traces of disturbance could be identified. This means that previously they may also have spent some time on the surface of the fill of the pit.

Grave 11
A few human bones were discovered in a primary position in Squares 37/1988 and 38/1988, that is in the part of Pit 13 that falls within Square 38. This location is north of Grave 10 at about a depth of 100 cm.

Grave 12
This burial was found in Squares 25/1987, 41/1989 and 43/1989 at a depth of 110 cm. The grave was in part dug into the Körös Period strata. It was a 10th to 11th centuries burial with a horse skull and horse extremity bones along and above the thigh of the corpse. Grave goods included a pair of iron stirrups and iron cheekbits. Orientation: north-northwest – south-southeast, the grave Pit could not be observed (*Plate 43,2*).

Grave 13
This burial located in Square 41/1989 has a northeast-southwest orientation and contained the remains of a contracted adult skeleton. The body lay on its right side. Originally, this person was buried (or perhaps rather tossed as may be judged from the skeleton's position) close by a concentration of sherds deposited on top of the sterile soil. No grave goods were found.

Grave 14
Late Bronze Age inhumation in Square 52/1989. Orientation: S-N, depth: 80 cm. Three artifacts were included as grave goods: a small cup, a bowl and a larger cup (*Plate 36,2-4, Plate 43,4*).

Grave 15
A small infant skeleton was found in Square 54/1989 at the very bottom of the black humus. Orientation: northwest-southeast. It dates perhaps, to the 10th to 11th centuries.

It must be added that all of the Körös culture refuse pits contained sporadic human bones which did not belong to formal burials. In the Square 29 part of Pit 12 remains of a very small, perhaps one year old infant were discovered. The exact position of these bones, however, could not be worked out.

Among the *ovens* Oven 1 (on the border of Square 18/1986 and Square 49/1989), Oven 3 (on the border of Square 21/1987 and Square 25/1987) and Oven 6 (in Pit 13) belong to the Period of the Árpád Dynasty. Oven 1 was opened, drawn and reburied in order to preserve it since the adjacent part of Square 49 was not excavated in 1989. The parts of Oven 6, located south of Square 37/1988 remained similarly unexcavated. Körös Period ovens include Oven 2 in Pit 9, Oven 4 by Pit 10, Oven 5 (on the border between Square 28/1987 and Square 55/1989), Oven 7 in Pit 13 and Oven 8 in Pit 14. This latter refuse Pit also served as an ash deposition spot for Oven 8 in the later part of its existence. Oven 5 had a dual division, with one half being almost a half metre higher than the southern section. The walls were all burnt red and the whole structure was dug into the sterile soil. Oven 8 has a very complicated structure so that it did not even appear typical of the Körös Period at first sight. Later it became clear that its opening led to the upper layer of Pit 14 and its ash was scraped into that feature. The body of the oven was arched which could have been achieved by digging under the Körös Period surface, although it is also possible that the cavity was scooped sideways with access from Pit 14. The main cavity was in sterile soil while the upper cavity extended into the subhumus layer. It could be clearly observed that prior to plastering, roughly cut slabs of wood were horizontally wedged into the soil from the direction of the opening in order to support the plastered arch. Negative imprints of this wood were preserved in the well fired arch of the oven.

The pieces of wood were placed in a radial form, pointing to the oven's opening. A shaft of oval cross-section and approximately 50 cm in diameter joined the oven's arch at its western part. The uppermost preserved point of this shaft could be observed in the subhumus and the greyish Körös Period settlement layer respectively. The dating of this complex and rare oven is unambiguous, since a Körös type bowl (broken into several pieces) was discovered in it. It was, in a way, stuck under the oven's side arch in the eastern part of the feature.

Oven 4 was small but arched as well. It had a simple bun shape and was built on the sterile soil. The walls of this oven were well fired. Ovens 7 and 2 were similar, perhaps somewhat larger, around 80 cm in diameter. Both of them were built on the edge of adjacent refuse pits.

The fireplaces all essentially belonged to the Körös Period and no plastered firing surfaces could be observed. They rather resembled round surfaces dug into the sterile soil or the actual floor level. They had a burnt layer and well-fired lumps of clay scattered over them. One of them (Fireplace 1) was associated with a Pit (Pit 4), the other (Fireplace 2) stood by itself in Square 25/1987. Neither fireplace contained restorable vessels or had foundations containing sherds.

Other features such as the late Avar Period or Period of the Árpád Dynasty ditches are not described here. They will be discussed in detail together with the elaborate presentation of Körös culture features such as Pit stratigraphies and house structures in the final publication.

The position of most important (predominantly Neolithic) features shows very well that large refuse pits with rich find materials were located along the southern edge of the oval elevation oriented northeast-southwest. Of the two Körös Period buildings, House 1 was located on the ridge of this mound near its northeastern end while House 2 was built more on its southern slope around the mid-line. Only smaller features were to the north while the middle area of this elevation contains no Körös features at all. Considering that this central surface marks the highest point of this otherwise very low mound, it is likely that this small area served as a communal ground for social life at this small settlement, which at most could have consisted of two 10 to 12 metres long houses at any given time. The fact that the area available for house construction was limited to these dimensions is shown by the late Avar Period and Period of the Árpád Dynasty situation as well. In these times a maximum of three buildings (Houses 3, 4 and 5) may have stood on the same elevation. These, however, were significantly smaller than the Körös Period houses. The very small size of population and short occupation are also indicated by the small number of Neolithic burials: no more than ten people were interred here, the majority of whom were children.

A short description and identification of illustrations as well as a brief inventory of the published artifacts

Plate 1: So-called carinated bowl whose belly has a sharp profile line and pedestalled bowls (5, and 1, 3, 6), pedestalled bowl of unusual shape (4), "metallic" ware with shiny black-brown surface and knobs on the belly's sharp edge (2), clover-form pedestal (6) covered with brownish red slip,

1: Pit 5, 90-130 cm
2: House 2, 50-60 cm
3: Pit 10, 60-80 cm
4: Pit 12 ?, 30-60 cm
5: Pit 12, 80-100 cm
6: Pit 10, 30-90 cm

Plate 2: Vessels with neck: with a medium tall neck leaning slightly outward (1-6), or sometimes having a cylindrical shape (7),

1: House 2, vessel 2
2: Square 43, between 60-90 cm
3: Pit 4, 70-90 cm
4: Pit 12, 60-120 cm, under House 2
5: Pit 5, 45-55 cm
6: Pit 4, 90-130 cm
7: Pit 13, 90-130 cm

Plate 3: Pedestalled bowls, the pedestals of which are low and narrow relative to the bowl's dimensions. The bowls are usually high bodied, closing in an upward direction. The pedestal may be cylindrical (3), broadening downward in a slight or more expressed arch (2, 5-6). These pieces are moderately elaborate. Many of them are corroded (4-5). A thin-walled vessel (1) with a smoothly burnished surface and yellowish light brown in colour deserves attention. Sherds from this specimen were found not only in several parts of the same feature, but also came to light from other features,

1: Pit 12 and House 2, that is Square 29: 0-30 cm, 30-40 cm, 30-60 cm; Square 27: Pit 12; Square 30: 0-30 cm, 30-40 cm; Square 31: 50-85 cm
2: Pit 13, 80-150 cm
3: Pit 5, 50-80 cm
4: Pit 12, 50-80 cm
5: Pit 4, 90-160 cm
6: Pit 13, 80-120 cm

Plate 4: Spherical bowls standing on a low pedestal or ring with a tall or medium tall body. The sides usually close inward without marked rims or with a barely discernable rim curving slightly outward (4). They are usually dark (brown or greyish brown) in colour. This medium fine ware is often heavily tempered with chaff (5). They are commonly corroded (6). The bowl (3) is worthy of mention: it stands on two elongated human legs (the other view is shown in *Plate 11,5)*,

1: Pit 13, 80-100 cm
2: Pit 12, 160-180 cm
3: Pit 13, 60-80 cm
4: House 1, 30-50 cm
5: Pit 5, 40-70 cm
6: Pit 12 and House 2, 0-30 cm (Square 27, 60-90 cm under the house and Square 29, 0-30 cm, House 2, Sherd concentration 3)

Plate 5: Round bellied cups standing on a heavy flat base or low pedestal ring (1), four clover form small legs (6). Their low and short rims are clearly sectioned and may be vertical or slightly inward leaning (2-3). A variety with a broad, extended belly may look flat (1 and to some extent 4). Their surface is medium fine and corroded depending on the extent to which they were exposed to weathering (6),

1: Pit 12, 105-220 cm
2: Pit 12, 100-130 cm
3: Pit 12, 30-60 cm, 90-110 cm
4: Pit 13, 60-100 cm
5: Pit 12, 60-90 cm
6: Pit 10, 60-90 cm, 90-120 cm

Plate 6: Vessels usually standing on low, sectioned bottoms similar to the specimens shown in *Plates 4* and *5*. Some of them have a clearly semi-spherical bowl shape (6-7), others have a cylindrical or globular bowl (4) or an upward closing side with (2-3) or without a sectioned rim (1). Their surface is medium fine displaying occasional corrosion (3),

1: Pit 5, 90-110 cm
2: Pit 13, 60-125 cm
3: Pit 13, 80-100 cm
4: Pit 13, 100-125 cm
5: Pit 13, 80-200 cm
6: Square 32 west, 50-80 cm (House 2 ?)
7: House 1, 30-50 cm

Plate 7: Closed bowl types standing on a low and sectioned bottom, pedestal disc (3) or four low legs (4). An exception (6) has a cylinrical upper part and an outward leaning rim. As a consequence of their closed form, the profile of their belly line is relatively sharp and occurs symmetrically at the mid-height (1-4, 7) or somewhat beyond this line (6, 8). Some of these artifacts are remarkably well executed and characterized by thin walls, as well as a finely smoothed and burnished surface which may be blackish (4) or dark brownish (1-2). It is a relatively rare form and category, although a number of such fragments from other vessels were found, sometimes even with an impressed line decoration,

1: Pit 12, 80-150 cm
2: Pit 13, 125-130 cm
3: Pit 12, 60-90 cm
4: Pit 13, 60-90 cm, 100-180 cm
5: Pit 12, 65-90 cm
6: Pit 13, 60-100 cm
7: Pit 13, 100-120 cm
8: Pit 13, 80-120 cm

Plate 8: Bowls standing on a taller pedestal disc or on a low tubular pedestal. Some of these have a cylindrical bowl (7-8), or outward leaning rim (7), as well as an outwardly expanding truncated conical form (1-6). Some of the pedestals are cylindrical (4, 5-7). Others broaden in a downward arch (3 and especially 5). They are characterized by medium fine execution. None of the broadly arched types have straight walls. One specimen has a semi-spherical bowl (6),

1: Pit 12, 50-100 cm
2: Pit 12, 125-150 cm
3: Pit 13, 85-110 cm
4: House 2, 30-60 cm

5: Pit 12 ?, 30-60 cm
6: Pit Square 15, 30-60 cm (House 1 ?)
7: House 2, 50-60 cm
8: Pit 12, 60-90 cm

Plate 9: Large, thick walled vessels which are of spherical or upside-down truncated cone shape. They usually stand on four legs which may be either round or angular in cross-section. Some of these vessels have an uneven rim formation. This group is characterized by medium or poor quality execution. Usually only the larger specimens are decorated with pinched patterns, barbotine and deeply cut lines,
1: Pit 13, 100-120 cm
2: Pit 13, 60-125 cm
3: Pit 13, 120-160 cm
4: Pit 13, 100-120 cm
5: Pit 9, 110-130 cm
6: Square 21, 30-60 cm
7: Pit 9, 80-110 cm
8: Pit 9, 80-130 cm

Plate 10: Large, thick-walled bowls of spherical shape standing on four legs of either round or angular cross-section. Larger varieties of the bowls shown in *Plate 9* with totally similar characteristics,
1: Square 26, 30 cm and 60-90 cm, House 2 ?
2: Pit 9, 130-170 cm
3: Pit 10, 60-90 cm

Plate 11: Usually small bowls of an upside-down truncated cone shape with a arched wall (1) or a sectioned bottom part (2-3). The rim has a lacey decoration produced by finger imprinting although a pinched pattern is possible as well (1). The other vessels are round bowls or cups standing on legs. Their shape is in every aspect similar to the vessels standing on sectioned bottoms or pedestal discs (see *Plates 3 to 7*!). Here a specimen standing on very low legs may be seen (4), as well as another sitting on two low longitudinal legs (5; the other view of this vessel is shown in *Plate 4,3*). There are forms with a medium fine surface finish (7-8),
1: Pit 12,130-160 cm
2: Pit 12, 150-200 cm
3: Pit 12, 80-110 cm
4: House 1, 40-50 cm
5: Pit 13, 60-80 cm
6: Pit 10, under 120 cm
7: Pit 9, 130-175 cm
8: Pit 5, 50-70 cm

Plate 12: Vessels standing on legs including open bowls which broaden upwardly (2-4, 6), or narrow upwardly without a sectioned rim (1). Other types have a cylindrical vertical shape (7 and perhaps 8) and an outward leaning, funnel-like rim (5). In a few cases diverging (2, 5-8) and stump-like legs have been noted (3; see also *Plate 11,4!*). The execution of these usually undecorated vessels is medium poor,
1: Pit 13, 60-120 cm
2: Pit 13, 60-200 cm
3: Pit 12, 60-145 cm

4: Pit 5, 160-120 cm
5: Pit 9, 110-170 cm
6: Pit 12, 60-90 cm
7: Pit 9, 130-175 cm
8: Pit 13, 100-120 cm

Plate 13: This plate shows vessels with legs. All of these are cups with a spherical and a closed, upwardly narrowing form. Body proportions may vary, but the upper parts in all cases lack profiling and decoration. The legs may sometimes be quite small (8) or tall, and either round or angular with a conical form (5) as well as flat and angular (2, 7). There are always four such legs, however,

1: Pit 9, 80-100 cm
2: Pit 9, 60-110 cm
3: Pit 9, 130-175 cm
4: Pit 9, 80-110 cm
5: Pit 13, 80-125 cm
6: Pit 13, 125-150 cm
7: Pit probably House 1 (Squares 7 and 11, 40-50 cm)
8: House 1

Plate 14: Similarly to those shown in the previous plate these are also spherical vessels standing on four legs. Some of them (1-2) have an upwardly narrowing form (as in *Plate 13*), while others have a lower or taller sectioned rim. The rim may lean slightly outward (3 and 5) or stand vertically (4, 7-8). These vessels have a medium quality execution with smoothed, sometimes heavily worn surfaces (7). One of the types (7) is characterized by a sharply broken – carinated – belly profile line and corresponds to similar spherical forms standing on sectioned bottoms or pedestal discs (*Plate 7,1-2, 7*),

1: Pit 12, 30-90 cm
2: Pit 12, 100-130 cm
3: Pit 12, 90-120 cm
4: House 1, 55-60 cm
5: Pit 13, 80-125 cm
6: Pit 13, 100-125 cm
7: House 1, 40-50 cm
8: Pit 13, 120-125 cm

Plate 15: Small, spehrical form similar to cups standing on sectioned bottoms or low pedestal discs (1-2) which is characterized by a spherical (2-3) or sectioned belly. This carinated line may even have a slight edge (1, 8). Some forms may have an ovoid shape (5). Sometimes relatively large knobs or horizontal handles with openings may be found on the belly edge or at the broadest part of the belly. These occur as single features or in pairs (1-2, 6-8 as well as 3). The majority are decorated, especially with a network (4), nail impressions (6), barbotine (4) with a line of nail impressions on the shoulder and other similar motifs (7). In general, the execution of this pottery is worse than average. In the case of forms with rims, these rims may be vertical (4, 7) or lean outward to varying degrees (3, 6, 8) and may even be sectioned by finger imprints,

1: House 2, 30-40 cm
2: Pit 4, 40-130 cm
3: Pit 13, 100-120 cm
4: Square 13, 50-70 cm

5: Pit 13, 100-120 cm
6: Pit 13, 100-150 cm
7: Pit 9, 110-130 cm
8: Pit 12, 90-140 cm

Plate 16: Larger varieties of the type shown in the previous plate. The only exception is a specimen with a rather highly located edge on the belly's profile line and a rim sectioned by incisions (4). Other types have a slightly outward (1-3, 5) or inward (6) leaning rim. They are decorated by pinched patterns (1-2, 6) or barbotine (3, 5). They are of medium or even poorer quality with the exception of the vessel shown in 1. Although this specimen is decorated only by pinched patterns, its walls are burnished to a shiny brown,
1: Pit 5, 90-110 cm
2: Pit 12, 80-160 cm
3: Pit 5, 40-110 cm
4: Pit 4, 90-130 cm
5: Pit 12 and House 2, 30-40 cm, 60-90 cm
6: Pit 12 and House 2, 100-120 cm, 30-60 cm

Plate 17: Large size vessel standing on four low legs from the artifactual assemblage of House 2 (1), and the inside bottom surface of the large storage vessel (2) also shown in *Plate 25,1* showing signs of prehistoric repair work (slightly lighter in color). This bottom part probably cracked during the drying process. The repair was unsuccessfully attempted by applying a soft clay plastering on the vessel's bottom. It is possible that this crack was one of the reasons why this piece was deposited in Sacrificial Pit 4,
1: House 2, vessel "b"
2: Sacrificial Pit 4 = Square 31, Pit 3

Plate 18: Storage vessels with a spherical body. Some of these have a sectioned bottom (1) others do not (2-3). In addition to undecorated forms and knobs on the wall, a net decoration sometimes occurs. These vessels are characterized by the usual simple execution with some smoothing and polishing (2),
1: Pit 13, 100-140 cm
2: Pit 12, 120-160 cm
3: Pit 12, 100-160 cm

Plate 19: Spherical storage vessels with a relatively narrow opening similar to the specimen in *Plate 18,1*. The rim is clearly divided, leaning slightly outward (1,3-4) or vertically. It is decorated by pinched patterns and single, paired or divided knobs (2-4) located symmetrically on the lower third of the body on four sides of the vessel. A relatively rare, tall form also occurs (4). An intentional blackening (painting?) inside the rim is common with this type. The paint sometimes leaked on the outside of the vessel leaving black strips (this latter is not shown in the photographs),
1: Pit 13, 60-150 cm
2: Pit 13, 60-125 cm
3: Pit 5, 110-140 cm
4: Pit 4, 110-130 cm

Plate 20: Varieties of the large vessels with spherical bodies shown in Plates 18 and 19: both broad (2) and narrow (1, 3) mouth openings, rims vary from low (4) to high and outward (5) or slightly inward leaning (1, 3-4). These vessels may be undecorated or decorated with barbotine (1, 3-4) and single as well as divided knobs placed symmetrically in the lowest region of the vessels' body,

1: House 1, 40-50 cm
2: House 2 and Pit 12, Square 29, 0-60 cm ; Square 32, 60-90 cm
3: Pit 12, 80-105 cm
4: Pit 4, 130-170 cm
5: Pit 9?, 0-90 cm

Plate 21: large storage vessels similar to the previous types (*Plates 19* and *20*) on strongly sectioned bottoms with pinched and barbotine patterns. The semi-spherical, bomb-shaped storage vessel with knobs and ridges but without decoration is a new form,

1: House 1, 60-70 cm
2: Pit 9 ?, 0-90 cm
3: Pit 9, 130-175 cm

Plate 22: Broad, large dishes on clearly sectioned bottoms or even pedestal discs with a smooth (2, 3) or pinch decorated rim (1, 4), with barbotine ornamentation on the lowest part of the body (1-4) or decorated by large knobs especially on the lower parts. A more open (1-3) and a relatively closed form both occur (2),

1: Pit 12, 100-120 cm
2: Pit 9, 90-110 cm
3: House 2, vessel 8, 65-90 cm
4: House 2, vessel 13 and Pit 12, 65-90 cm

Plate 23: Additional forms of large storage vessels: broad dishes standing either on sectioned bottoms, pedestal discs or four legs (5). Some have a broad body (2-5), others have a relatively closed or almost cylindrical shape (1). The rim is decorated by pinching in all five cases. Simple coarsening (1) and all sorts of pinched patterns occur (2-5) on these vessels. Large handles, knobs and even large handles with horizontal openings may be found on their lower sections,

1: Pit 12, 105-125 cm
2: Pit 12, 60-90 cm
3: Pit 12, 130-160 cm
4: Pit 9, 90-120 cm
5: Pit 12, 60-160 cm

Plate 24: This butt-shaped vessel is another form of storage jars typically with an arrangement of four large handles on one side, all of them having horizontal openings. The three main types shown here include the pear-shaped (1-2), cylindrical (3) and spherical body formation with a tall neck (4). They are decorated by rows of pinched patterns (1) or nail incisions. They sit on strongly sectioned pedestal discs (3) or four legs (1). Find 2 is a small bowl. It is special for its set of unusually large handles with horizontal openings placed on its belly,

1: House 1 40-50 cm, 50-70 cm
2: Pit 5, 70-90 cm
3: Pit 5, 110-130 cm
4: Pit 12, 60-110 cm

Plate 25: Two large storage vessels found in a ritual context. One of them stood at the bottom of Sacrificial Pit 4 and was found in this position even at the time of excavation. A completely intact, standard shape net weight was found at the bottom of this vessel. Its rim, as has been mentioned, was probably dislocated by tillage. A crack seen on the inside of this vessel was probably repaired before firing (1; see *Plate 17,2* as well). The other piece was probably dug into a thin yellow clay layer near to or in the northeastern post hole of House 2. Its upper part was destroyed by tillage. The direction of this disturbance shows that the vessel was placed by the post hole in a slightly oblique position. As may be seen on the lower part of the picture published here, additional motifs occur below the barbotine decoration. These include a pattern of straight incised lines and a line of nail pinched decoration below it. This row of pinched pattern forms a curving line with the first unit of the series being slightly bigger and longer. This is probably the representation of a snake. These signs provide proof that the vessel may have been part of a special "post-sacrifice" ritual(2),[15]

1: Sacrificial Pit 4
2: Square 25, northwest, by the post hole

Plate 26: A set of miniature or tiny vessels. In practice these represent the complete ceramic repertoire. The only exception may be the group of flat bowls (1, 4) which do not occur among the usual large forms in our material. The side of object 1 has two drilled holes in it, although the function of this artifact is unknown. In addition to a variety of other bowls (2-3, 4, 6-7, 11-12), and cups (5, 9-10) a few special forms are worthy of mention. One of them has an ovoid form with pinched decoration (16), while the other has a body form sectioned in an oblique plane (17). A cup of amorphous shape (15) and two small vessels perhaps depicting human feet (13, 14) also deserve attention. The representation of a human foot is particularly likely in the case of the piece shown under 14,

1: House 1, 40-70 cm
2: Pit 5, 50-60 cm
3: Pit 12, 65-90 cm
4: House 2, 30-60 cm
5: Pit 5, 70-90 cm
6: Pit 12, 80-90 cm
7: House 2?, 30-50 cm
8: Pit 5, 70-90 cm
9: Pit 12, 50-80 cm
10: Pit 5, 40-70 cm
11: Pit 12, 60-90 cm
12: Pit 12, 50-80 cm
13: Pit 4, 70-160 cm
14: Pit 5, 110-130 cm
15: Pit 12, 60-160 cm
16: Pit 4, 50-70 cm
17: Pit 10, 30-60 cm and Square 25, south 40-50 cm corresponding probably to House 2

---

[15] See footnote 7

Plate 27: Special types including handled cups with spherical or semi-spherical bodies (6, 7) which are sometimes compressed (5) or even have a belly with a sharp profile line (4). The handle is usually placed in an oblique position on the vessels' body always starting at the rim and ending usually at the bottom (6). Sometimes the handle is attached to the body's lower section or to the bottom of the edge shown in the profile line (4, 5, 7). Some of them are of medium fine quality (4, 5), while others are rougher with pinched decoration (6-7). An upward pointing knob on the handle's top or two laterally placed knobs are among the common decorative motifs.[16] In one case, horizontally placed human hands decorate the vessel (not represented). A small vessel of an upside-down truncated cone shape with a small nipple reconstructed onto it may have served as a baby dish used in "bottle feeding" (1). Two other pieces (2, 3) first seem to be tall and slender necks or deformed tubular pedestals. In fact, however, they may have been spouts, of large ascoidale vessels because their narrow inner ends (as shown in the upper part of pictures) are asymmetric. They were in part closed and opened. Unfortunately, neither of these fragments could be refitted with any of the other sherds found at the site,

1: Pit 13, 150-200 cm
2: Pit 9, 140-170 cm
3: Pit 9, 40-80 cm
4: Pit 5, 110-130 cm
5: Pit 5, 110-130 cm
6: Pit 4, 40 cm
7: Pit 13, 80-100 cm

Plate 28: Human figurines. Of these, the single complete (only one of the buttocks is damaged: it probably separated and was lost during recovery) specimen found at this excavation deserves special attention. This idol is roughly executed and composed of two matching parts. Only the nose is marked (1). As was mentioned previously, it is possible that this figurine belonged to Sacrificial Pit 5. The two halves of a body were found in the case of another specimen as well (7), however, the neck and head of this idol are missing and it was found secondarily fired without additional contextual (e.g. cultic) evidence. ALP decorated sherds were also found in the Körös Period material recovered in the immediate neighborhood of this piece.[17] An additional three pieces (2, 4, 5) include the lower part of a trunk, a right (2, 5) and a left leg (4). Unfortunately, only one (the right side) half of the largest figurine was found with the representations of the knee and toes. It cannot be excluded that a matching head covered by a completely similar grey slip, made of a comparable material and of a proportional size (*Plate 29,5*) in fact belongs to this figurine, although it came to light from another feature and cannot be fitted to the body fragment directly (6). The flat faced figurine (3) may represent a male.

1: Pit 12, or perhaps Sacrificial Pit 5, 160-200 cm
2: Pit House 1, Square 11 south, 0-50 cm
3: Pit Pit 13, 100-120 cm
4: Pit Grave 7 pit, secondary position, Square 11
5: Pit House 1, by Grave 6, 30-40 cm
6: Pit 9
7: House 1, 40-70 cm

---

[16] Compare with J. Makkay: Endrőd - Szujókereszt. *MittArchInst* 8-9 (1981), 209-213.

[17] See footnote 14

Plate 29: Human figurines and an object of unknown function. Idol fragments (1-5, 7-9) show that with the exception of the previously mentioned piece (*Plate 28,1*) neither complete nor restorable pieces were found and even matching fragments are missing (the only association may be hypothesized between the fragments shown on *Plate 28,6* and *Plate 29,5*). This plate shows thre heads of different types (1-3), one of them with a characteristic hair design (2, the face is damaged). Another specimen damaged on the top of its head (4) is different from the previously described threee idols in as much as it does not belong to the steatopygic type. It has a simple columnar shape, characterized by a schematic genital represented by an upward pointing triangle with a hole[18] and a separate umbilicus. Another fragment (7) must have belonged to a large steatopygous figurine, while (9) shows the side view of the artifact shown on *Plate 29,7*. It is important to mention a piece (8) which was completely different from the rest of the idols both in terms of its material and execution. It is made of a very finely levigated clay without any organic ingredients or temper. It is of a light terracotta colour and is beautifully polished. It is the right side of the body of a steatopygous idol. The function of the object (6) in this plate, having five perforations of different sizes, is unknown. Its tip (head?) has been broken off from its narrow, upward pointing end. It is possible that this object was a tool used during the course of weaving, although it shows no marks of use or other wear. See also *Plate 32,3*,

1: Pit 10
2: Pit 10
3: Pit 4
4: Pit 9
5: Pit 10
6: Pit 10
7: Pit 12
8: Pit 12
9: Square 4, 40-70 cm perhaps by House 1

Plate 30: Human representations: reliefs. Two fragments similar to each other originate from the neck part of two large vessels (one cannot rule out the possibility that they in fact belong to the same storage jar). On each of these fragments a schematic human figure with an oval face (no facial characteristics were shown) and uplifted hands are depicted (1, 3). Another fragment originates from the rim of a female-shaped vessel with the face simply marked, that is, with nose and eyes (2). A right hand (4) was probably located on the belly part of a large, non-spherical vessel. Under the neck of another large, thick-walled storage vessel an unusually expressive human face and the beginning of his neck may be seen (5). In the current treatise this representation is regarded as a male,

1: Pit 5, 40-70 cm
2: Pit 12, 130-160 cm
3: Pit House 2, 30-40 cm
4: Pit House 1, 50 cm
5: Pit 4, 50-90 cm

[18] In relation to this type of representation see M. Tulok: A Late Neolithic idol of conical type. *ActaArchHung* 23 (1971), 3-17.

Plate 31: Human representations: reliefs. Relief (1) shows a woman and is placed on the shoulder of a large vessel. The lower trunk is seen from the back, while the top of the body is shown from the front. The head is missing. We may consider this as an effort to show the body from all main directions and, thus, the principle may be called "total frontality". It is noteworthy, that according to our present knowledge, reliefs following this principle show predominantly (more exactly exclusively) females. Currently, they have come only from this site (four specimens) and the nearby Endrőd 39 settlement (2 specimens) in such great numbers. The other object (2) is the wall fragment of a large vessel showing two schematic human forms side by side. They resemble two female bodies with marked steatopygia,

1: Stray find, 1986, collected by Dénes Virágh before excavations.
2: Pit 12 or perhaps Sacrificial Pit 5, 160-200 cm

Plate 32: Small finds such as lamps (1, 4, 5) include three and four legged specimens as well. No complete lamp was found at this site. The first (1) was found in Sacrificial Pit 1 (as described previously). It was placed in the uppermost layer together with a mass of fish bone and scales, among the remains of the last burnt offering. There is a small, spherical vessel standing on two human feet with the neck unfortunately missing: this may have been the part which had a human face on it, assuming that this artifact was in fact an anthropomorphic (female shaped) vessel (2). The third object (3) is identical with that on *Plate 29,6,*

1: Sacrificial Pit 1
2: Pit 4, 70-100 cm
3: Pit 10, 80-100 cm
4: Grave 7 pit, secondary position
5: Pit 13, 85-110 cm

Plate 33: Bone tools.[19] It is characteristic of the relatively rich bone tool material of this site that only a few complete specimens were found (for example, no complete spoons were recovered). The great number of bone artifacts may also be related to the unusually small number of stone tools at this site. This plate shows a few selected pieces. Some spoons are heavily eroded (10) or originally fragmented (9, 11). All of them are made from cattle long bones with carving and polishing. The find material contains a complete series of artifacts illustrating spoon manufacturing. This ranges from the raw material through blanks (carved but unpolished forms) to the end product. Two other implements (massive bodied points) represent a type of which only complete examples were found. Their function is unknown. One side of these points is slightly concave, while the other side has a shouldered cross-section (8). All of these specimens were heavily used and show high use polish. There are simple small points (2-3), fine points (4), asymmetric pegs (pointed on one side at one end and on the other side at the opposite end), and an absolutely intact fine needle (1, 5 and 6 respectively),

1: Pit 13, 150-200 cm
2: Square 18, 50-70 cm
3: Pit 13, 125-150 cm
4: Pit 10, 120-140 cm
5: Pit 13, 120-150 cm
6: Pit 12, 80-100 cm

---

[19] Cf. J. Makkay: Knochen-, Geweih-, und Eberzahngegenstände der frühneolithischen Körös-Kultur. *ComArchHung* 1990, 23-58.

Plate 34: Fine fragments decorated with stroke burnished or finely incised lines. This decoration has three main variations. One of them is a pattern incised with a sharp tool prior to, or rarely after, firing (1 in this plate shows such a piece incised before firing). The other relatively deeply impressed line pattern is created by several strong pressures and rubbing (stroke burnishing). In the third case, no surface relief is left where the tool was moved around, and the pattern may be seen only in the differential colour and shine of the surface (pattern burnishing). The object shown under number 2 belongs to the deeply smoothed category. The first piece (1) is important because its two associated fragments came to light from two different features in two different years,

1: Pit 13 = Square 38, 60-90 cm and Pit 12 = Square 29 middle, 65-90 cm
2:House 2 = Square 32, 30-40 cm and Pit 12 = Square 30, 100-160 cm

Plate 35: Fragments of a simple or pedestalled bowl decoreted with a stroke burnished net pattern both on its outer and inner surfaces. The significance of this artifact is that its fragments were also found in two different features, House 2, 30-40 cm, Pit 12, 60-90 cm

Plate 36: Large bowl of a rare form, otherwise unknown at this site. It is decorated by plastic and engraved patterns and large, disc shaped knobs. Fragments of this piece were found in two different features: the deepest part of Pit 12 and Post Hole 2 in Square 34 which probably belonged to House 2 (1). Objects 2-4 are grave goods from the Bronze Age burial (Grave 14).

Plate 37: Excavation photographs

1: Pit 5 shown from a north-northwestern direction. Graves 5 and 6 (left to right) are shown in front. The opening of Sacrificial Pit 1 may be seen in the centre of Pit 5, to the left: Grave 7
2: Fireplace 1 in Square 17 shown from the northeast
3: The large storage vessel with snake decoration in the northwestern corner of Square 25. The spot of a post hole may be seen directly to the left

Plate 38: Excavation photographs

1: Pit 13 shown from a northwestern direction. Oven 6 from the Period of the Árpád Dynasty may be seen in the background
2: Pit 13 shown from a southeastern direction. To the left: Pit 2 in Square 38

Plate 39: Excavation photographs

1: Section of Sacrificial Pit 5 shown from a southwestern direction, that is, the inside northern end of Pit 12 with the section wall
2: Same as the previous picture (1) at an earlier stage of excavation before the large cattle horn core was removed from the section's edge
3: The excavation of Sacrificial Pit 2: outline on the sterile soil, inner section seen from a southwestern direction
4: The same as Picture 3 in an earlier stage: the section at a depth of 30 cm (plough zone). As may be seen, a dark Bronze Age layer almost seals the opening of this sacrificial shaft. Round, light spots in the profile section are rodent burrows

Plate 40: Excavation photographs

1: The section of Sacrificial Pit 6 in the profile wall between Squares 44 and 41 and the section inside this sacrificial shaft seen from a northwestern direction. A light sterile layer of clay deposit may be clearly seen. This filled the shaft along its inner surface and results in a hyperbolic cross-section

2: Sacrificial Pit 6 shown in Picture 1 in an earlier stage of the shaft's excavation and the discoloration observed at the bottom of its section
3: Sacrificial Pit 7 in Square 52 seen from the southwest. The section is located in front of the eastern wall of Square 52
4: Pit 12 and Sacrificial Pit 5 (left) in Squares 33 and 36 (above) seen from the northwest
5: Sacrificial Pit 8 in Square 40 seen from the east, with collapsed fragments of the large vessel. To the right:: an upside down four legged vessel and animal bones

Plate 41: Excavation photographs
1: Grave 13 seen from the southwest
2: Grave 13 seen from the northwest
3: Grave 9 seen from the west
4: Detail from House 2 with net weights in Square 29

Plate 42: Excavation photographs
1: The shaft or chimney of Oven 8 (left) and its shaved off arch at the depth of the plow zone (left) seen from above from a slightly northwesterly direction
2: Pit 14 with Oven 8 at one end seen from the south-southeast
3: Close up view of the same features. The four legged vessel mentioned in the text was found to the right of the assumed mouth of the oven, under the almost horizontal arch on the eastern side of the oven chamber
4: Oblique planes seen at the bottom of the oven's arch which display the imprint of the slabs of wood which supported the plastering of this horizontal arch

Plate 43: Excavation photographs
1: Grave 7 seen from the south-southeast
2: Grave 12 seen from the south-southeast with a horse skull in the foreground
3: House 5 from the Period of the Árpád Dynasty or the late Avar Period in the northern part of Squares 29 and 31 seen from the west-northwest
4: Late Bronze Age Grave 14 with grave goods seen from the north

The distribution of find material by features as published in the Plates

Pit 4 in Plates
2,3; 2,6; 3,5; 15,2; 16,4; 19,4; 20,4; 26,13; 26,16; 27,6; 29,3; 30,5; 32,2

Pit 5 in Plates
1,1; 2,5; 3,3; 4,5; 6,1; 11,8; 12,4; 16,1; 16,3; 19,3; 24,2; 24,3; 26,2; 26,5; 26,8; 26,10; 26,14; 27,4; 27,5; 30,1; 37,1

Pit 9 in Plates
9,5; 9,7; 9,8; 11,7; 12,5; 12,7; 13,1; 13,2; 13,3; 13,4; 15,7; 20,5?; 21,2?; 21,3; 22,2; 23,4; 27,2; 27,3; 28,6; 29,4

Pit 10 in Plates
1,3; 1,6; 5,6; 10,3; 11,6; 26,17; 29,1; 29,2; 29,6; 32,3; 33,4

Pit 12 in Plates
1,4?; 1,5; 2,4; 3,1; 3,4; 4,2; 4,6; 5,2; 5,3; 5,5; 7,1; 7,3; 7,5; 8,1; 8,2; 8,5?; 8,8; 10,2; 11,1; 11,2; 11,3; 12,3; 12,6; 14,1; 14,2; 14,3; 15,8; 16,2; 16,5; 16,6; 18,2; 18,3; 20,2; 20,3; 22,1; 22,4; 23,1; 23,2; 23,3; 23,5; 24,4; 26,3; 26,6; 26,9; 26,11; 26,12; 26,15; 29,7; 29,8; 30,2; 31,2?; 33,6; 34,1; 34,2; 35; 36,1; 40,4

Pit 13 in Plates
2,7; 3,2; 3,6; 4,1; 4,3; 5,4; 6,2; 6,3; 6,4; 6,5; 7,2; 7,4; 7,6; 7,7; 7,8; 8,3; 9,1; 9,2; 9,3; 9,4; 11,5; 12,1; 12,2; 12,8; 13,5; 13,6; 14,5; 14,6; 14,8; 15,3; 15,5; 15,6; 18,1; 19,1; 19,2; 27,1; 27,7; 28,1; 28,3; 32,5; 33,1; 33,3; 33,5; 34,1; 38,1; 38,2;

Pit 14 in Plates
42,2; 42,3

House 1 in Plates
4,4; 6,7; 8,6; 11,4; 13,8?; 14,4; 14,7; 20,1; 21,1; 24,1; 26,1; 28,2; 28,7; 30,4

House 2 in Plates
1,2; 2,1; 6,6?; 8,4; 8,7; 10,1?; 15,1; 16,2; 16,5; 17,1, vessel "b"; 20,2; 22,3, vessel 8; 22,4, vessel 13; 26,4; 26,7?; 26,17; 30,3; 34,2; 35; 36,1; 41,4

House 5 in Plate
43,3

Sacrificial Pit 1 in Plates 32,1; 37,2
Sacrificial Pit 2 in Plate 39,3-4
Sacrificial Pit 4 in Plates 17,2; 25,1
Sacrificial Pit 5 in Plates 31,2?; 39,1-2?; 40,4
Sacrificial Pit 6 in Plate 40,1-2
Sacrificial Pit 7 in Plate 40,3
Sacrificial Pit 8 in Plate 40,5
Square 4, 40-70 cm in Plate 29,9
Square 7+11, 40-50 cm = House 1 in Plate 13,8
Square 11 south = House 1 in Plate 28,2
Square 11, 70-80 cm = Grave 7 fill in Plates 28,1; 32,4
Square 13, 50-70 cm in Plate 15,4
Square 15, 30-60 cm (perhaps House 1?) in Plate 8,6
Square 18, 50-70 cm in Plate 33,2
Square 21, 30-60 cm in Plate 9,6
Square 25 northwest, by the post hole in Plates 25,2; 37,3
Square 25 south, 45-50 cm = House 2 in Plate 26,17
Square 26+30, 60-90 cm (House 2?) in Plate 10,1
Square 31, Pit 3 = Sacrificial Pit 4 in Plates 17,2; 25,1
Square 32 west, 50-80 cm (House 2?) in Plate 6,6
Square 34, 60-90 cm in Plate 2,2
Square 34, Pit 2 in Plate 36,1
Stray find in Plate 31,1

Reconstructed-completed vessels recovered from different features

Plate 3,1: Pit 12 and House 2
Plate 4,5: Pit 12 and House 2
Plate 16,5: Pit 12 and House 2
Plate 16,6: Pit 12 and House 2
Plate 20,2: Pit 12 and House 2
Plate 22,4: Pit 12 and House 2
Plate 26,17: Pit 10 and House 2
Plate 34,1: Pit 12 and Pit 13
Plate 34,2: Pit 12 and House 2
Plate 35: Pit 12 and House 2
Plate 36,1: Pit 12 and House 2

Data on associated sherds found at very different depths of the same refuse pit

Plate 2,4: Pit 12, 60-120 cm
Plate 3,1: Pit 12, and House 2
Plate 5,3: Pit 12, 30-110 cm
Plate 5,6: Pit 10, 60-120 cm
Plate 6,13: Pit 13, 80-220 cm
Plate 7,1: Pit 12, 80-150 cm
Plate 7,4: Pit 13, 60-180 cm
Plate 12,2: Pit 13, 80-200 cm
Plate 12,3: Pit 12, 60-145 cm
Plate 15,2: Pit 4, 40-130 cm
Plate 16,1: Pit 12, 80-160 cm
Plate 19,1: Pit 13, 60-150 cm
Plate 23,5: Pit 12, 60-160 cm
Plate 26,13: Pit 4, 70-160 cm
Plate 26,15: Pit 12, 60-160 cm
Plate 32,2: Pit 4, 70-160 cm
Plate 34,2: House 2 + Pit 12, 30-160 cm

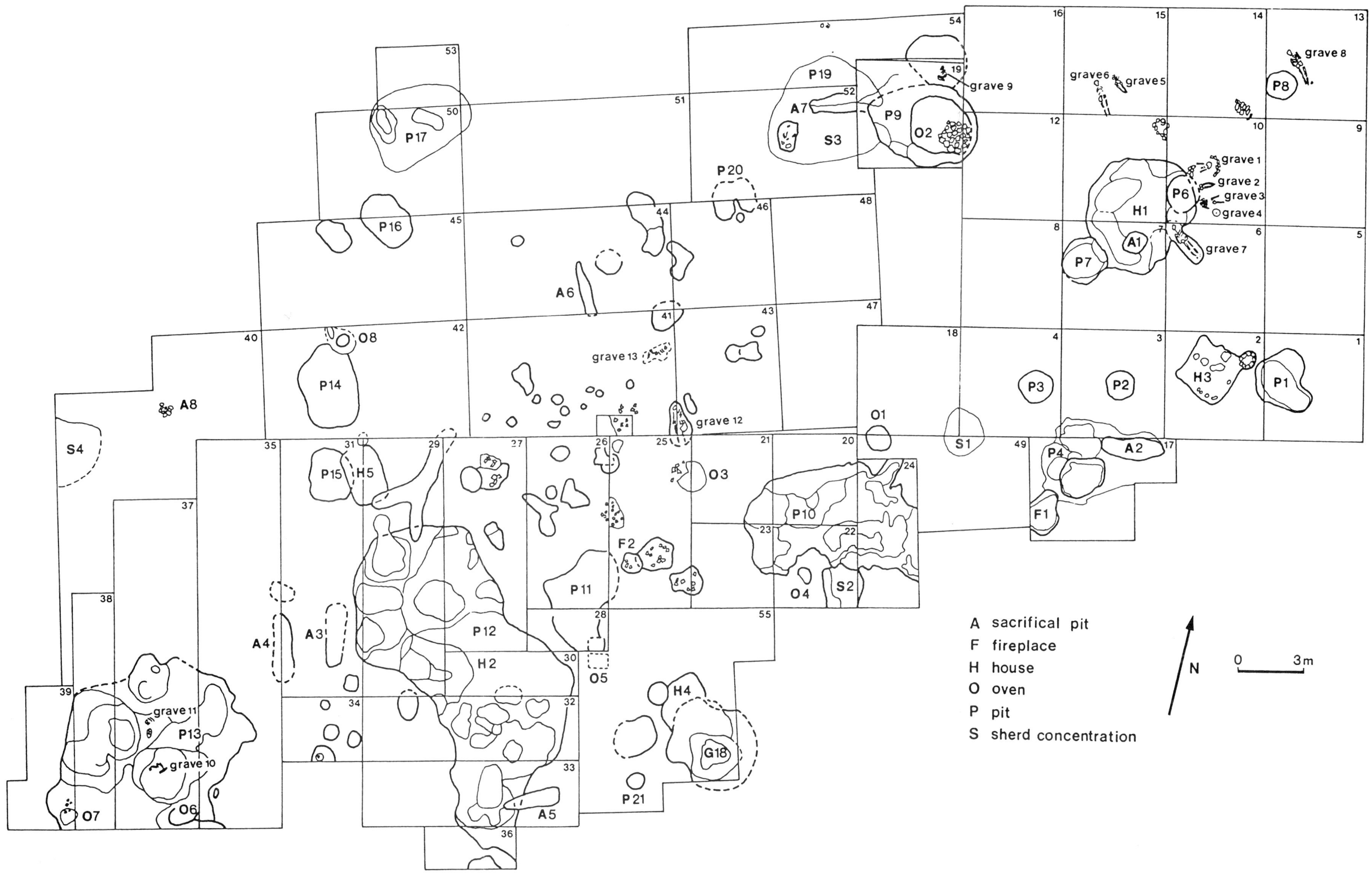
A sacrifical pit
F fireplace
H house
O oven
P pit
S sherd concentration
N
0 3m
grave 1
grave 2
grave 3
grave 4
grave 5
grave 6
grave 7
grave 8
grave 9
grave 10
grave 11
grave 12
grave 13
H1
H2
H3
H4
H5
P1
P2
P3
P4
P6
P7
P8
P9
P10
P11
P12
P13
P14
P15
P16
P17
P19
P20
P 21
A1
A2
A3
A4
A5
A6
A7
A8
O1
O2
O3
O4
O5
O6
O7
O8
S1
S2
S3
S4
F1
F2
G18

General map

Plate 1.

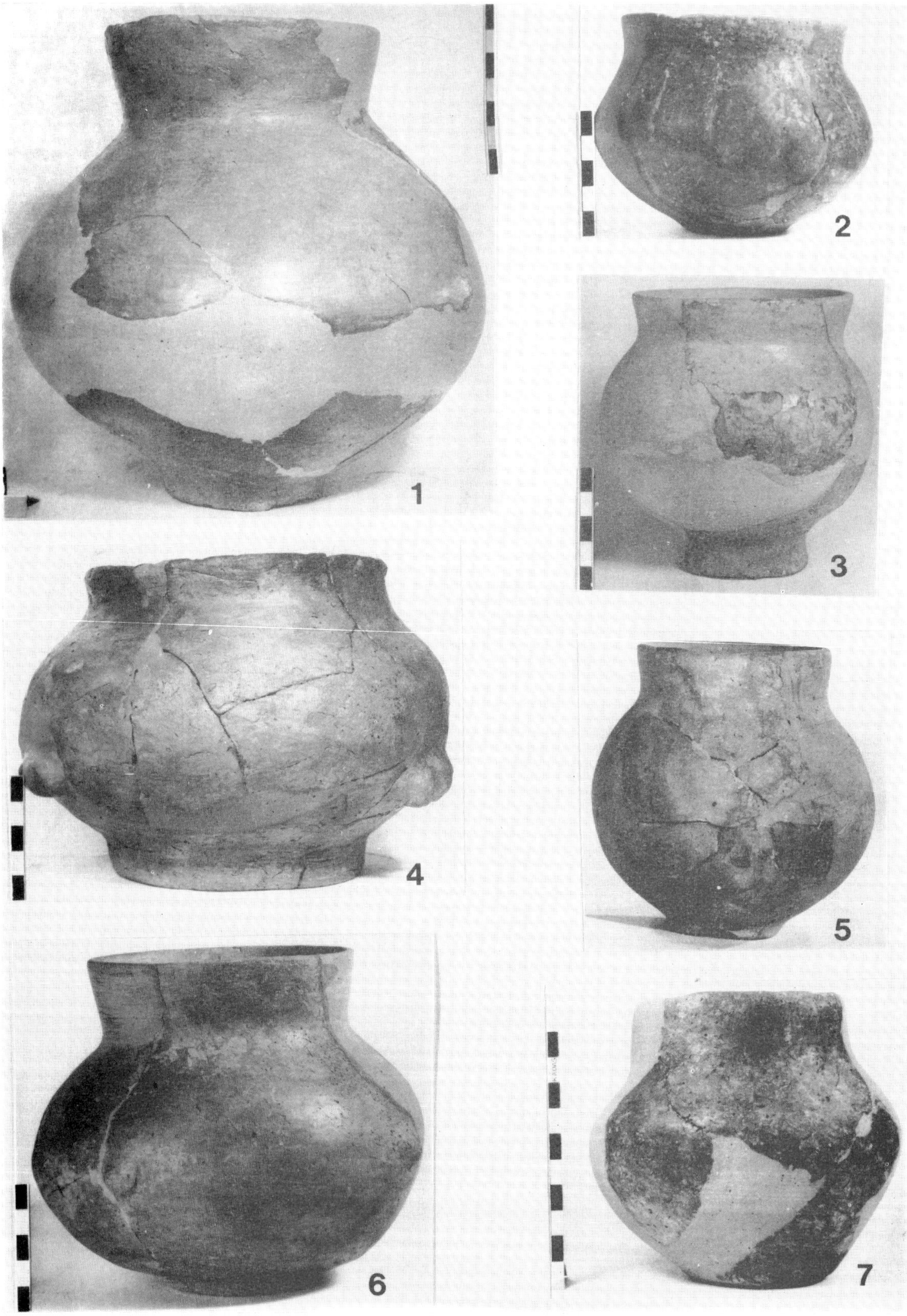

Plate 2.

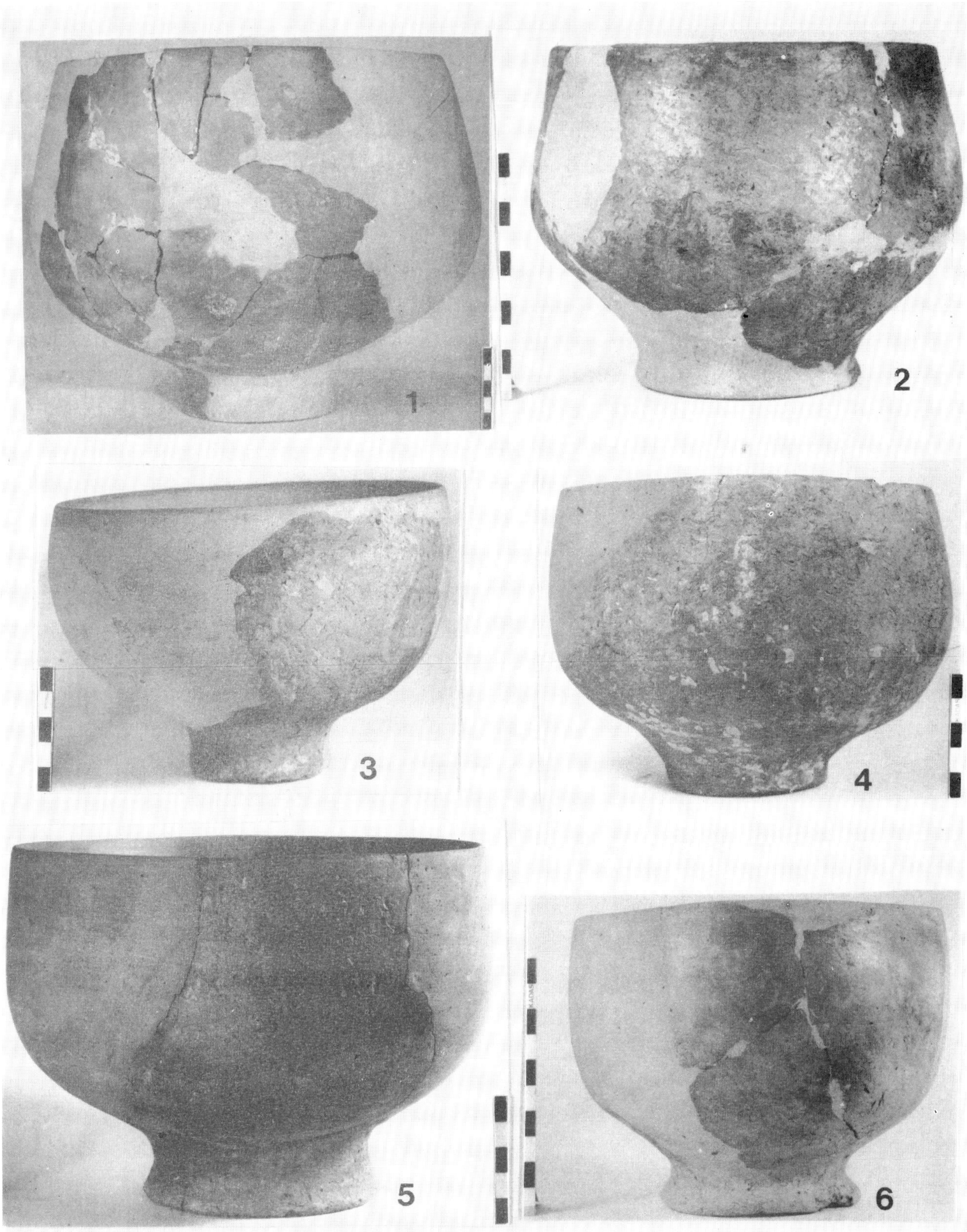

Plate 3.

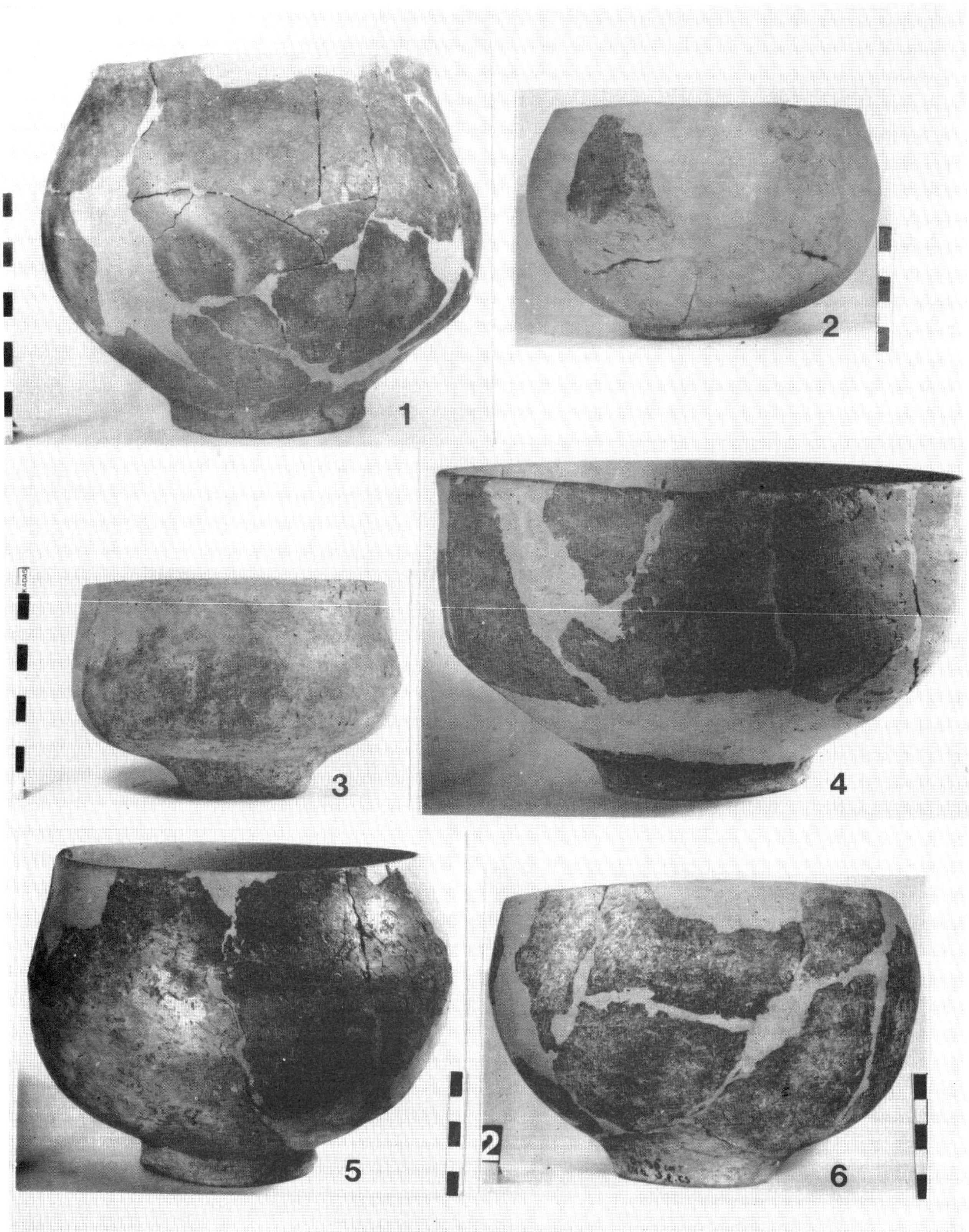

Plate 4.

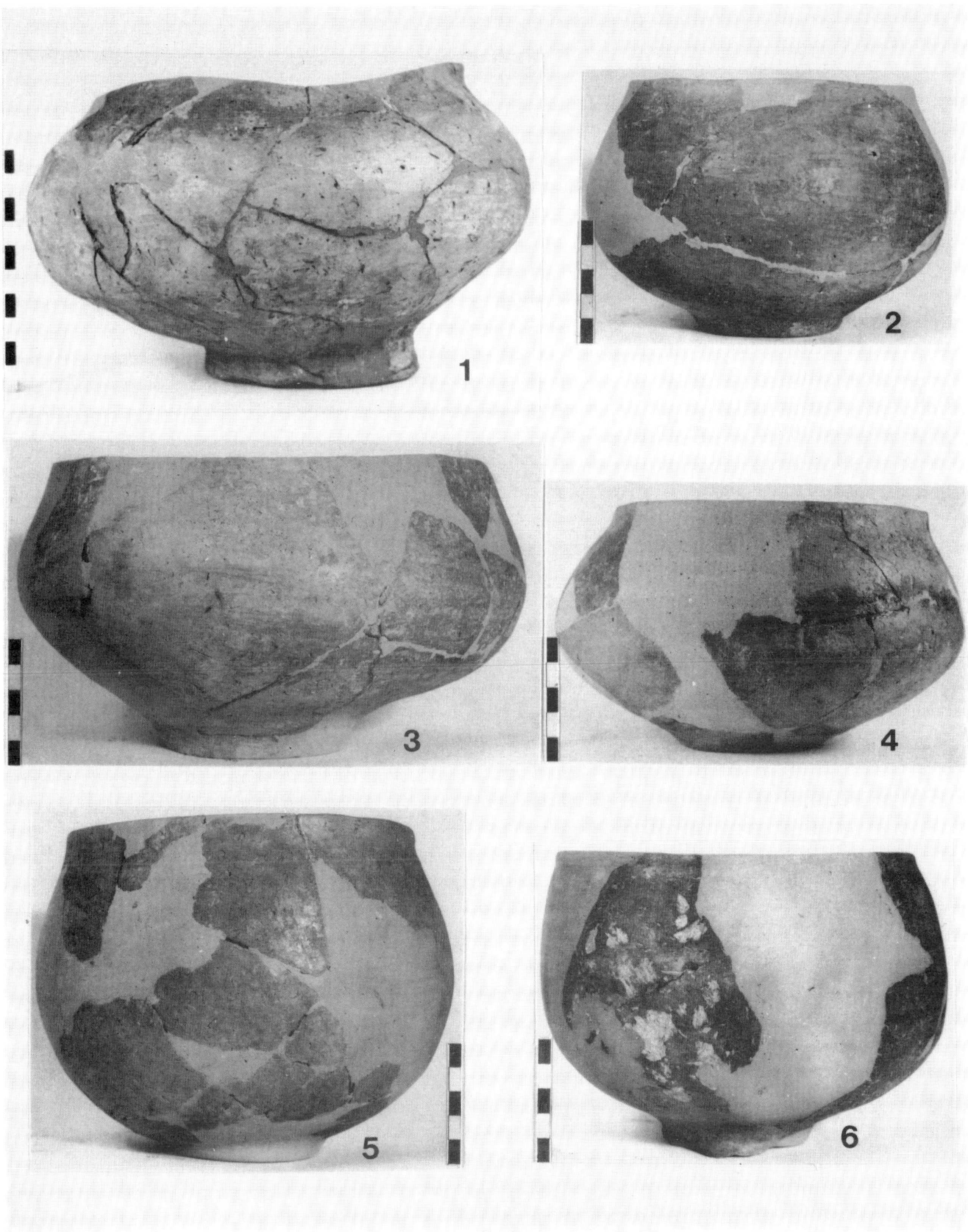

Plate 5.

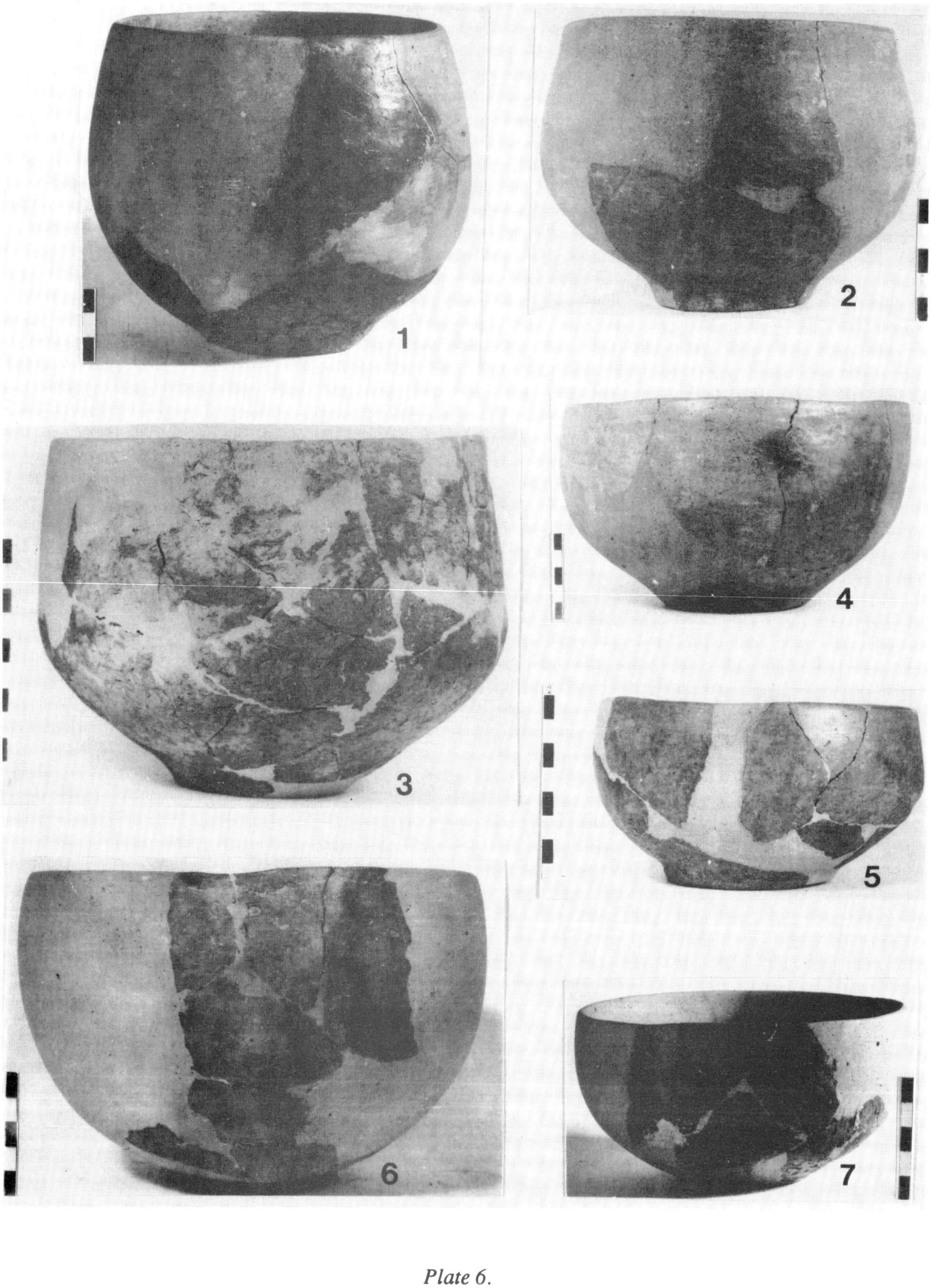

Plate 6.

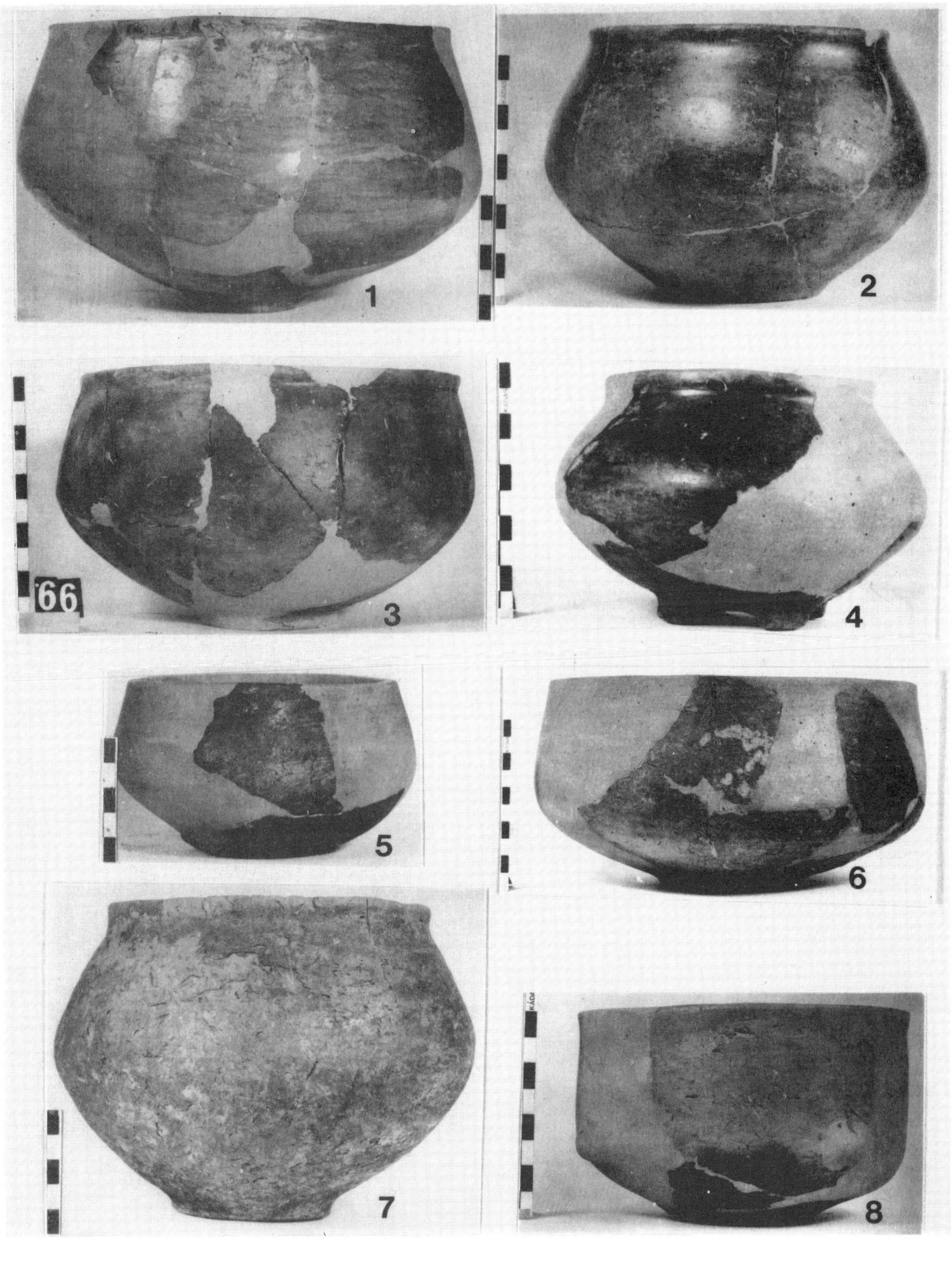

Plate 7.

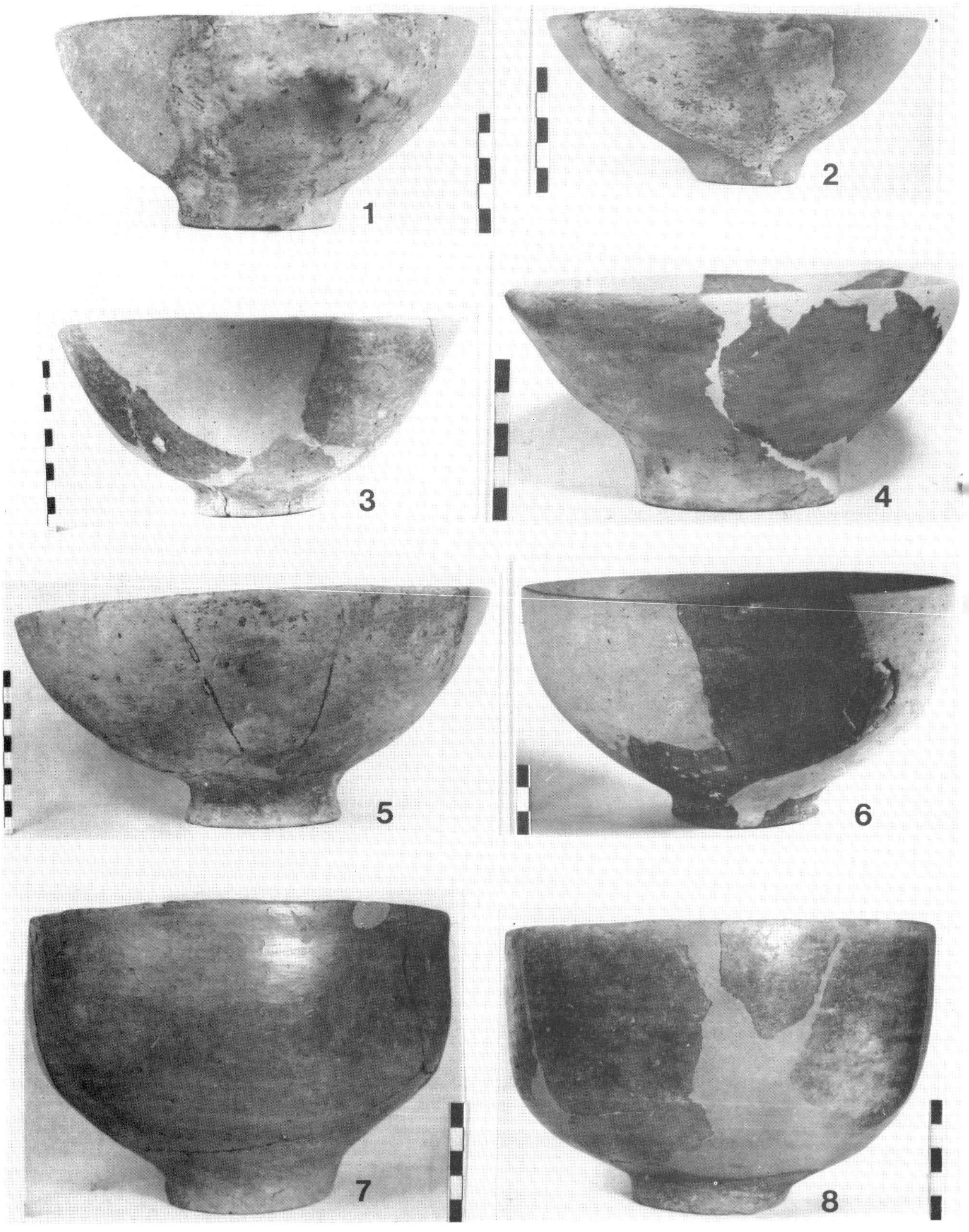

Plate 8.

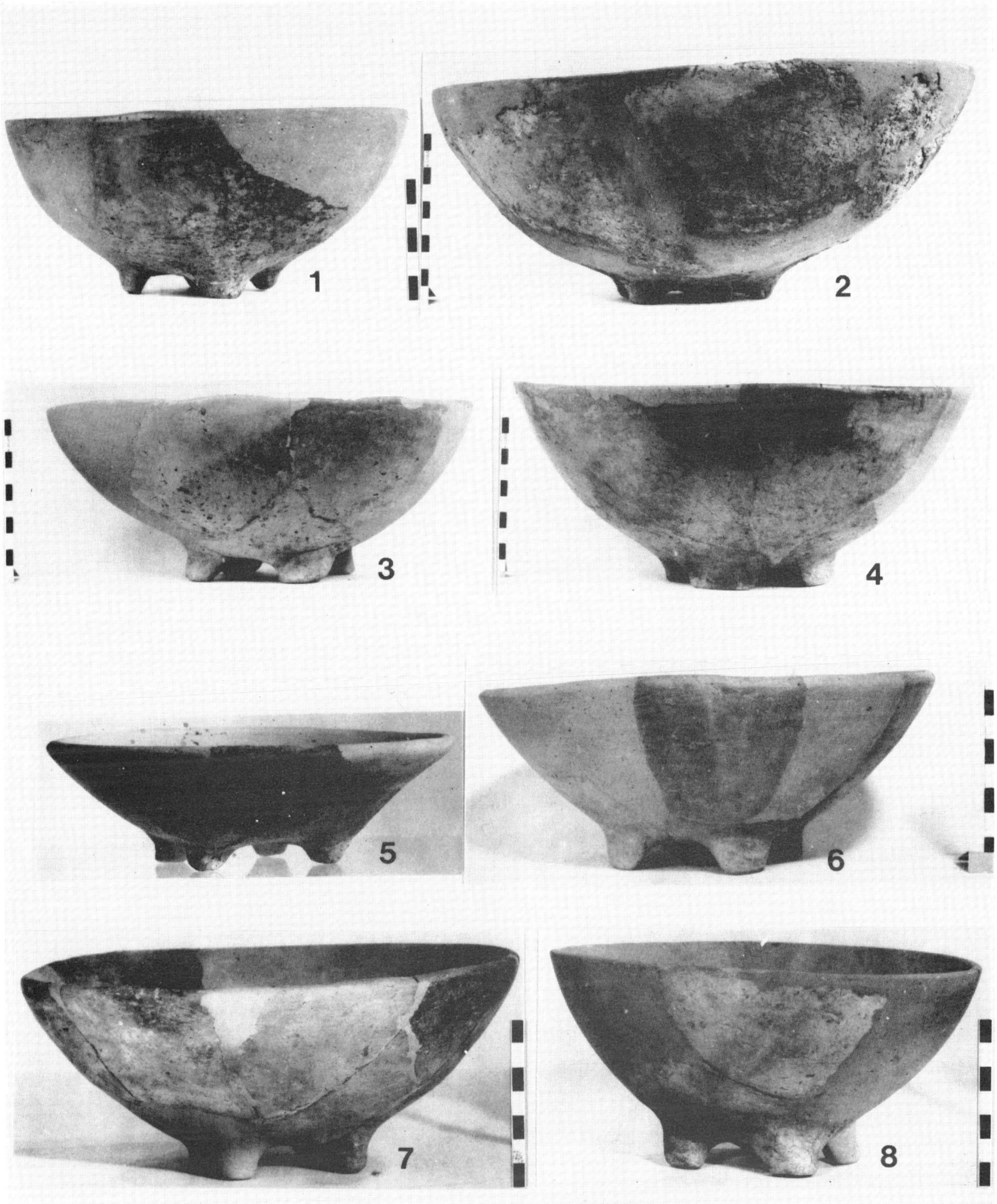

Plate 9.

Plate 10.

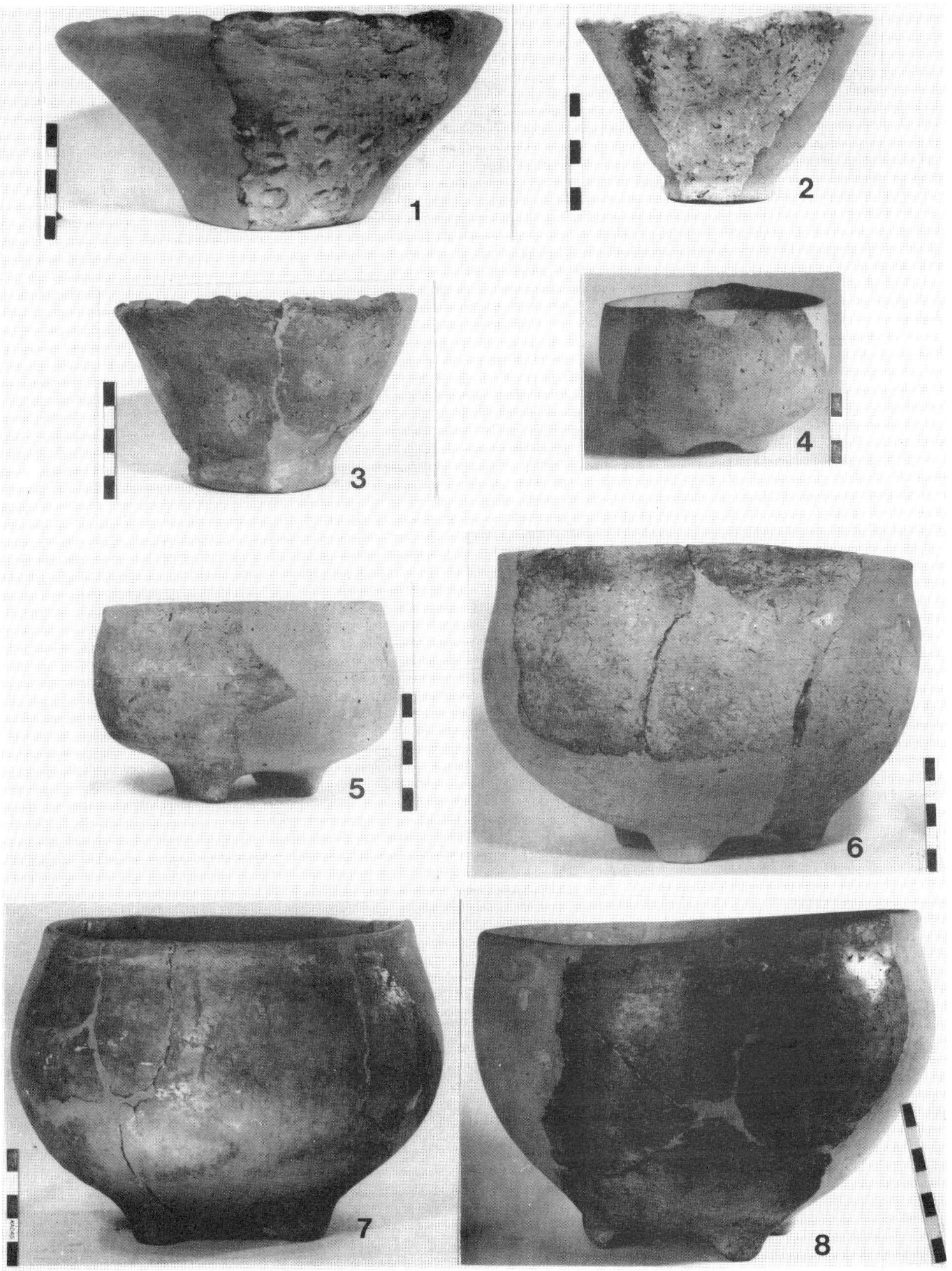

Plate 11.

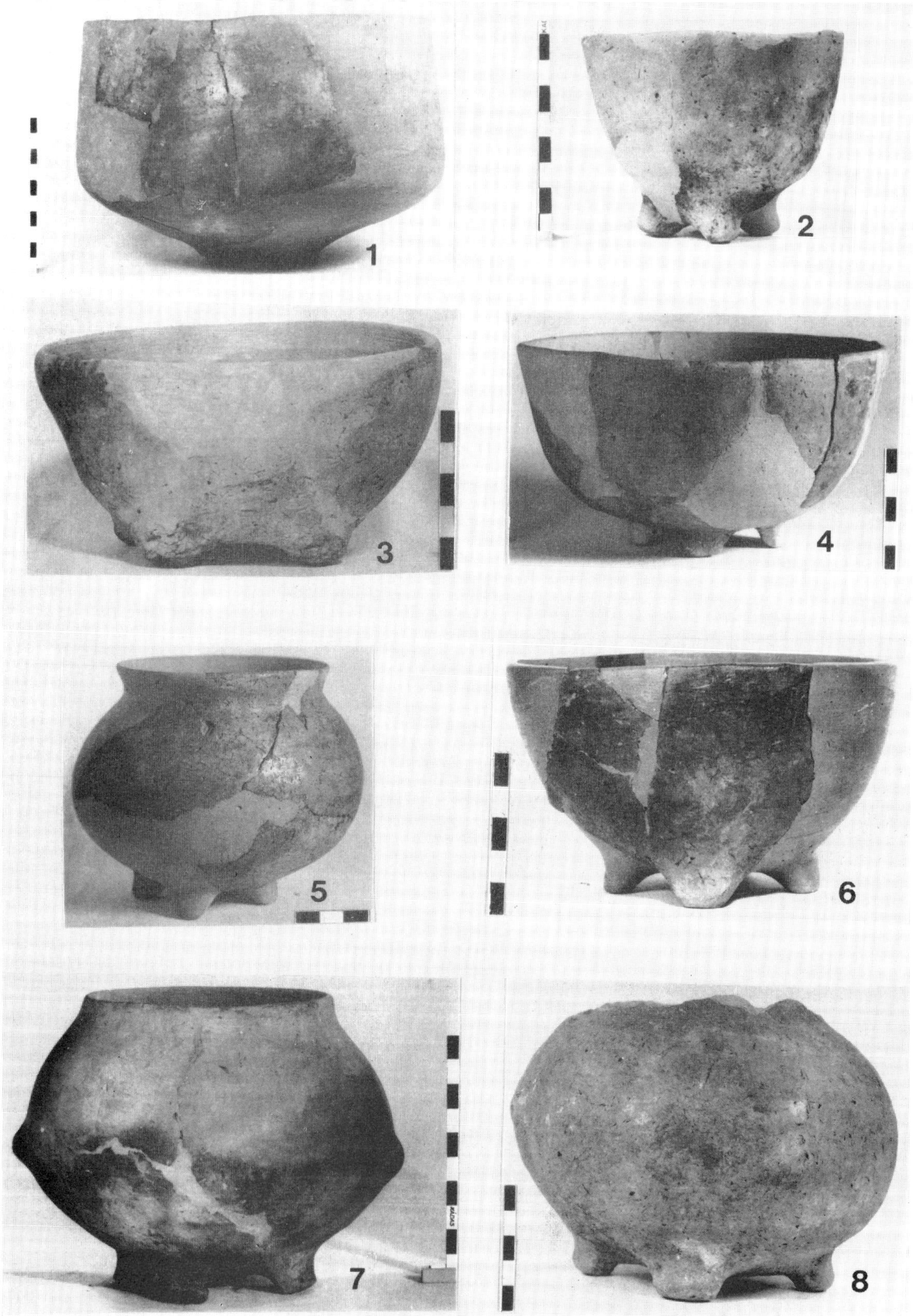

Plate 12.

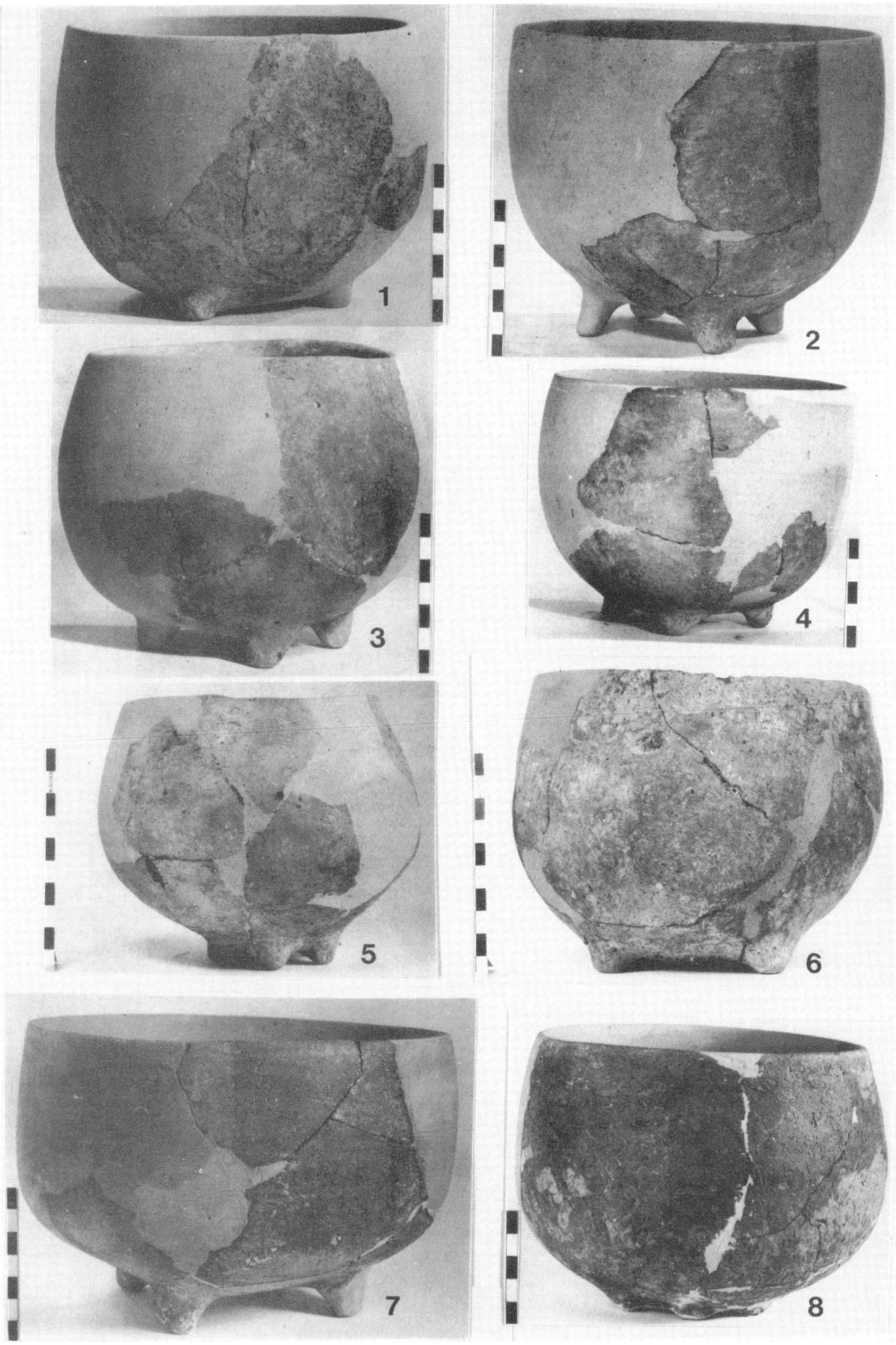

Plate 13.

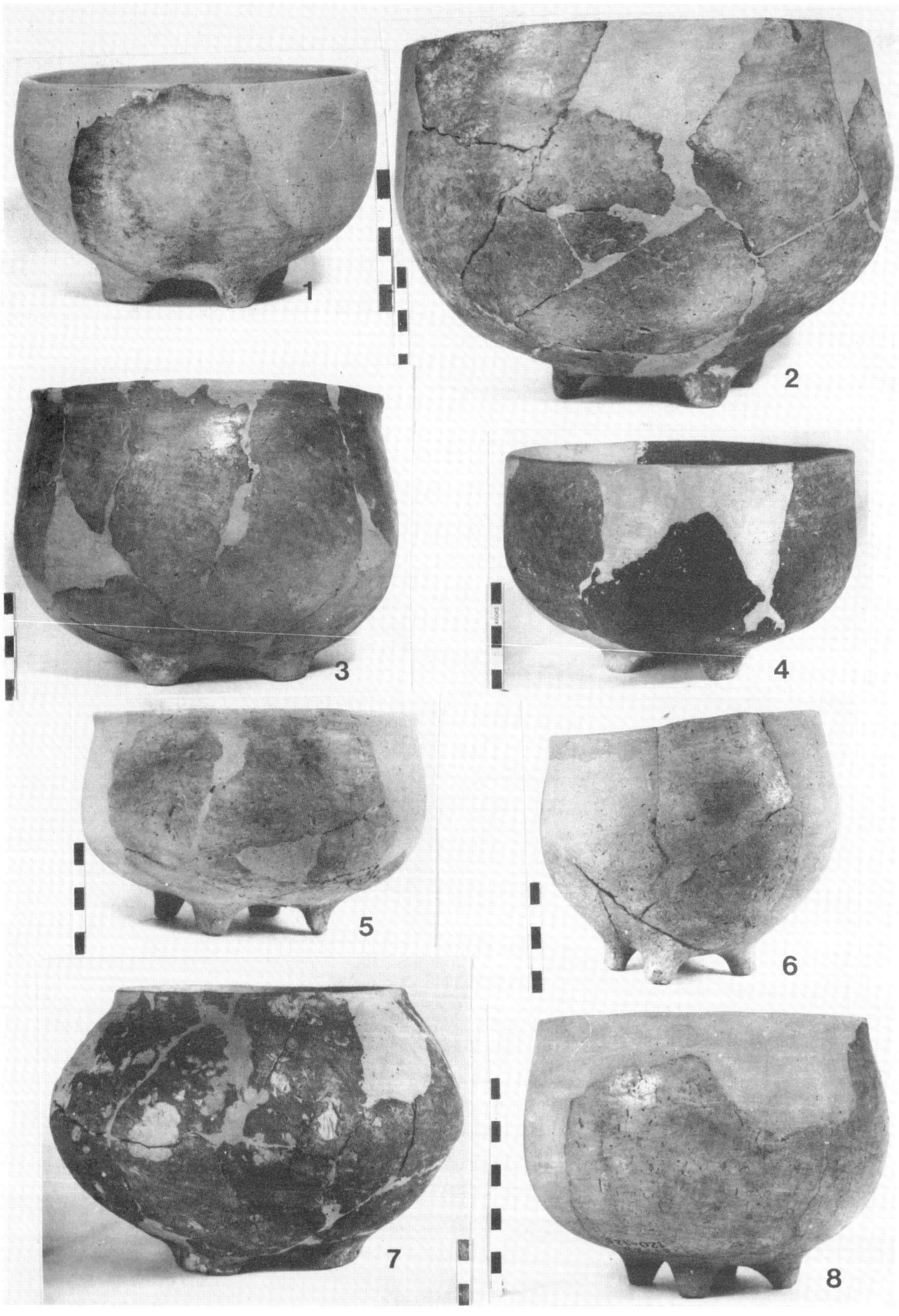

Plate 14.

Plate 15.

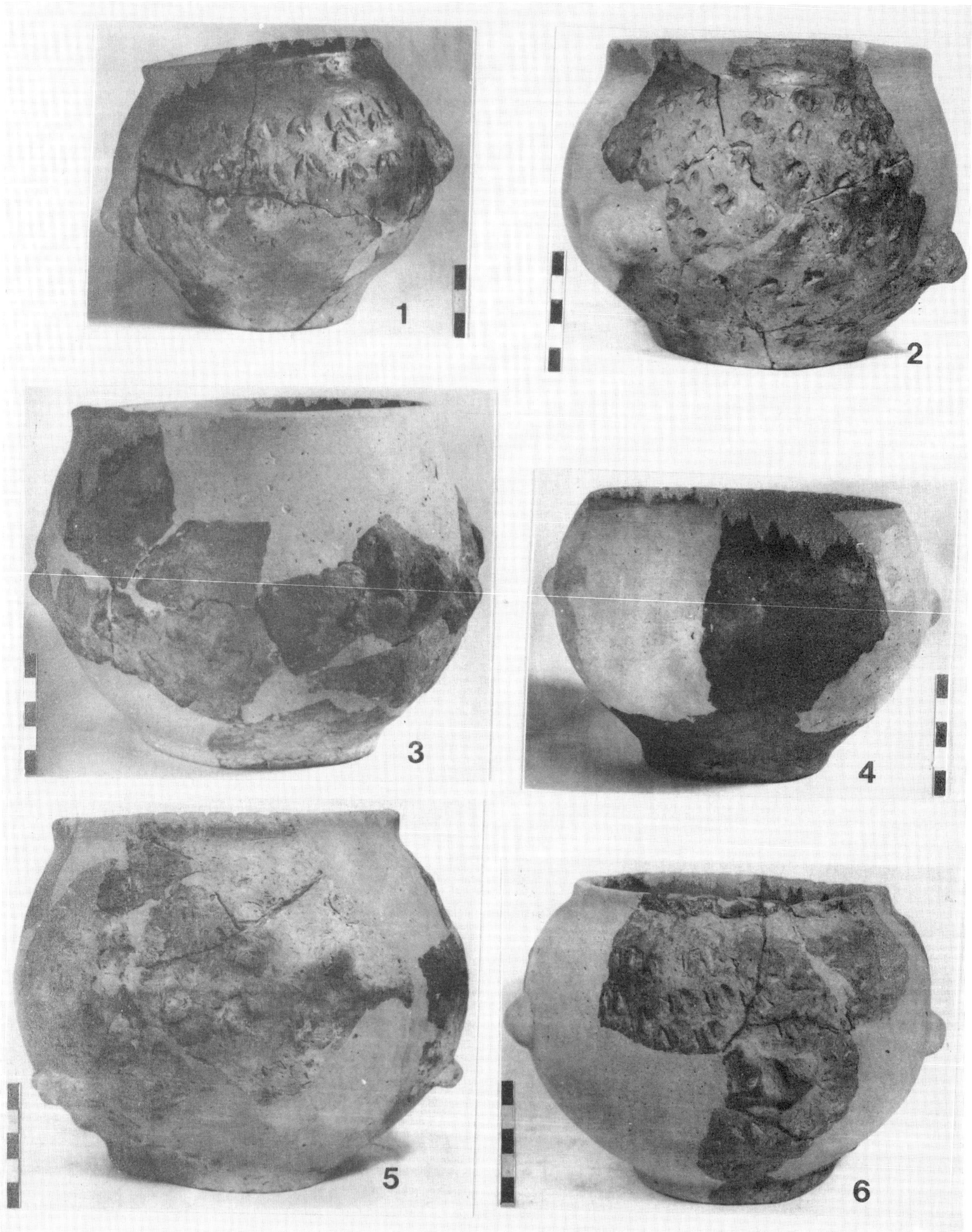

Plate 16.

1

2

Plate 17.

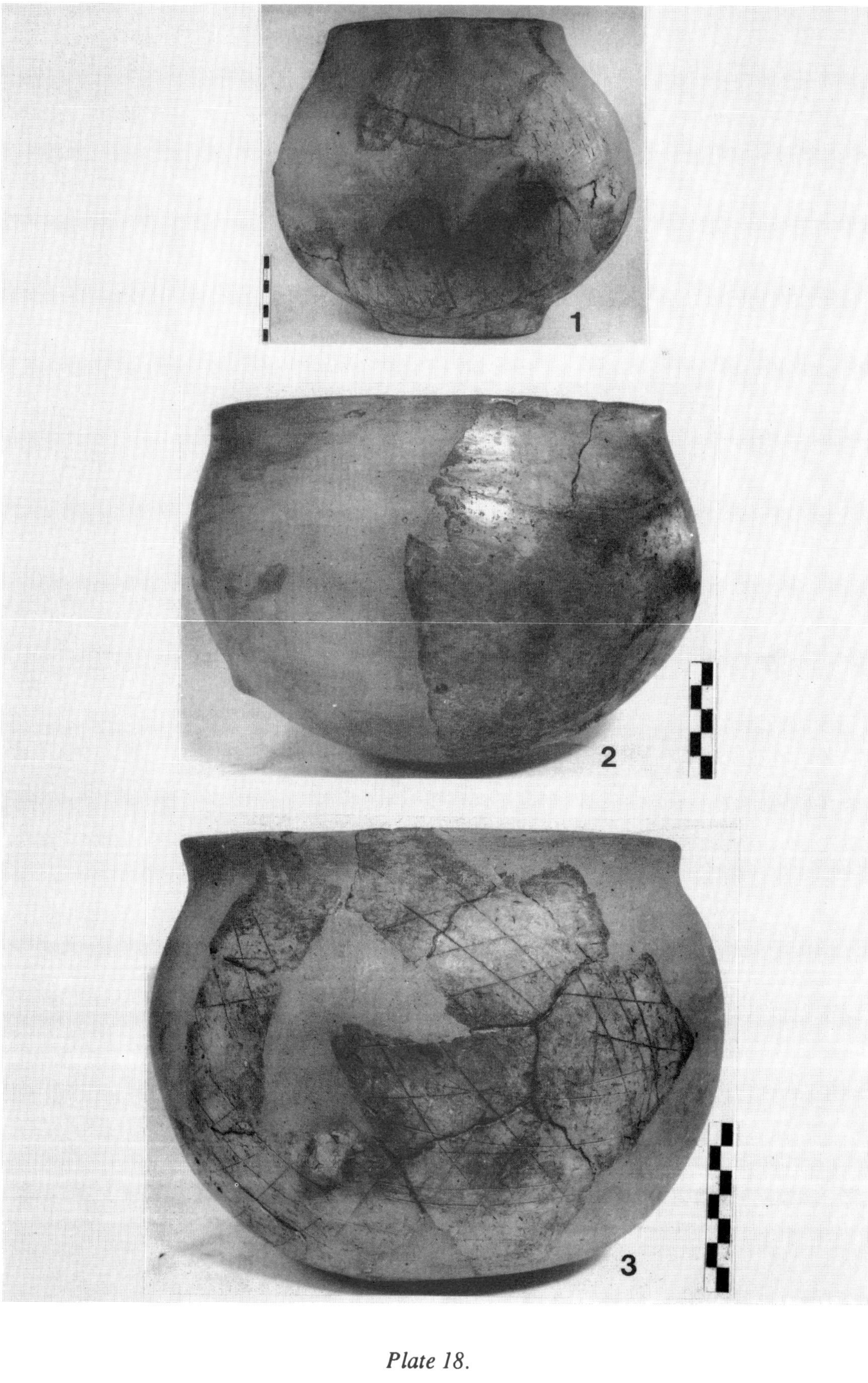

Plate 18.

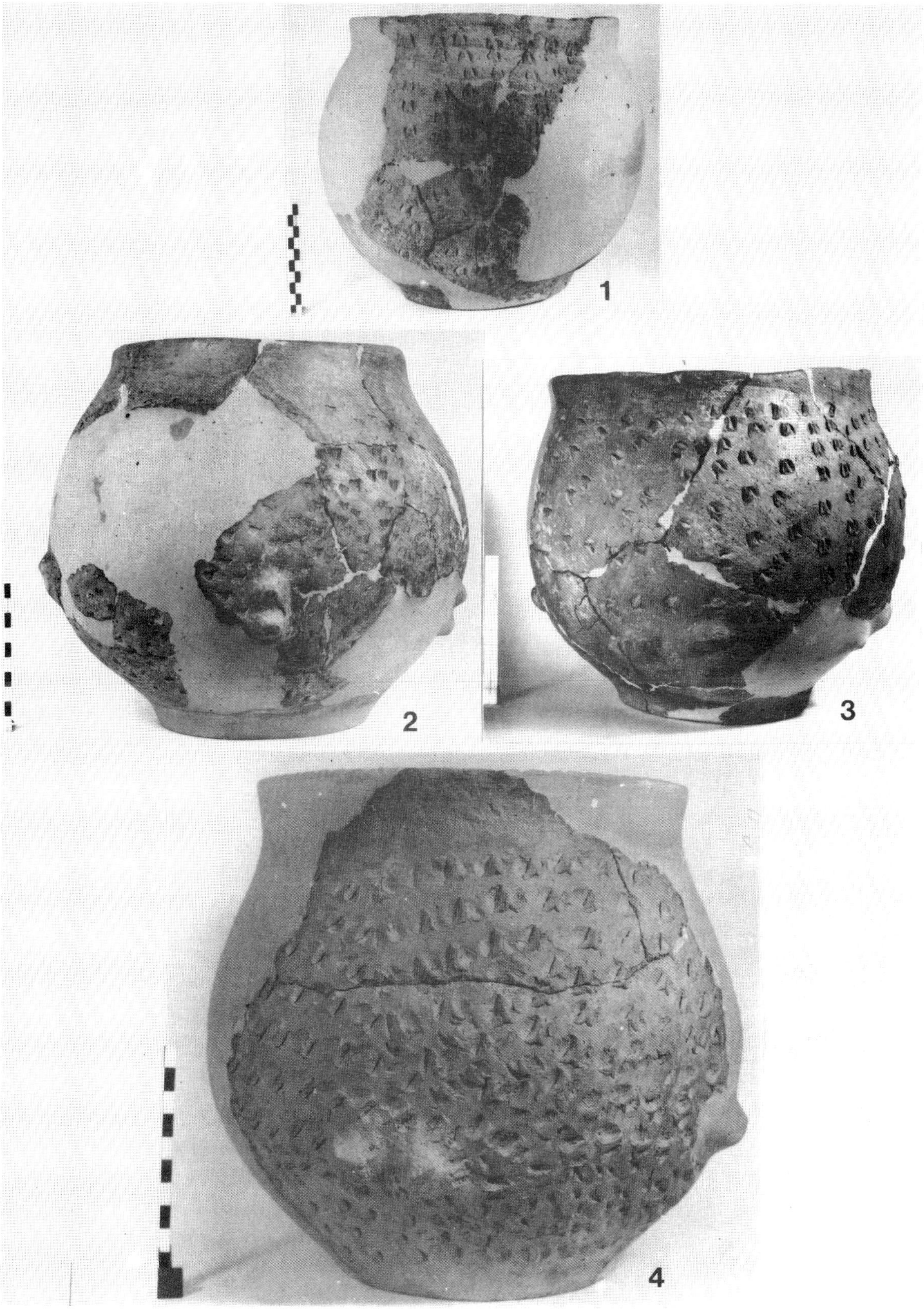

Plate 19.

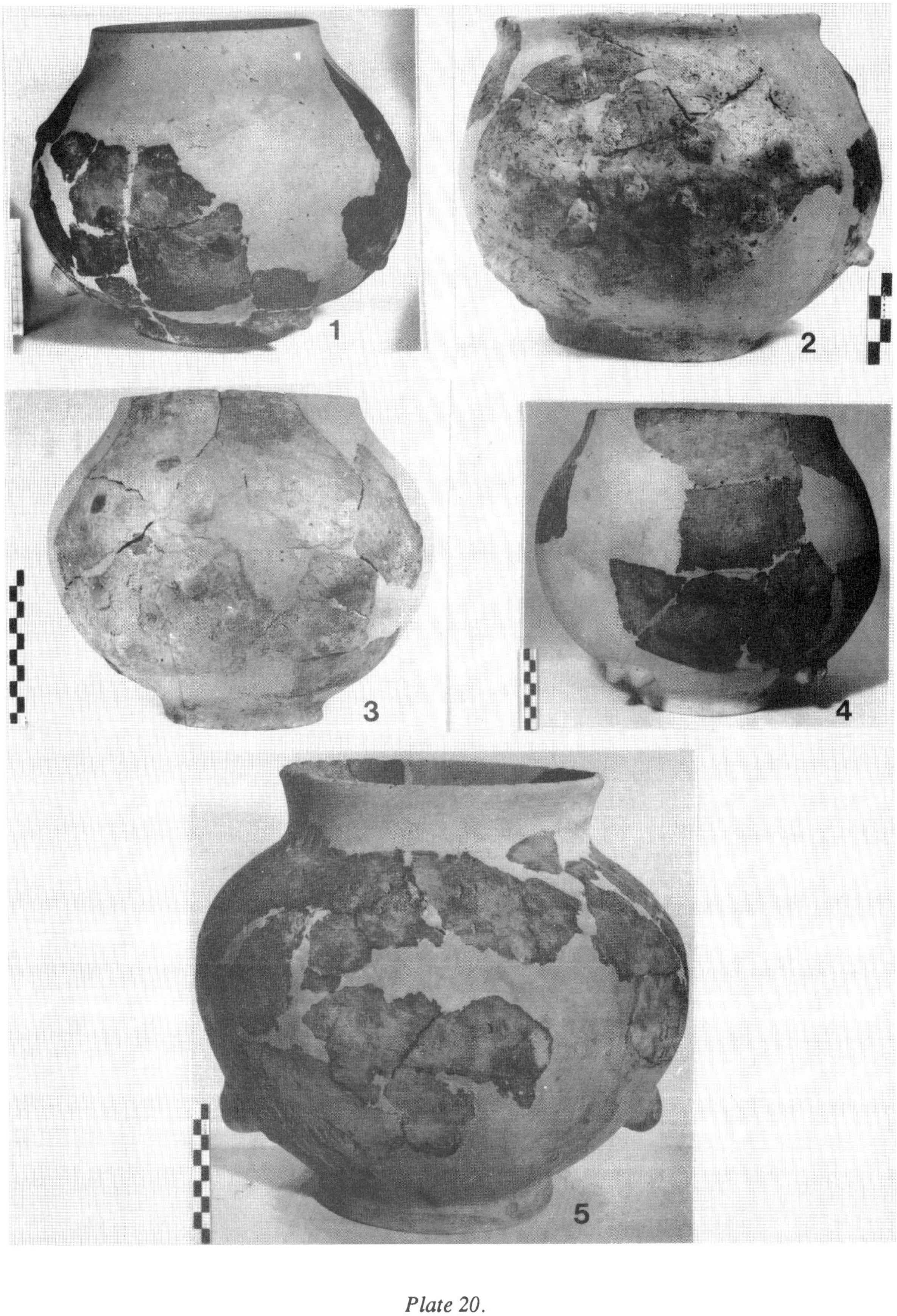

Plate 20.

Plate 21.

Plate 22.

Plate 23.

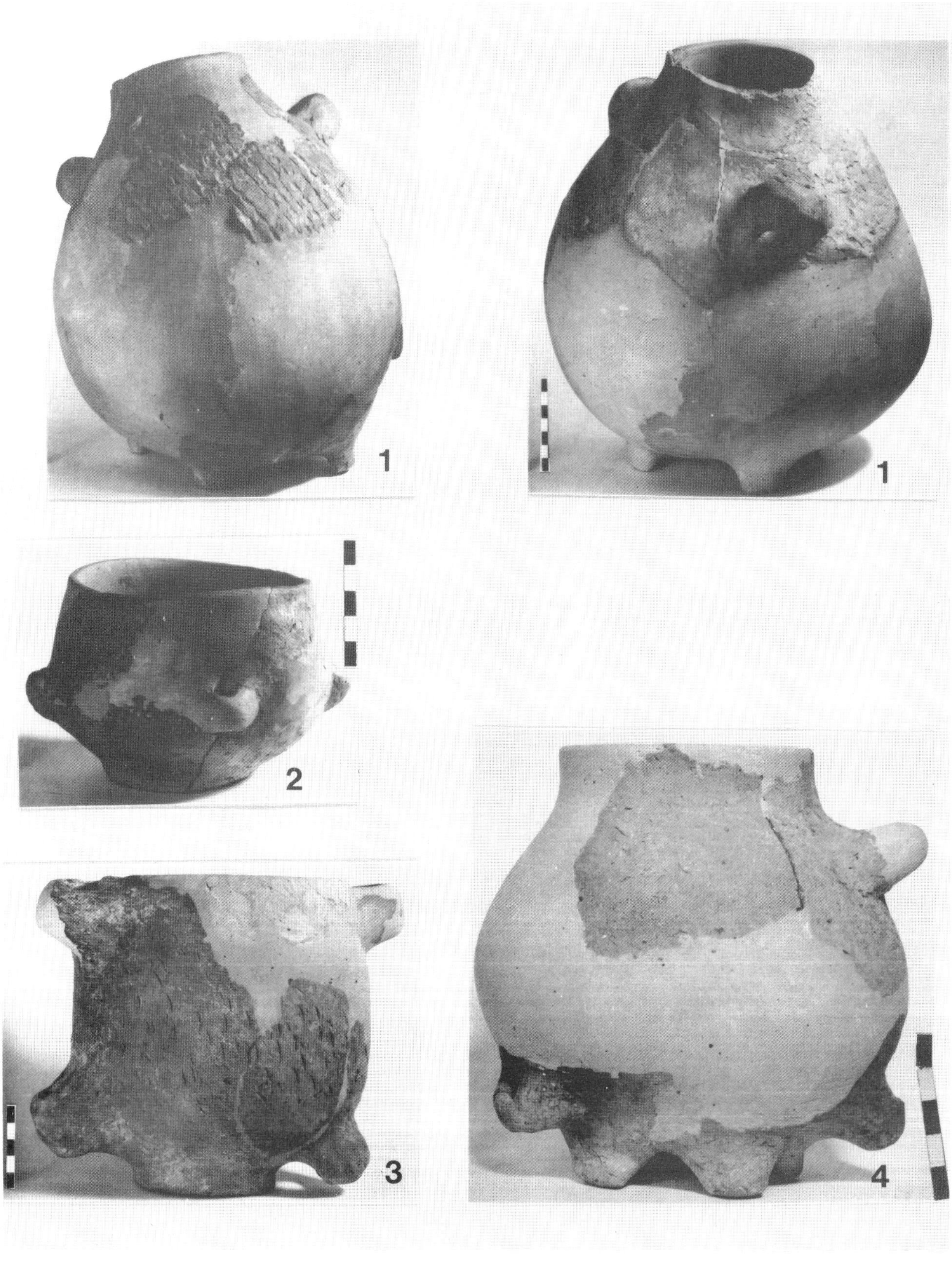

Plate 24.

Plate 25.

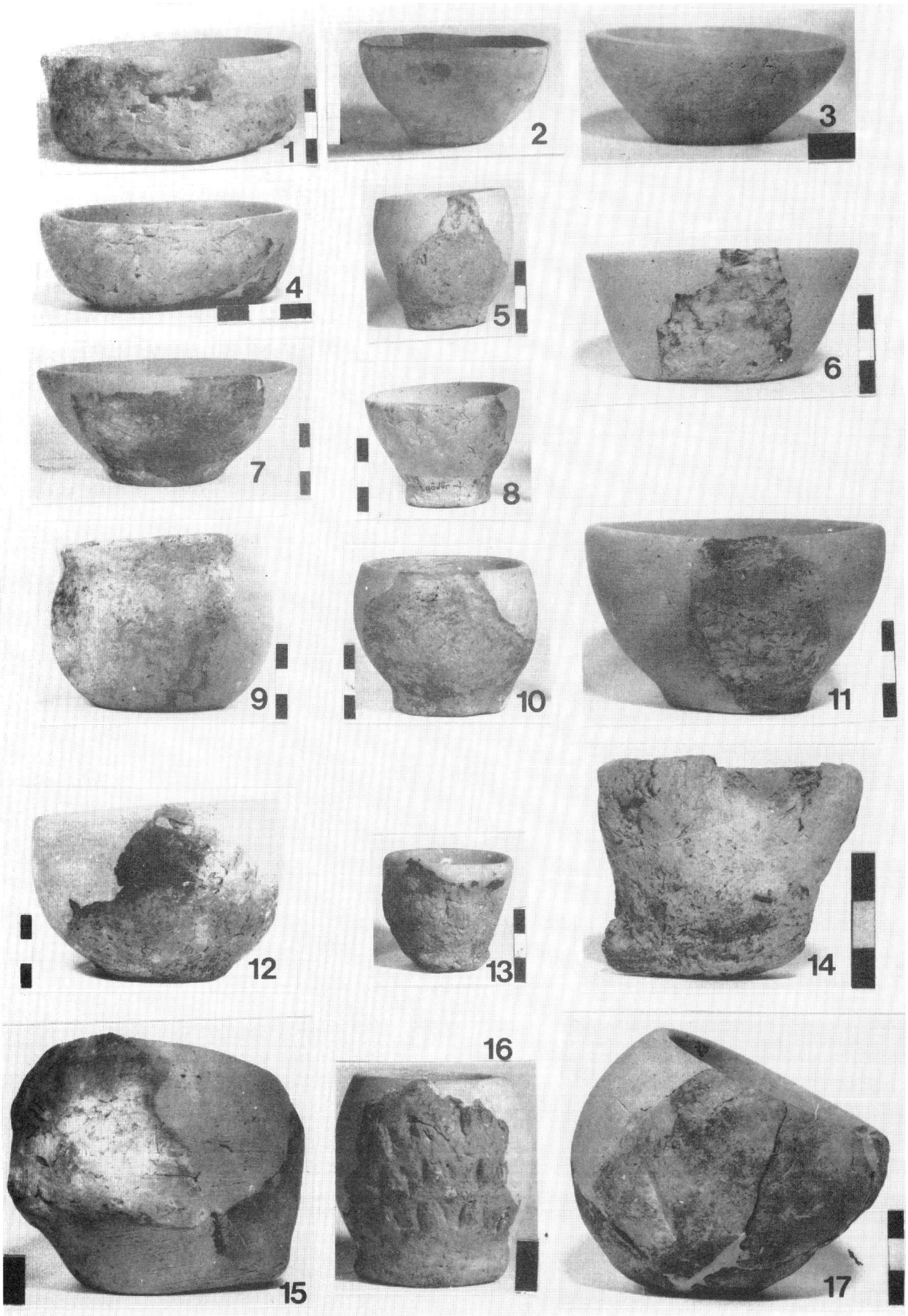

Plate 26.

Plate 27.

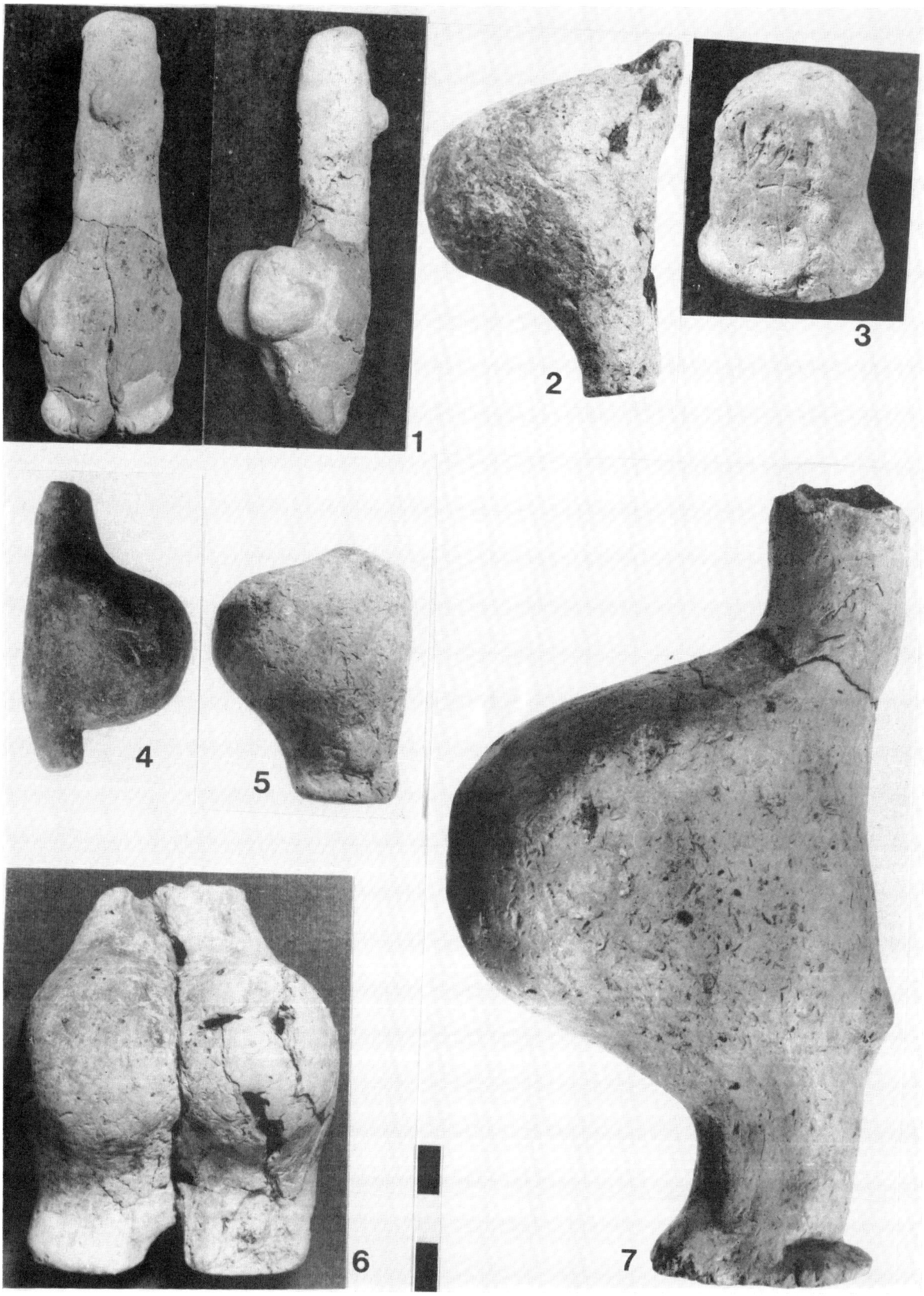

Plate 28.

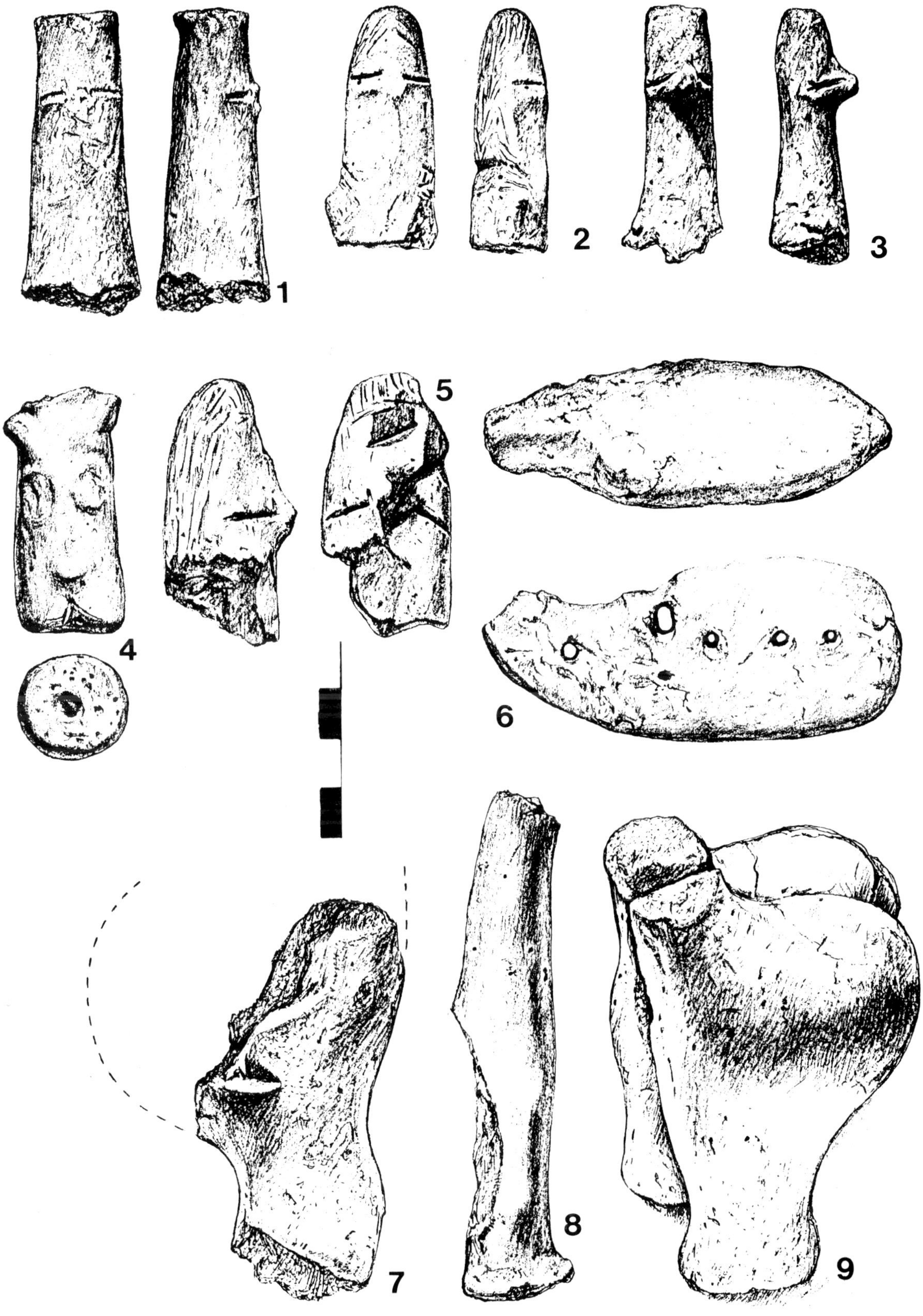

Plate 29.

Plate 30.

1

2

Plate 31.

Plate 32.

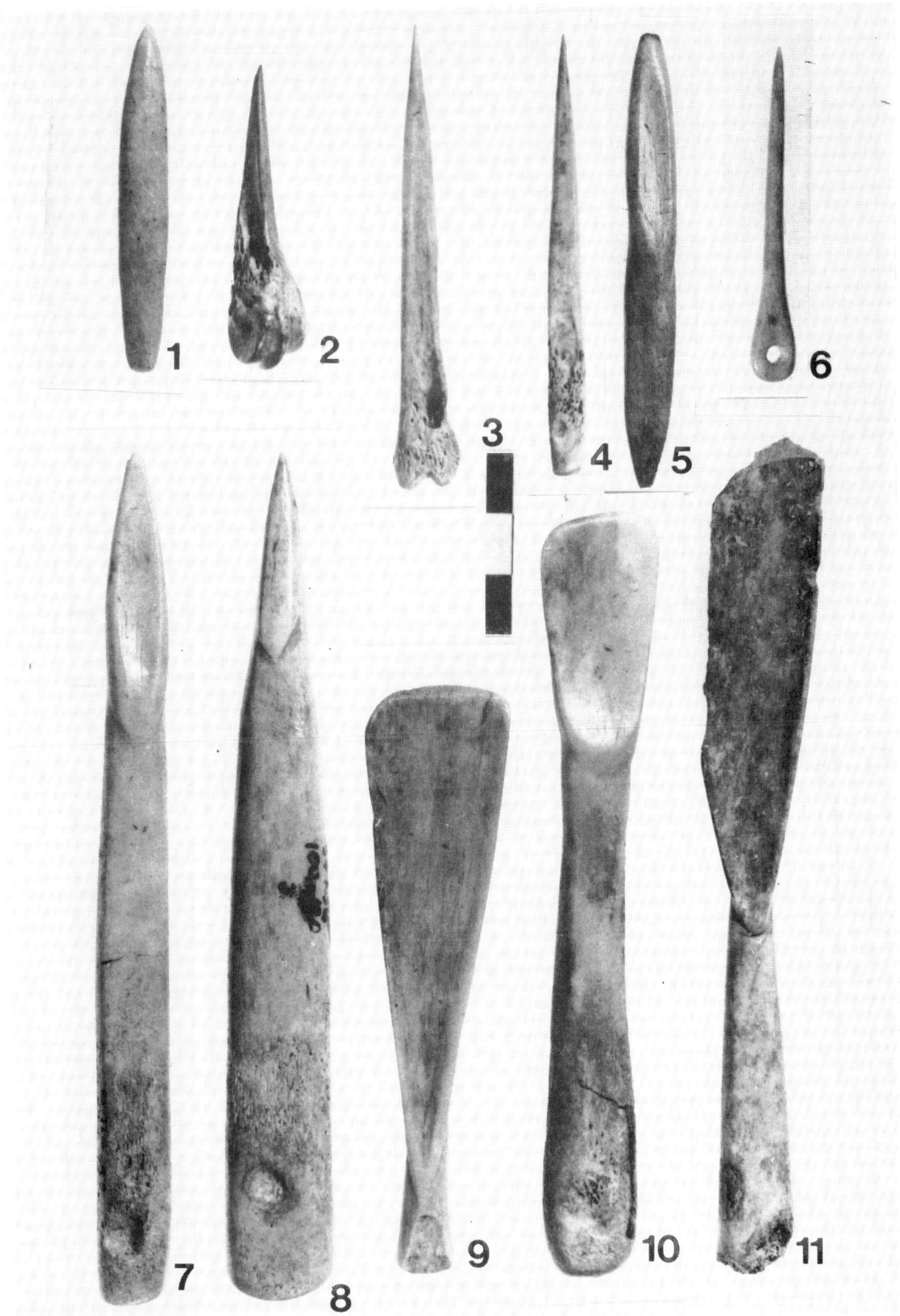

Plate 33.

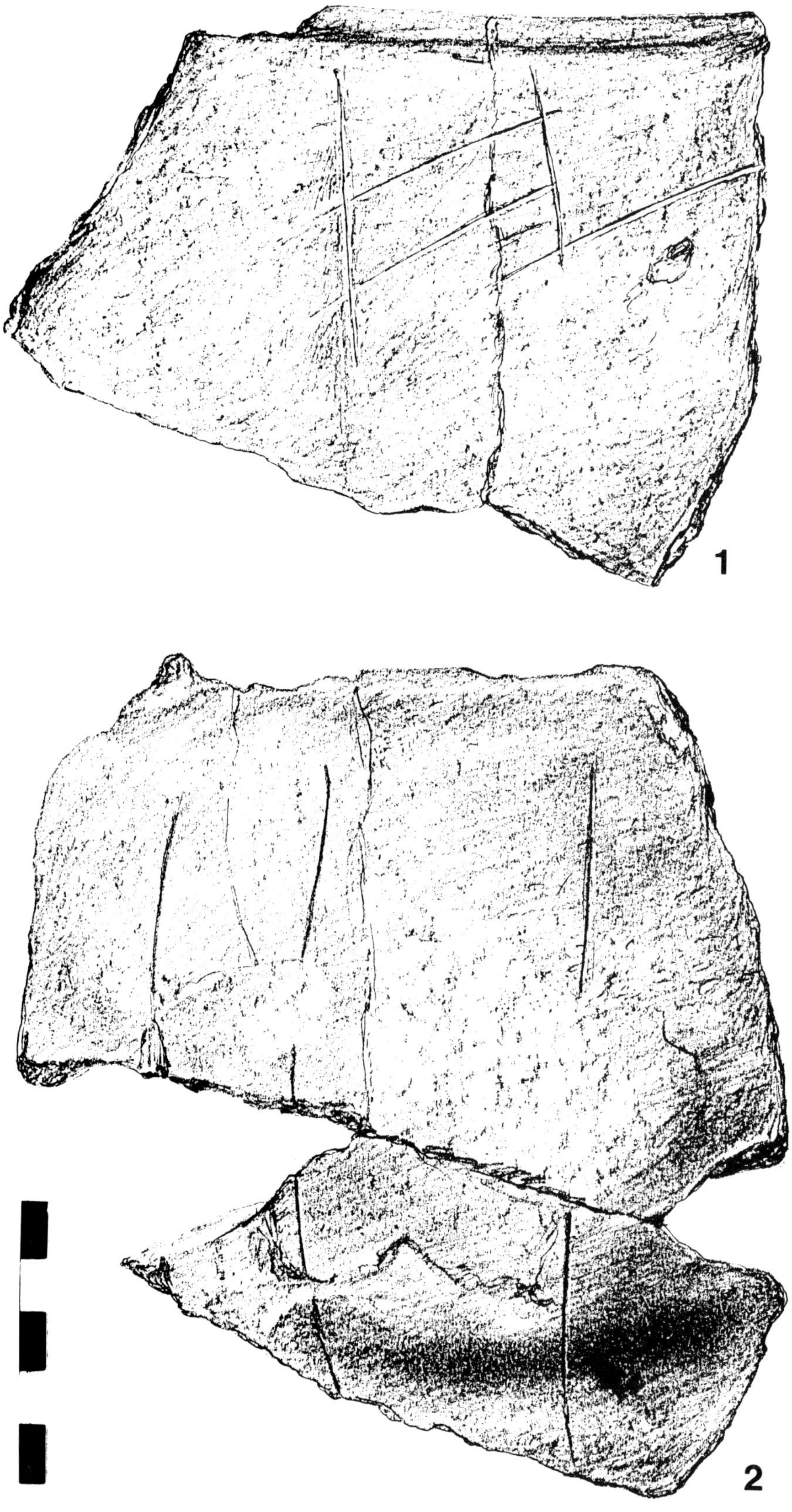

Plate 34.

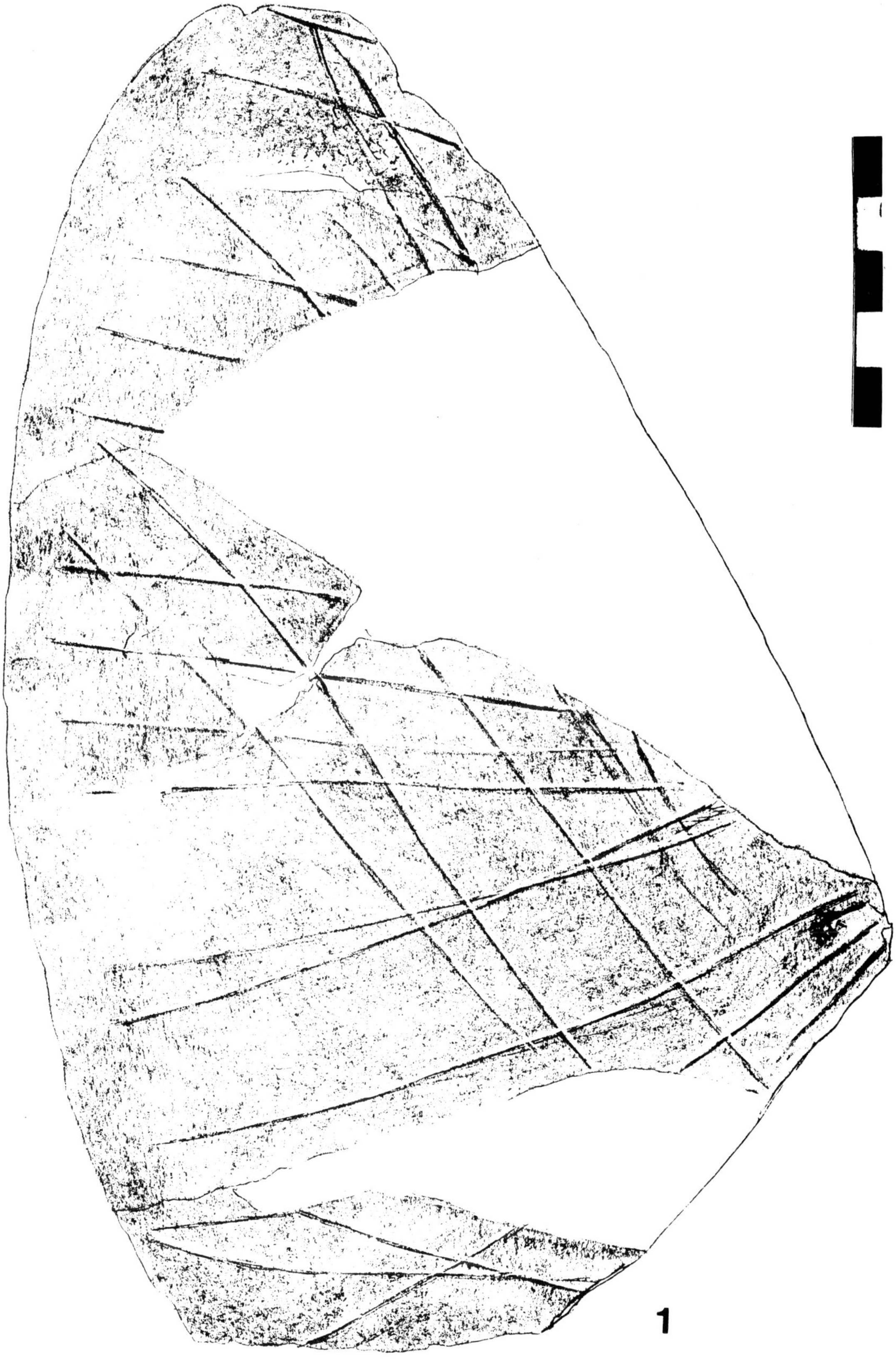

Plate 35.

Plate 36.

Plate 37.

1

2

Plate 38.

Plate 39.

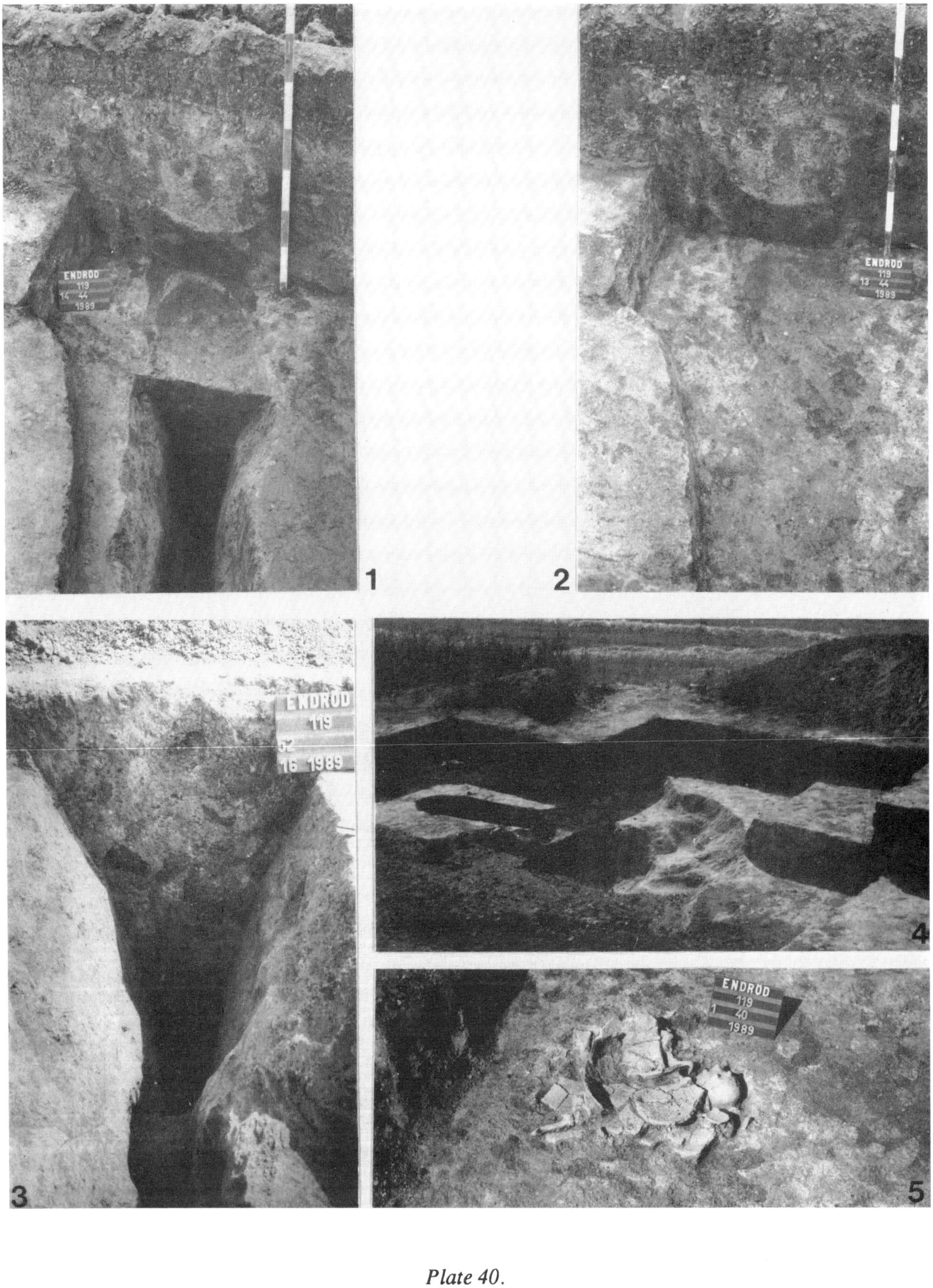
ENDROD
119
14 44
1989
1
ENDROD
119
13 44
1989
2
ENDROD
119
52
16 1989
3
4
ENDROD
119
1 40
1989
5

Plate 40.

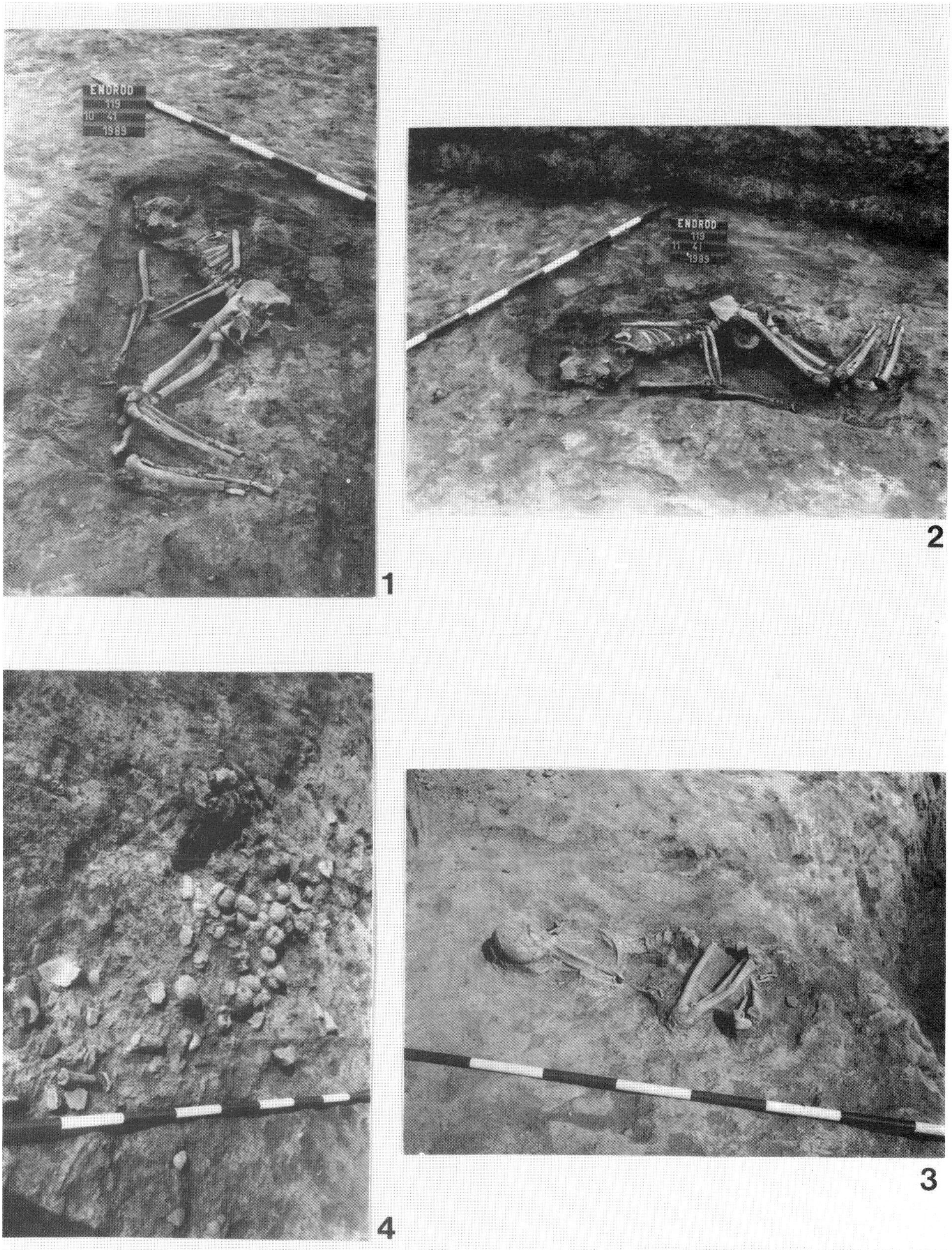

Plate 41.

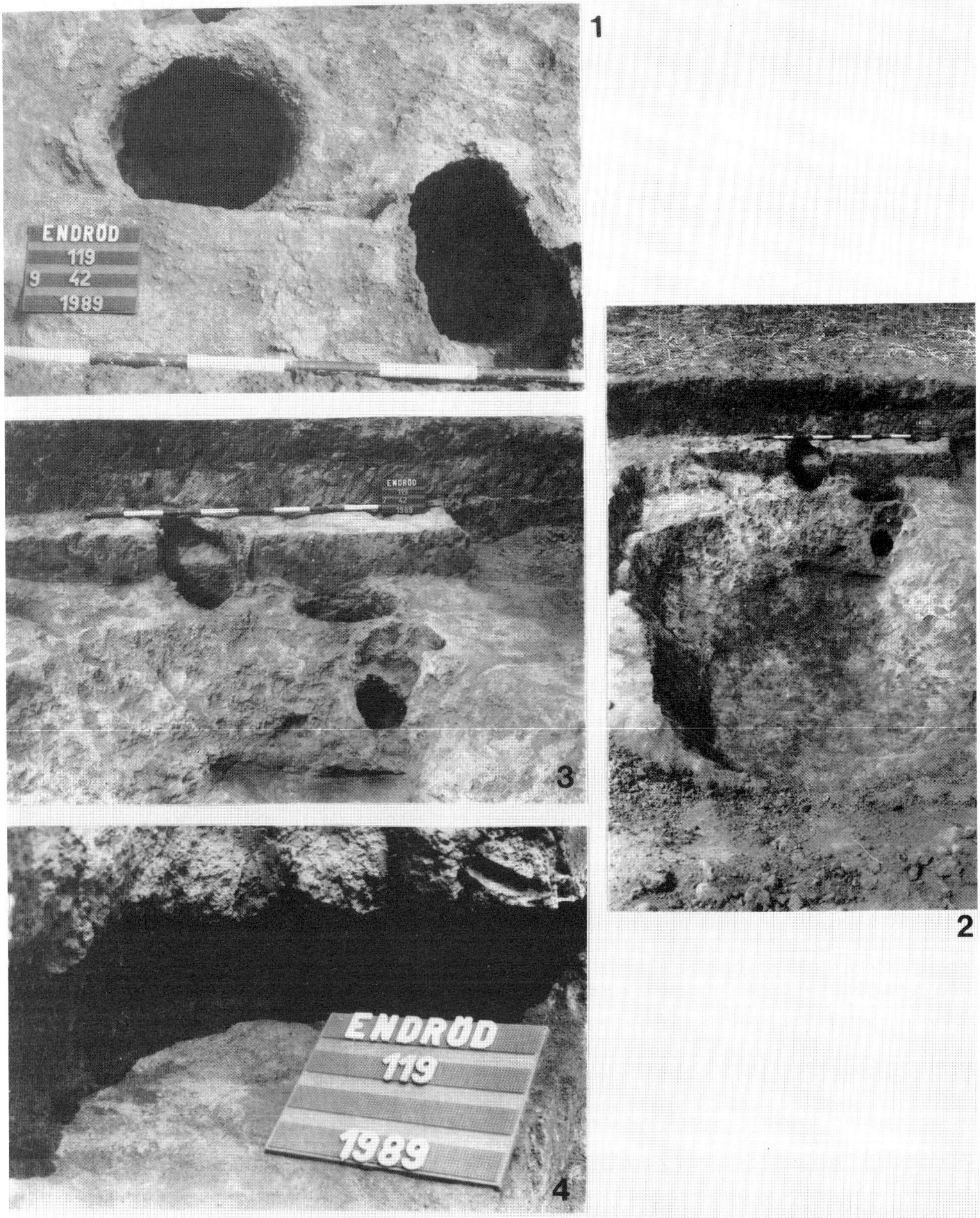

Plate 42.

1

2

3

4

Plate 43.

The Early Neolithic vertebrate fauna of Endrőd 119

SÁNDOR BÖKÖNYI

Introduction

The main aim of the excavation of the Endrőd 119 site was to uncover, as completely as possible, a small but complete Körös culture settlement. We are of the opinion that a fully excavated site provides incomparably more information about the structure of the settlement, the various activities conducted there and details of the everyday life of its inhabitants than small-scale excavations of several sites would provide even if there actually constituted a larger excavated area. The complete excavation of whole settlements does not result in information linearly comparable to small scale digs but can answer questions which could not even be posed in the case of a small excavation over a limited area of a settlement even if that produced some attractive finds, by chance.

The excavation of the site of Endrőd 119, carefully and precisely carried out by J. Makkay, not only revealed a number of houses, unfortunately badly damaged by deep ploughing, and tens of thousands of potsherds, but also yielded a considerable quantity of animal remains as well; in fact the largest animal bone sample ever collected from the excavation of a Körös settlement.

As a result, the author is in the lucky position of being able, to cooperate with the excavating archaeologist, in answering questions concerning the animal stock of a whole settlement, its age structure, the kill-off pattern and the development of the different domestic species as well as possible seasonal wanderings (transhumance), the role of hunting, fowling and fishing as secondary sources of food and raw materials of animal origin, and the seasonality of these activities as well. In this respect one must solicit essential help from the ornithologist and also the ichthyologist who identify and evaluate the bird and fish remains. If one was lucky enough to be able to ascertain the number of individuals in a household and at the length of habitation, one could attempt to reconstruct the diet of the inhabitants, not only qualitatively but also quantitatively: in other words, one could try to determine the quantity of consumed meat and through it possibly the calorie intake from animal sources of the inhabitants. Consequently, one could also get some idea of what the ratio of foodstuffs of plant origin occurred in human consumption.

The sample

The sample is a typical example of kitchen waste with two exceptions. The first exception is an animal sacrifice consisting of a cattle rib fragment, an ovicaprine left proximal radius fragment and a dog's skull, two mandible halves, atlas, axis, 5 cervical, 7 thoracic, and 4 lumbar vertebrae, the os sacrum, and four ribs as well as a wild duck's os coracoideum. The other is a, more or less, complete skeleton (without skull and mandible) of a juvenile roe deer that may have been dug into the archaeological levels at a later date.

All other remains are composed of food remnants in addition to debris thrown away when the carcasses were processed (e.g. horn cores, antlers, toe bones, particularly third phalanges, etc.). Typical kitchen waste comprises no whole skeletons or large articulated components. Whole skulls are also missing, and where skulls were found it may be suspected that they were intrusive (e.g. some fox and badger skull fragments because these animals prefer digging underground burrows into light soils, and the loose layers of archaeological sites can certainly be considered sort of this type). The larger skull fragments found include two cattle frontal bones with complete horn cores, a goat cranium fragment with the proximal parts of both horn cores present and several similar ones from sheep representing mainly hornless skulls (*Fig. 11*), although some have rudimentary (*Fig. 10*) or "palustris" (*Fig. 8, 9*) horn cores.

The number of collected horn cores is conspicuously high: not less than 194 sheep, 45 cattle, 13 goat and 3 aurochs horn cores or horn core fragments were recovered. From this point of view Endrőd 119 is one of the richest sites ever uncovered.

We are particulary fortunate in the number of long bones which survive on the site. By measuring the full extent of these bones it is possible to ascertain the height at the animals withers, that is its absolute size, and sometimes the constitution of the animals can be determined. *Table 1*, presenting the numbers and distributions of these complete long bones in the different mammal species from the site reveals a lot of interesting data in this respect. For example among the domestic mammals all but pig were represented by such bones. (The lack of complete pig bones can easily be explained on the following grounds: 1/ the pig bones were very rare, 2/ the pig, as a typical "meat" animal, was mainly killed when immature and, thus, whole adult bones would not be found, 3/ its bones were chopped up in such a crude way that there was no chance for any remaining long bone to escape this fate).

Table 1. Complete long bones in the sample

	humerus	radius	metacarpus	femur	tibia	etatarsus	total
cattle	1	1	8	-	-	5	15
sheep	6	54	37	2	6	54	159
goat	-	4	1	-	-	6	11
dog	-	2	-	-	1	-	3
aurochs	-	-	1	-	-	-	1
roe deer	-	5	3	-	1	3	12
wild cat	-	-	-	1	-	-	1
wolf	-	1	-	-	-	-	1
hare	2	2	-	1	2	-	7
total	9	69	50	4	10	68	210

We have no explanation for the numbers of long bones of wild mammals. However, it is a fact that aurochs bones were so large that they had to be chopped up in order to be put into the cooking pots (the survival of a single metacarpal is mere chance: this particular bone yields little meat or marrow). The small (roe deer) and very small (brown hare) "meat species" yielded larger quantities of whole long bones. There is no way to explain the presence of two complete carnivore long bones.

Still refering to *Table 1*, it is not surprising that among the herbivores, the metapodials were the most common of the unbroken bones, representing 56.73 per cent of all herbivore bones (not counting the hare because the metapodials of carnivores and hare are not included in *Table 1*). It is not clear, however, why so many sheep radii occurred (25.96 per cent of the whole herbivore sample). In addition, both in the sheep and goat samples subadult and juvenile radii were very numerous. They do not occur in *Table 1* because their distal epiphyses were missing. To make the picture complete, is should be pointed out that radii are also conspicuously frequent in the comparatively small goat and roe deer samples.

Those bones in the sample which do not occur in *Table 1* were mainly badly broken, and in spite of the fact that virtually no post-deposition destructive forces seem to have been in operation on them, the number of measurable bones is rather low in comparison to the largeness of the mammal sample as a whole. Most of the bones were broken into small fragments or splinters before being discarded, presumably in order to facilitate extraction of the edible material. Few bones show cut marks from butchering, while the gnaw marks by dogs are extremely rare. The case is not the same with burning marks that originated during food preparation or fires in the settlement.

In the course of excavations at the site, traces of activities from more recent periods than the Early Neolithic (a Copper Age pit, several Bronze Age pits, some Migration period graves, and even a quite recent grave) were recognised. Since the animal remains of these later periods could not always be differentiated from those of the Körös culture, where there was the slightest doubt all bones from the affected area were excluded from the investigation in order to insure as precisely as possible the authenticity of the material.

The fauna

This chapter can be divided into two clearly distinct, but, nevertheless, in certain ways related parts: the domestic and the wild fauna. (The species and their frequencies are shown in *Table 2*. For the identification of the bird and fish remains the author expresses his gratitude to Professor D. Jánossy and Mr. I. Takács respectively.)

Table 2. The fauna list

		specimens	individuals
cattle - Bos taurus L.		5139	183
sheep – Ovis aries L.	2332		
goat - Capra hircus L.	298	15357	788
sheep/goat - Ovis/Capra	12717		
pig - Sus scrofa dom. L.		140	20
dog - Canis familiaris L.		87	12
domestic animals		20723	1003
aurochs - Bos primigenius Boj.		243	23
red deer - Cervus elaphus L.		138	16
roe deer - Capreolus capreolus L.		671	37
wild swine - Sus scrofa fer. L.		376	29
wild ass - Asinus hydruntinus Reg.		16	4
wild cat - Felis silvestris Schreb.		2	1
polecat - Putorius putorius L.		2	1
marten - Martes sp.		1	1
badger - Meles meles L.		13	5
fox - Vulpes vulpes L.		14	4
wolf - Canis lupus L.		29	5
carnivore - Carnivora		1	1
vole-rat - Arvicola terrestris L		1	1
lesser mole rat - Spalax leucodon Nordmann		1	1
brown hare - Lepus europaeus Pall.		137	8
wild mammals		1645	137

Slavonian grebe - Podiceps of auritus L.	1
black stork - Ciconia of nigra L.	2
great white egret - Egretta alba L.	3
little egret - Egretta garzetta L.	1
grey heron - Ardea cinerea L.	1
spoonbill - Platalea leucorodia L.	1
mallard - Anas platyrhynchos L.	12
ferrugious duck - Aythya nyroca Güld.	3
pochard - Aythya ferina L.	6
shoveler - Spatula cf. clypeata L.	1
grey-lag /?/ goose - Anser cf. anser L.	1
white-fronted /?/ goose - Anser cf. albifrons Soop.	2
mute swan - Cygnus olor Gm.	2
black grouse - Lyrurus tetrix L.	7
partridge - Perdix perdix L.	1
crane - Grus grus L.	6
little bustard - Otis tetrax L.	2
great bustard - Otis tarda L.	8
coot - Fulica atra L.	1
tawny owl - Strix aluco L.	1
not yet identified birds	80
unidentifiable birds	465
pond tortoise - Emys orbicularis L.	1
catfish - Sirurus glanis L.	311
carp - Cyprinus carpio L.	201
pike - Esox lucius L.	27
pike perch – Lucioperca lucioperca L.	1
not yet identified fishes	5
unidentifiable fishes	126
other wild animals	1279
Total	23647

The domestic fauna present the picture typical for Early Neolithic sites in Southeast Europe (Boessneck 1962, 27ff.; Higgs 1962, 272; Bökönyi 1964, 87ff.; 1969, 222; 1971, 643; 1973b, 167f.; 1974, 21, 26, 51; 1976, 21; 1984a, 30f.; 1984b, 27ff.; 1989, Table 132; Jarman and Jarman 1968, 242ff.; Gejvall 1969, Table 5; Dennel 1972, 61; Payne 1973, 59; Schwartz 1976, Table LII; 1981, Fig. 3, etc.) and Southern Europe (Whitehouse 1971, 239ff.; Poulain-Josien 1975, 409ff.; Ducos 1976, 165f.; Geddes 1980, 121ff.; 1981, 371ff.; Bökönyi 1977, 82, Table 1; 1983, 237ff.; Bökönyi – Kretzoi 1983, 129ff.; etc.). Its essence is that a/ domestic animals represent an overwhelming majority in the fauna and are generally far more common than their wild counterparts, b/ among the domestic animals caprovines represent the absolute, sometimes enormous majority, c/ they are followed, in lesser numbers, by cattle or alternatively pig, and d/ the group of domestic animals brought up by the dog, found extremely rarely. There are some exceptions to this general rule, e.g. in Transylvanian sites (Necrasov 1961, 205ff.; 1964, 167ff.) and at Divostin, Serbia (Bökönyi 1988, 420, 429f.). However, in the Transylvanian sites the small sample size and at Divostin probable collection problems, were the cause of the divergance from the above picture. (Since then, an excavation has been carried out at the Iron Gate gorge of the Danube where emphasis was put on precise bone collection. This site yielded a high caprovine frequency, similar to other Early Neolithic settlements in the Balkans.)

This sort of animal husbandry came over in this form from Southwest Asia, most probably from West Anatolia, where the wild forms of the caprovines lived in a great abundance (Bökönyi 1976, 21; etc.). Since its leading species, sheep and goat are animals of warm climate and dry environment they survived well in the southern regions of Europe. Nevertheless, this fauna in an unchanged form reached northwards to the Carpathian Basin but never crossed the northern chain of the Carpathians. Only individual animals reached territories north of the Carpathian Basin, but never this type of animal husbandry. In fact, this latter also had to change under the pressure of a colder climate and moister soil in favour of the locally domesticable and environmentally perfectly adapted cattle and pig (Bökönyi 1971, 644; 1974, 26, 56; 1984, 30ff.).

At Endrőd 119, the domestic fauna was very typical for Hungary: the ratio of caprovines was high (74.11 % of the number of specimens and 78.56 % of the number of individuals) while cattle occupied second place (with 24.8 % of the number of specimens and 18.25 % of the number of individuals) as in every Körös site in Hungary. The number of pig and dog was exceptionally low showing that these two species did not have any serious economic importance in the settlement.

In comparison to the domestic fauna that in an "obligatory" way only consisted of the five usual species of the Neolithic of the Near and Middle East and the temperate belt of Europe as well, the wild fauna was very rich in species. In the mammalian fauna alone 15 wild species (5 ungulates, 7 carnivores, 2 rodents and 1 leporide) occured, and added to them were the species in the very rich avifauna (with 20 species) and ichtyofauna (with 4 species). Thus, altogether 39 wild species could be counted in this undoubtedly large bone assemblage.

Large numbers of species are not unusual in sites of the Early Neolithic if the collecting work had been good at the excavations. One does not even need a large sample to have this, e.g. in the small sample (279 specimens) of Maroslele Pana no less than 24 wild species, among them 14 bird species, were present (Bökönyi 1964, 87f.). This is caused by two things: 1/ The inhabitants exploited all environmental zones available in the close vicinity of the settlements. 2/ Most of the settlements were situated close to watercourses and lakes or on emerging islands of inundated areas. The water and the shorelines had an uncomparable richness in animal species, particularly birds and fishes, all excellent protein and fat sources for fowlers and fishermen.

Among the wild animals of Endrőd 119, the five ungulate species (aurochs, red deer, roe deer, wild swine and European wild ass) made up nearly half of the whole wild sample (48.29 per cent). If only the mammals are considered, they represented 87.69 % of the wild sample. If the rodents and the brown hare are also included as animals clearly hunted for their meat, their ratio reached 96 per cent. This clearly shows that the main aim of hunting just as that of fowling and fishing was to get meat either in order to complement their diet or to relieve the domestic stock from the pressure of being the only meat source of the increasing human population. Their exploitation gave to domestic species the chance to increase their numbers through breeding.

As for the wild species occurring here, the presence of the Hydruntinus ass is no longer surprising. Just the opposite: it would be surprising if this "leading" fossil of the Early Neolithic and occasional survivor into the Middle Neolithic would not appear in such a large Early Neolithic sample.

The other four ungulate species can be considered as common prey in all Early Neolithic sites. Out of the ordinary is the unusually high occurrence of roe deer in the sample. This could be observed in other Békés County Neolithic sites too and it may be connected with some environmental factor such as the greater extent of light forests or forested steppe. The final answer must come from palaeobotanical results, and it is hoped that the microregional project will provide it.

The seven carnivore species make up a good list, in fact one of the largest in an Early Neolithic site in Central, Southeast or South Europe. The otter should only be mentioned as a species which should have occurred among them because of the well-watered environment and the waters rich in fishes. At the same time, neither the brown bear nor the lynx should be expected because this environment was not really an ideal habitat for them at all.

The rodents are represented by micromammals: the vole-rat and the lesser mole rat. Both finds appear to be contemporary with the other bones from the site.

There is, however, a rodent species that is noticably absent. This is the beaver. It is a quite common species in the hunted fauna of Early Neolithic sites in Hungary and the North Balkans. In these areas, it had perfect environmental situations, and strangely enough it is absolutely missing in this large bone assemblage. It cannot be supposed that a large otter population killed off the beavers in the region because otters were not present or at least were not hunted either. The only explanation is that the soft-wood trees might not have been common in the region and consequently, this species was not able to build its dams forming its special home.

The 15 wild mammalian species compose a wild mammalian assemblage that is the richest in species. Such a large number of wild mammal species has not been found in any other Early Neolithic site in the three larger geographical units dicussed here. There is one site, however, Grotte IV de St. Pierre de la Fage (Hérault) in France (Bökönyi and Kretzoi 1983, 129) that was far richer in wild mammals. Its wild mammal collection, however, derived from four sources: human activity, prey of predatory birds living in the cave, and bats cohabiting with men and predatory birds living in the cave which occasionally died there.

The high number of wild mammal species in the the animal bone sample can be excellently demonstrated by the fact that in an article (Vörös 1980, 47) with a summarizing aim using the animal remains from 18 excavations of Körös sites (out these only two of which were identified by him while another had been published in detail [Maroslele-Pana; Bökönyi 1954, 87, 93]; the remainder were either old publications from the beginning of the century or had been partially published in a summarizing monograph [Bökönyi 1974], the 18 sites altogether produced only 15 wild mammal species. As a matter of fact, one out of the 15 wild species of that article (the wild horse) simply could not have occurred in a bone assemblage from the Körös culture because it had only been described in early reports when the existence of Asinus hydruntinus was not yet known in Hungary (in modern excavations not a single horse bone has been uncovered from Körös sites in Hungary). Another wild species, the "wild sheep" (in all probability Vörös does not seriously suppose the presence of a wild sheep species in the Carpathian Basin and this is why he puts its name in quotation marks) did not exist anywhere on the European mainland either. Thus, eighteen Körös sites did not yield as many wild mammal species as Endrőd 119 did. It is true, however, that the number of the identified wild animal remains from this site alone surpasses the quantity of animal bones from all the eighteen other sites together.

There is another species that is conspicuously underrepresented. It is, however, a batracian, the pond tortoise. Its shell fragments generally occur in large numbers in prehistoric sites, particularly Neolithic sites. Nevertheless, some of these finds can be intrusive: tortoises dig themselves into dry soil for hibernation and the layers of archaeological sites are among their favourite because, in addition to being dry, they are also soft. One can therefore only accept a tortoise carapax or plastron fragment as contemporaneous if they show cut marks or burns. The presence of one single carapax fragment is strange and can only be compared to the absence of otter and beaver bones.

Every wild animal species has its own habitat that is well determined by the soil, contour, climate, water, food, competing species, etc., and it is a general rule in archaeozoology that if the remains of a given species appear in the bone assemblage of a site it is sure that its habitat can be found as well within two days walking distance. Habitat identification is not so strict for carnivores because food, an important determining factor of the habitat, is basically the same everywhere.

Keeping the wild fauna of Endrőd 119 in mind, the following habitat types must have been present in the vicinity of the site.

1. Lightly forested grassland ("Parklandschaft"). This was the typical habitat of the aurochs, roe deer, Hydruntinus ass (however, this species could easily live on treeless grassland too) and brown hare. Of the birds, only the small and the great busstard and, to some extent, the partridge are grassland species.

Since roe deer was by far the most frequent wild mammal species while aurochs needed the largest grazing land, it is highly probable that this habitat type covered most of the area where the inhabitants of the settlement used to hunt.

2. Large dense forest rich in undergrowth. This is the favourite habitat of the red deer. However, since the number of red deer killed is very low, one can reasonably suppose that dense forests were not abundant in the area. At any rate, this type of habitat must have yielded some of the wild swine too. Several of the wild carnivores preferred it as well. Among the birds, only the tawny owl lives in this habitat type.

3. Gallery forests along watercourses and around lakes. This was the favourite habitat of the wild swine because of the plentiful food and the vicinity of water. The roe deer may also have been present here, although they mainly used it as cover in daylight. Nevertheless, roe deer which make frequent small "meals" could have found food there too.

4. The treeless, although rich in reed and plant, shores of lakes or creeks of various size are closely linked to the last habitat type. In fact, the water with its rich vegetation and large number of invertebrate fauna, also belonged to this habitat type providing food and shelter to most members of the site's avifauna. The water itself contained the plentiful fish fauna too, offering proper conditions accomodating different needs in variable conditions. The two latter watery habitat types provide homes to the remaining 16 birds species and thus, 80 per cent of the avifauna at this site.

5. Agricultural land provided good feeding grounds for roe deer and wild swine but other wild ungulates also visited it, sometimes causing extensive damage to the crops. This led to the more intensive hunting of these "crop destroyers" (Zeuner 1963, 199). Among the birds, several species also visited it for food.

The domestic species and their wild forms

Cattle and aurochs

Not much is known about the domestic cattle population of the Körös culture. The reason for this is that a/ cattle bones have not been found in large enough quantities at Körös sites, and b/ their state of preservation has not been particularly good (they have usually been badly fragmented) and the number of whole long bones has been quite low.

Despite the very low number of Körös animal bone finds (Bökönyi 1954), two cattle metatarsals were found in 1954 at the site of Hódmezvásárhely Bodzáspart. These metatarsals indicated that the cattle were rather large. This is not surprising because such Early Neolithic cattle had been domesticated only a few generations previously and were close both in form and size to their wild ancestor, the aurochs. Now, a fair number of cattle bones have appeared in the animal bone sample of Endrőd 119. The cattle sample was qualitatively rather good as well, there are altogether four large cranium fragments (unfortunately with their horn cores broken off) which clearly reveal that the shape of the cranium that was very similar to that of the aurochs. The forehead was flat or even a little concave while its intercornual ridge was wide and straight. There were two horn cores attached to about half of the forehead and the intercornual ridge (*Fig. 1, 3-4*). They display similar features. The horn cores themselves are large, with a form extremely similar to that of, the aurochs (see *Fig. 1, 1* and *Fig. 2*). In fact, all the six cattle horn cores in *Fig. 1* are long and closely resemble aurochs horn cores.

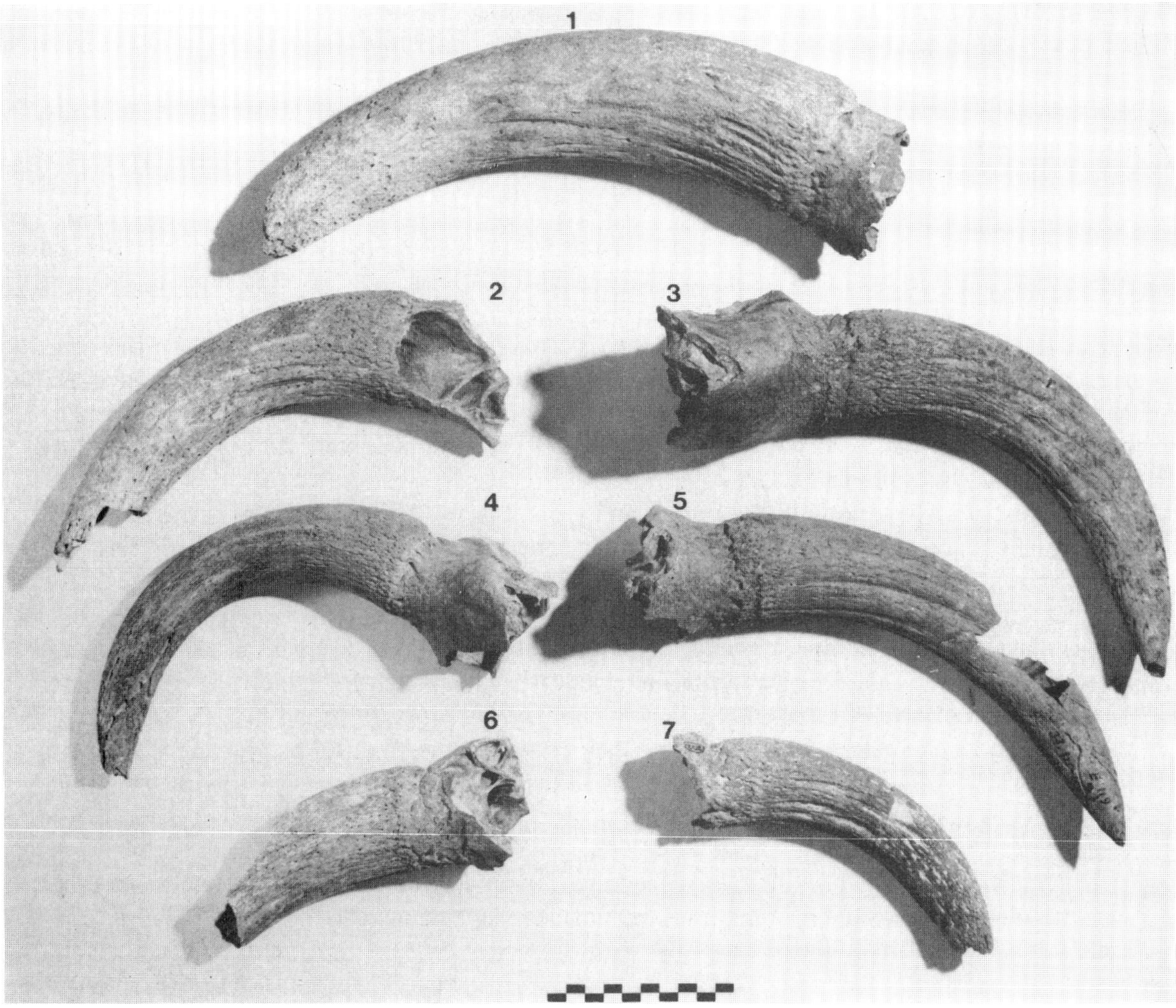

Fig. 1. One aurochs (1) and six domestic cattle (2–7) horn cores.

As *Table 3* shows, the great majority of cattle horn cores belong to this group (it is called the primigenius type because of its similarity to the wild aurochs) while medium sized and small horn cores represent a small minority. The small fragments with unidentifiable forms also occur in very low numbers. In summary: the form of horn cores and horn core fragments clearly demonstrate that the domestic cattle population of the Early Neolithic Endrőd 119 was widely variable, showing that after domestication these variants were able to survive under human protection.

Table 3. Cattle horn core types from Endrőd 119

	short	medium sized	large	unidentified	total
juvenile	-	-	2	3	5
subadult	-	-	-	-	-
adult	-	-	13	-	13
unidentified	2	3	11	11	27
total	2	3	26	14	45

The appearance of short horned cattle (the so called brachyceros type) is quite interesting. Such cattle undoubtedly developed not long after domestication. At Çatal Hüyük, for example, such horn cores were found in the 7th millenium B.C. (Perkins 1969, 178). The earliest find in Europe has until now been a right horn core from Nosa in northeast Yugoslavia, not far from the Hungarian border (Bökönyi 1974, 110; 1984a, 32, Abb.1).

As for the size of the domestic cattle at Endrőd 119, the eight whole metacarpals and five metatarsals in the sample provide absolute size data. Through their studies not only size (withers height) but also their sex may be determined. The results are shown in *Table 4*.

Table 4. Withers height and sex groups at Endrőd 119 domestic cattle based on the metapodial measurements

bone	greatest length	proximal width (cm)	smallest width	withers height	Nobis index	sex	slenderness index
metacarpus	196	57	29.5	121.13	29.08	♀	15.05
metacarpus	201	58	32	124.22	28.86	♀	15.92
metacarpus	202	62	31	124.84	30.69	♀	15.35
metacarpus	205	60	31	126.69	29.27	♀	15.12
metacarpus	205.5	60.5	31	127.00	29.44	♀	15.09
metacarpus	207	60	32	127.93	28.99	♀	15.46
metacarpus	213	60	35	131.63	28.17	♂	16.43
metacarpus	213	62	32.5	131.63	29.11	♀	15.26
metatarsus	216^{x}	46^{x}	27	118.15	21.30	♀	12.50
metatarsus	220	48	26	120.34	21.82	♀	11.82
metatarsus	228	49	23.5	124.72	21.49	♀	10.31
metatarsus	245	50	27	134.02	20.41	♀	11.02
metatarsus	246	53	27	134.56	21.54	♀	10.98

As *Table 4* indicates the average withers height is 126.88 cm based on the metatarsals and 126.36 cm based on the metacarpals and 126.68 cm if one takes both metapodials into consideration. It is about equivalent to the average withers height of the Neolithic cattle from Hungary (Bökönyi 1974, *Table 2*). The size variation of the cattle of Endrőd 119 was rather large, about 16.5 cm, as is customary in primitive populations where there was no selective breeding and any variant could survive under human protection. In addition, the number of cattle must have been be an indicator of wealth or status as has been suggested for Early Neolithic Divostin in, Serbia, (Bökönyi 1988b, 423) and as still exists in East African pastoral complexes (e.g. Watussi) even today.

There is however, one point that does not fit in with this hypothesis, and this is the low number of bulls. In societies in which cattle are indicators of status most of the bulls are usually kept. In Divostin the cow: bull ratio is nearly 1:1. At Endrőd 119, the high ratio of cows to bulls (12:1) suggests some secondary use of cows, probably for milk.

Some of the measurements of the postcranial bones also show the wide variation in the size of the Early Neolithic cattle of Endrőd 119, e.g. those of the proximal ends of radius or the astragalus, (*Figs. 3* and *6*) while others, e.g. the proximal metacarpus and the distal tibia (*Figs. 4* and *5*) are more uniform.

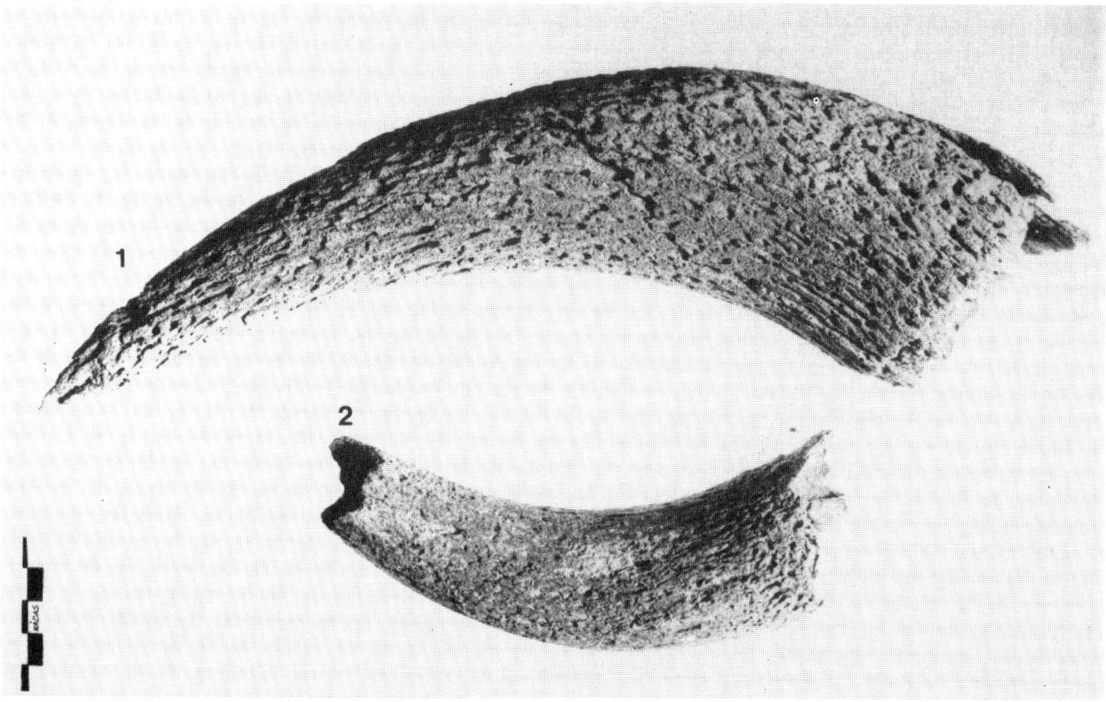

Fig. 2. Aurochs horn core (1) and horn core fragment (2).

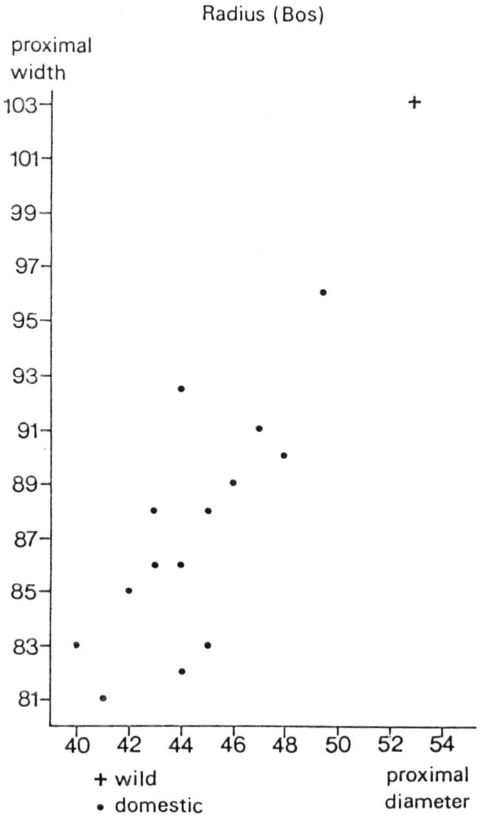

Fig. 3. Scatter diagram of proximal ends of radii. Bos.

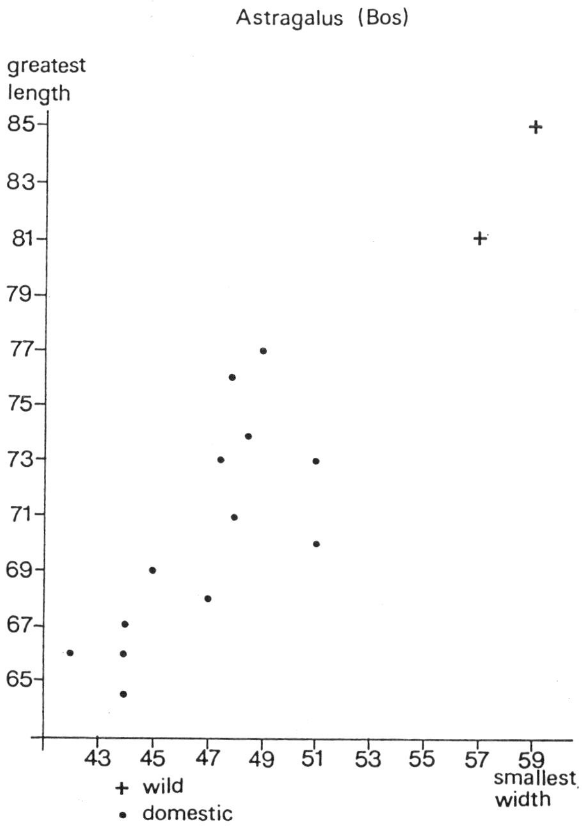

Fig. 6. Scatter diagram of astralagi. Bos.

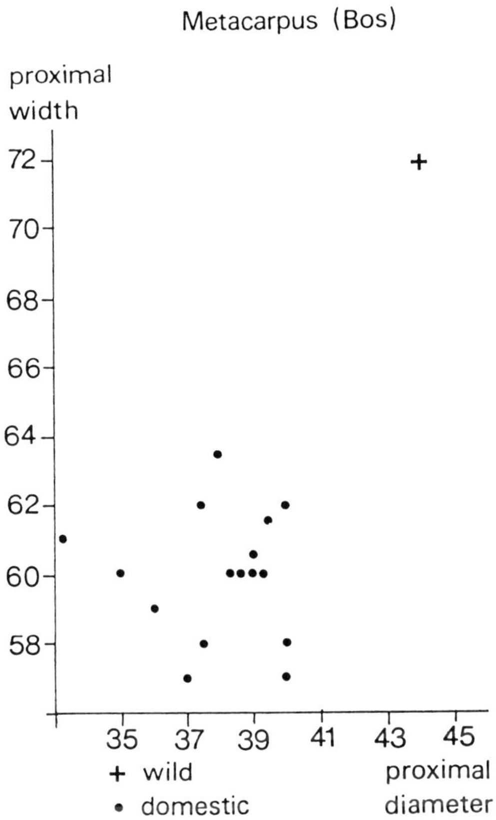

Fig. 4. Scatter diagram of proximal ends of metacarpals. Bos.

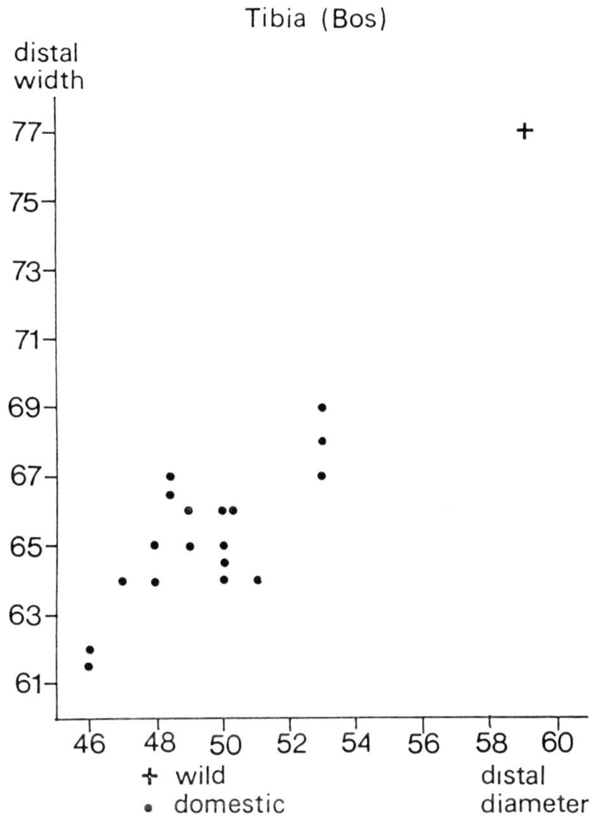

Fig. 5. Scatter diagram of distal ends of tibiae. Bos.

The above diagrams automatically raise the question of local cattle domestication on the site. No doubt, cattle had arrived in an already domesticated form in the Carpathian Basin by 5000 B.C. However the possibility of local cattle domestication cannot be excluded at any site once the principles of cattle domestication are known.

Table 5, which shows the distribution of the lower third molar (M3) as the most numerous measurable skeletal element (36 specimens) in the bone sample, demonstrates the existence of local domestication though the degree to which this occured is uncertain.

Table 5. Size distribution of the bovid lower M3

length	37	38	39	40	41	42	43	44	45
n	2	2	7	7	5	5	3	3	2

This Table undoubtedly shows a smooth transition between wild and domestic animals so unbroken that it is impossible to say where the measurements of domestic cattle end and those of aurochs begin. There seems to have been a wide area between 39 and 42 mm where the measurements of the two populations overlapped.

At the same time, out of the four scatter diagrams, that of the radius is the only one that produces one single transitional individual (similarly, there may be three or four transitional individuals in the measurement tables) while the other three show a wide gap between the domestic and wild populations.

This probably means that local cattle domestication existed. Its scale was rather small, however, and lagged far behind the large scale domestication of the Late Neolithic of the Great Hungarian Plain (Bökönyi 1988a, Figs. 7, 10, 13, 14, etc.).

The wild aurochs at the site were not particularly large. The right horn core (*Fig. 1, 1*) of the adult individual certainly falls into the lower half of the species's range of variation. The right horn core of a subadult male (*Fig. 2, 1*) would have reached a larger size. The third fragment probably came from a female on the basis of its small size and the lack of bone pearls on its base.

The bones of the extremities (in common with the lower M_3) are also of average dimensions. Only one single lower scapula fragment with the largest distal end of 90 mm came from a large, although not extremely large individual.

The only complete long bone – a metacarpal, with maximum length of 218 mm – points to a cow, while its 134.72 cm withers height falls far below the Hungarian average.

Caprovines

Sheep and goat represent an artificial unit and were handled in the faunal analysis as such. The reason for this is the great similarity of certain bones or bone fragments from these two small ruminants and because fragmented bones – in spite of several attempts to work out a reliable system for the distinction between sheep and goat bones, see Cornevin and Lesbre (1891), Gromova (1953), Boessneck, Müller and Teichert (1964), Payne (1969, 295f f.), Kratochwil (1969, 483f f.) – can in many case not yet be identified.

Among the caprovines of Endrőd 119 the sheep greatly outnumber the goat. Their ratio is 88.67:11.33% by specimens and 90.65:9.35 % by the number of individuals. This is the usual situation in the Körös culture: If a settlement has a statistically representative sample, one will always find a more or less pronounced dominance of sheep over the goat. It is uncertain, at present, whether this difference in proportionate numbers reflects ecological or economic circumstances.

a/ Sheep

After cattle, sheep were the most numerous and important group in the animal husbandry of Endrőd 119, although the species was not indigenous, that is, had no wild form in the Carpathian Basin. Its early domestication, the favourable climate and probably also its still continuing upswing pushed it ahead and only in the Middle Neolithic did the time arrive when it had to give over its leading place to cattle and pig and was pushed back to the rank of a third rate domestic species.

Nevertheless it had several revivals during the prehistoric period. Whenever an influx of population from the southseast or east reached the Carpathian Basin, it always brought with it a high number of sheep that certainly affected the local animal husbandry. Their effect was, however, not a long lasting one.

Among the domestic animals of Endrőd 119, sheep was the species which showed the widest variation in skull structure mainly through its horn cores and frontal bones, and which had the most complete long bones (mainly radii and metapodials). The skull structure demonstrated how numerous and large the changes in a species with high plasticity can be even in the early phase of its domesticated life. Sexing is possible in most of the cases. The second provided much useful data concerning its absolute and relative body size and allows a look into the trends of its development.

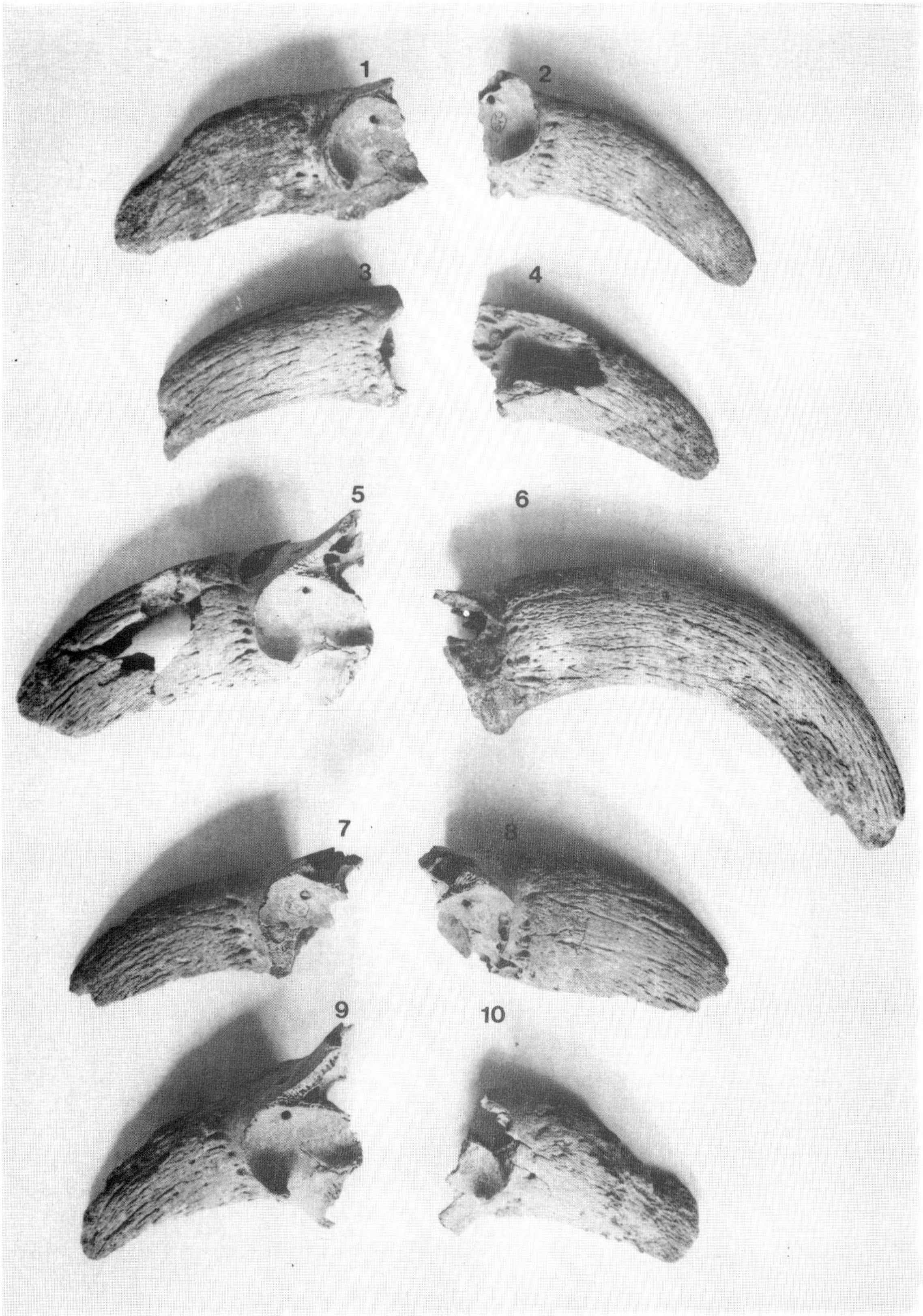

Fig. 7. Sheep horn cores.

Table 6. Distribution of different horn core types and hornlessness in sheep

age	hornless spec.	%	"palustris" spec.	%	rudimentary spec.	%	"copper sheep" spec.	%	horned spec.	%	total spec.
juvenile	0	0	10	12.82	1	11.11	13	32.50	3	27.27	27
subadult	11	19.64	27	34.62	2	22.22	15	37.50	4	36.36	59
adult	19	33.93	16	20.51	3	33.33	2	5.00	0	0	40
unidentified	26	46.43	25	32.05	3	33.33	10	25.00	4	36.36	68
total	56	100.00	78	100.00	9	99.99	40	100.00	11	99.99	194

This Table is not simply a documentation but also shows a kill-off pattern that will be evaluated in a later chapter. Nevertheless, it is a appropriate here to describe and introduce the different horn core types that had, beyond their morphological differences, other meanings too.

The horn core of the so called "copper sheep", whose best adult example is No.5 in *Fig. 7* and No.6 in *Fig. 8* is a heavy, outwardly twisted, long horn core with triangular cross section. It is a typical horn core for males, resembling that of wild males although not so flat and twisted. In fact, the name "copper sheep" does not refer to any group, type or breed, but is simply a short description of this horn core form, the horn core of the male sheep in the prehistoric European and southwest Asian sheep populations.

Fig. 8. Sheep horn cores (Nos. 1–5 are "palustris", No. 6 is "copper sheep").

The so-called "palustris" horn cores – the name was given by L. Rütimeyer, the great Swiss pioneer in animal domestication research (Rütimeyer 1961, 191) – are short, flat and goat-like although much shorter, pointing outwards and upwards, and are untwisted (*Fig. 8, 2,6; Fig. 9, 1,2*). They are one of the horn core forms of the females (Reitsma 1932, 45). The rudimentary horn cores (*Fig. 10* shows different stages of their development) and the hornless sheep (different forms of hornlessness can be clearly seen in *Fig. 11*) represent females.

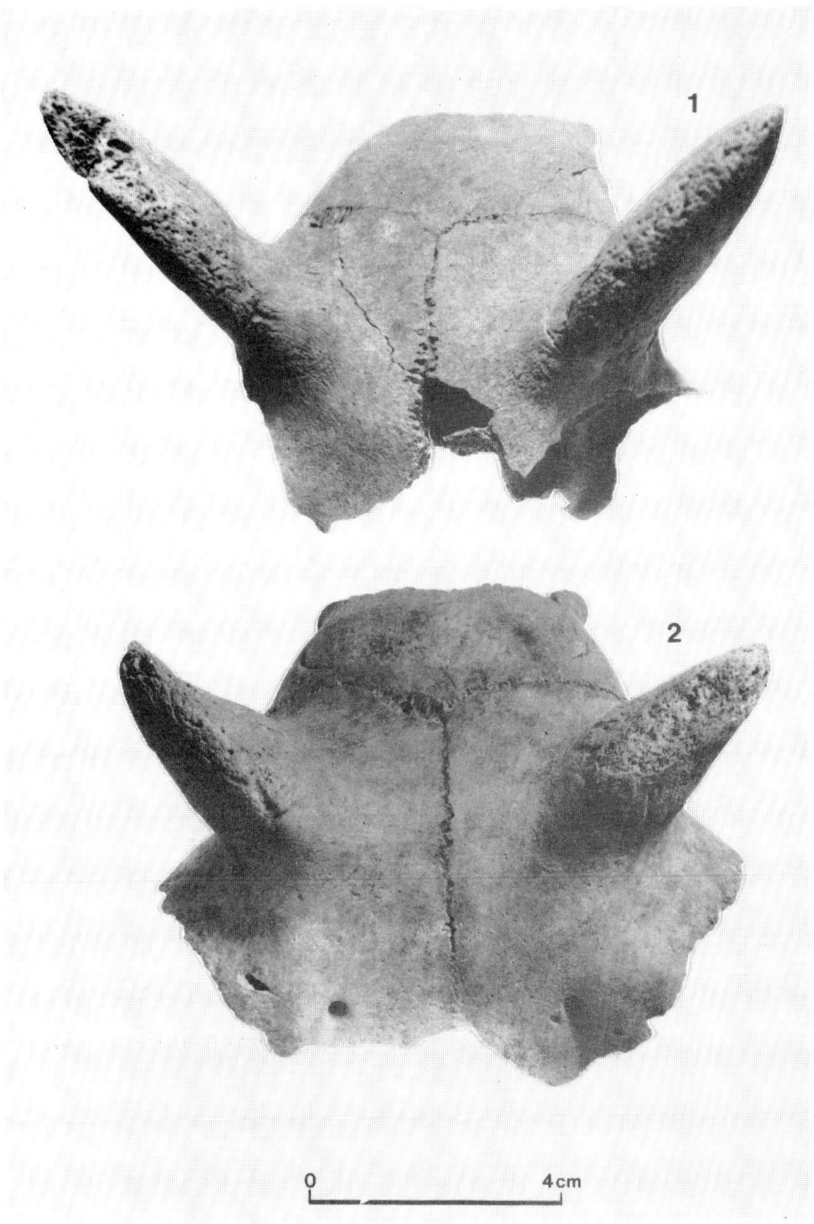

Fig. 9. Sheep cranium fragments with "palustris" horn cores.

The rudimentary horn cores may represent a developed phase – although not the final one – in the decrease of horn core size caused by domestication; it may however represent no more than a sudden mutation, such as occured in cattle and goat, in the C2 stratigraphic horizon (Bus Mordeh phase, ca. 7500 to 6500 B.C.) at Ali Kosh in the Deh Luran Plain in West Iran (Hole and Flannery 1967). After its first appearence it quickly spred all over Southwest Asia where it can be found in every larger pre-pottery or early pottery neolithic bone sample. It must have come to Europe with the first wave of Southwest Asian domestic fauna because, although it was not mentioned either at Nea Nikomedeia or at Argissa Magula, several hornless individuals could be found at Achilleion in Thessaly, Greece (Bökönyi 1989, 319). From that point on, such hornless individuals can be identified in every Early Neolithic site in the Balkans, the Carpathian Basin as well as in South Europe.

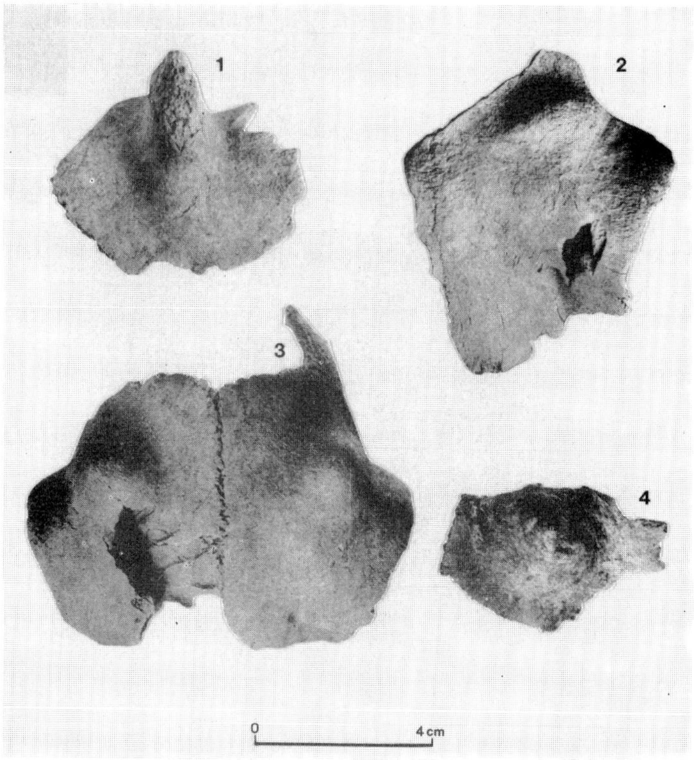

Fig. 10. Sheep cranium fragments with rudimentary horn-cores.

Fig. 11. Cranial fragments from hornless sheep.

Using these data, new light can be shed on the ratio of rams to ewes in the different age groups. The results are shown in *Table 7.*

Table 7. Ratios of rams to ewes in the different age groups

age	"palustris"+ rudimentary+ +hornless		"Copper sheep"	
	specimen	%	specimen	%
juvenile	11	12.36	13	43.33
subadult	40	44.94	15	50.00
adult	38	42.70	2	6.67
total	89	100.00	30	100.00
unidentified age	26		10	
Total	115		40	

Table 7 provides evidence that, even in this early period of animal husbandry, it was recognized that fewer males than, females were needed to sustain a breeding population. In fact, it was already customary by the time of specialized hunting and the beginning of domestication that the superfluous males were killed when meat was needed (Bökönyi 1969, 224f.). It is for this reason that the adult ewe: ram ratio is 86.49 : 13.51 %, in other words, one ram to 6.4 ewes. It is true that some of the subadults, that is two year olds, could be used for breeding in the same way as modern reindeer breeders also use mainly subadult males for breeding (Herre 1955, 180ff.). A part of the subadult females were used for breeding too.

The sheep from Endrőd 119 show a rather wide size variation even at first glance. It is also clear that they were not very large. Their build is somewhat variable too.

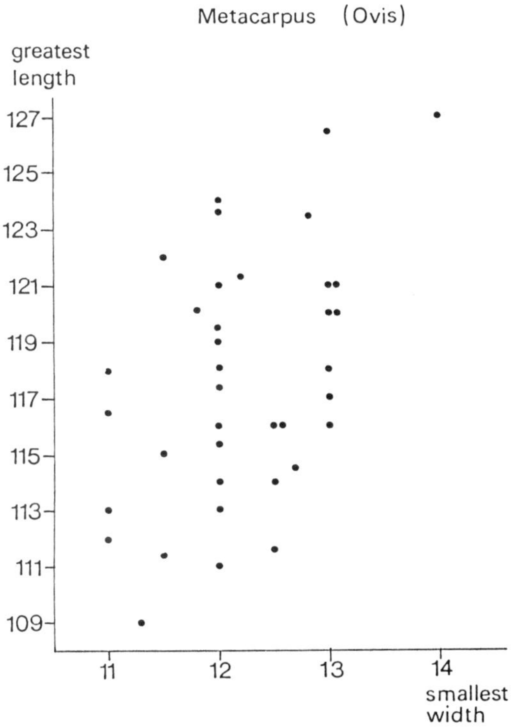

Fig. 12. Scatter diagram of sheep metacarpals.

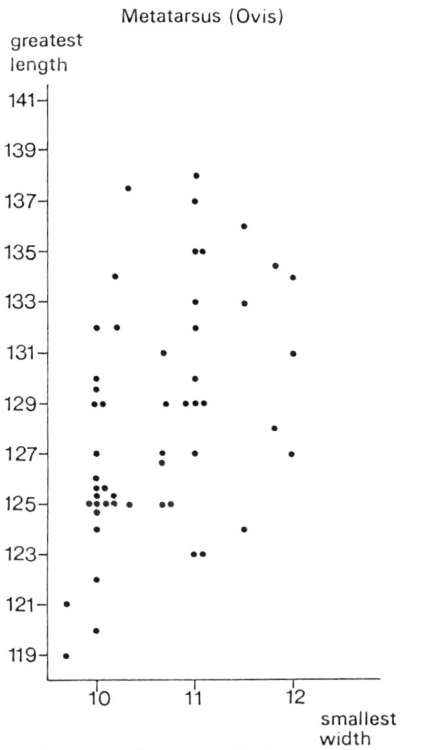

Fig. 13. Scatter diagram of sheep metatarsals.

To answer all these questions one has to study the complete long bones, particularly the metapodials, which can provide precise answers (see *Table 8* and *Figs. 12, 13*).

Table 8. Measurements and slenderness indexes with the determination of withers height and constitution of sheep

a/ *Humerus*

greatest length (mm)	smallest width (mm)	withers height (cm)	slenderness index
a/ *Humerus*			
127[x]	13.5	54.36	10.63
128	13	54.78	10.16
131[x]	13.5	56.07	10 31
133	13	56.92	9.77
133	13.5	56.92	10.15
143	16	61.20	11.35
b/ *Radius*			
128	15	51.20	11.72
130	16	52.00	12.31
130	16.5	52.00	12.69
132	15	52.80	11.36
135	15	54.00	11.11
135	15	54.00	11.11
135	16	54.00	10.85
135.5	15	54.20	11.07
136	13.5	54.40	9.93
137	14	54.80	10.22
137	14.5	54.80	10.58
137	15	54.80	10.95
137	15	54.80	10.95
137	15	54.80	10.95
137	16	54.80	11.68
138	14	55.20	10.14
138	14	55.20	10.14
138	14.3	55.20	10.36
138	15	55.20	10.87
138	16	55.20	11.59
138	17	55.20	12.32
139	19	55.60	13.67
140[x]	15	56.00	10.71
140	15	56.00	10.71
139	15.7	55.60	11.29
141	15	56.40	10.64
141	15.5	56.40	10.99

141.5	16	56.60	11.31
142	15	56.80	10.56
142	15	56.80	10.56
142	15.3	56.80	10.77
139	19	55.60	13.67
142	15.3	56.80	10.77
142	15.5	56.80	10.92
142	16.5	56.80	11.62
142	17	56.80	11.97
142.7	15	57.08	10.51
143	15	57.20	10.49
143.7	15.5	57.48	10.79
144	15.8	57.60	10.76
144	16	57.60	11.11
144	16	57.60	11.11
145	15	58.00	10.34
145	15	58.00	11.03
146	16	58.40	10.96
147	16	58.80	10.88
148	16	59.20	10.81
150	15.5	60.00	10.33
148.5	15	59.40	10.10
149	16	59.60	10.74
150	16	60.00	10.67
151.5	15	60.60	9.90
149	15.5	59.60	10.40
149	16	59.60	10.74
153	16	61.20	10.46
149	16	59.60	10.74

c/ *Metacarpus*

109	11.3	52.76	10.37
111	12	53.72	10.81
111.5	12.5	53.97	10.31
111.7	12.5	54.06	11.19
112^x	11	54.21	9.82
113	11	54.69	9.73
113^x	12	54.69	10.62
114	12	55.18	10.53
114	12.5	55.18	10.96
114.5	12.7	55.42	10.76
115	11.5	55.66	10.00
115.3	12	55.81	10.41
116	12	56.14	10.34
116^x	12.5	56.14	10.78
116	13	56.14	11.21
116	13	56.14	11.21
116.5	11	56.39	9.48

117	-	56.63	-
117.5	12	56.87	10.21
118	11	57.11	9.32
118	12	57.11	10.17
118	12.5	57.11	10.59
118	13	57.11	11.02
119.5	12	57.84	10.84
120	11.8	58.08	10.83
120	12	58.08	10.00
120	13	58.08	10.83
120[x]	13	58.08	10.83
121	12	58.56	9.92
121	13	58.56	10.74
121.2	13.2	58.66	10.89
122	11.5	59.05	9.43
123.5	12.8	59.77	10.36
123.7	12	59.87	9.70
124[x]	12	60.02	9.68
124.5	-	60.26	-
126.5	13	61.23	10.28

d/ *Femur*

160	14.5	54.68	9.06
166	15	58.60	9.04

e/ *Tibia*

189.5	14	57.04	7.39
194	13	58.39	6.70
194	13.2	58.39	6.80
198	14	59.60	7.07
199	14	59.90	7.04
204	14.7	61.40	7.21

f/ *Metatarsus*

119	9.7	53.67	8.15
120	10	54.12	8.33
121[x]	9.7	54.57	8.02
122[x]	10	55.02	8.20
123	11	55.47	8.94
123	11	55.47	8.94
124	10	55.92	8.06
124	10	55.92	8.06
124[x]	11.5	55.92	9.27
124.8	10	56.28	8.01

125^x	10	56.38	8.00
125	10	56.38	8.00
125	10.2	56.38	8.16
125^x	10.2	56.38	8.16
125^x	10.3	56.38	8.24
125	10.7	56.38	8.56
125	10.7	56.38	8.56
125.3	10	56.51	7.98
125.3	10	56.51	7.98
125.5	10	56.60	7.97
126	10	56.83	7.94
126.7	10.7	57.14	8.45
127	10	57.28	7.87
127	10.5	57.28	8.27
127	10.7	57.28	8.43
127	11	57.28	8.66
128	11.8	57.73	9.22
129	10	58.18	7.75
129^x	10	58.18	7.75
129	10.7	58.18	8.29
129	11	58.18	8.53
129	11	58.18	8.53
129	11	58.18	8.53
129.5	10	58.40	7.72
130	10	58.63	7.69
130^x	11	58.63	8.46
131	12	59.08	9.16
131	10.7	59.08	8.17
132	10	59.53	7.58
132	10.2	59.53	7.73
131	10.7	59.08	8.17
132	11	59.53	8.33
133	11	59.98	8.27
133^x	11.5	59.98	8.65
134	10.2	60.43	7.61
134	12	60.43	8.96
134.5	11.8	60.66	8.77
135	11	60.89	8.15
135	11	60.89	8.15
136	11.5	61.34	8.46
136.5	10.7	61.56	7.84
137	11	61.79	8.03
137.5	10.3	62.01	7.49
138	11	62.24	7.97
141	12	63.59	8.51

Withers height:

Summary:

	n	min.	max.	M
Humerus	6	54.36	61.20	56.71
Radius	54	51.20	61.20	56.40
Metacarpus	37	52.76	61.23	56.88
Femur	2	54.68	58.60	56.64
Tibia	6	57.04	61.40	59.12
Metatarsus	54	53.67	63.59	58.05
Total	159	51.20	63.59	57.19

Table 8 and *Figs. 12-13* reveal that the sheep of Endrőd 119 were small. This corresponds with the results of studies on other Neolithic sheep (Bökönyi 1978, 65f f.). Nevertheless, there were some middle sized individuals in the sample; these could either have been males that had reached full maturity or simply have originated from a later period in a disturbance which could not be recognized at the time of the excavation (such individuals were represented by two metacarpals and one metatarsal in *Figs. 12* and *13*.)

As for the slenderness of these sheep, the picture is rather variable. Nevertheless, it is impossible to distinguish sexual groups in this way because variations in the two sexes overlap considerably.

b/ Goat

The goat, far less numerous than sheep, yielded a much simpler bone sample than the sheep: the horn cores are less variable, hornless individuals are absent and whole bones are quite infrequent (though – as already mentioned – among the radii and metapodials whole bones of immature individuals are nevertheless quite common). Little can be inferred about the size of the adult population, however, because one does not know what size they would have attained if they could have reached their full maturity.

However, this is not a simple matter of smaller sample size. It is well-known that changes in goat horn cores are not as rapid as those of sheep, e.g. untwisted horn cores appear even now in areas where wild goats do not exist. Horn cores cannot therefore be used to argue recent domestication. Among goat horn cores, the number of variants is much lower than among sheep horn cores while hornless goats do not appear before ca. 2500 B.C. (Bökönyi 1974, 193). *Table 9* demonstrates the small number and low variability of goat horn cores at this site.

Table 9. Goat horn core variability at Endrőd 119

size	untwisted	twisted	unidentified	total
small	-	-	-	-
medium	1	3	4	8
large	1	-	1	2
very large	1	1	-	2
unidentified	1	-	-	1
Total	4	4	5	13

Although the sample size is small the equal number of untwisted ("aegagrus") and twisted ("prisca") horn cores is surprizing because twisting appeared and spred quickly immediately after the first wave of domestication (Bökönyi 1973a, 70). They certainly do not reflect sexual differences because there is a very large specimen among the untwisted (scimitar-like) horn cores and large horn cores are a characteristic of males. The comparatively high frequency of untwisted goat horn cores may perhaps be a local phenomenon in the Carpathian Basin. The Bronze Age site of Tószeg should be mentioned here, whose excavation in 1948 yielded only scimitar-shaped goat horn cores (Bökönyi 1952, 94f f.), a fact that also awoke Zeuner's interest (Zeuner 1963, 146).

Fig. 14. Two large goat horn core fragments (untwisted).

Fig. 14 shows two such untwisted horn cores. Nos. 2 and 3 in *Fig. 15*, are also untwisted. The others are twisted. A very large horn core fragment can be seen in *Figs. 2* and *3* each, one untwisted and the other one twisted.

Fig. 15. Goat horn core fragments.

The withers height can be determined on the basis of the greatest length of the metapodials. (Schramm 1967, 103). See *Table 10.*

Table 10. Determination of the withers height of goats using Schramm's method (Schramm 1967, Tab.8)

bone	greatest length (mm)	withers height (cm)
metacarpus	119.5	68.71
metatarsus	110.5	59.01
metatarsus	113	60.34
metatarsus	114.5	61.14
metatarsus	116	61.94
metatarsus	118.2	63.12
metatarsus	128^{x}	68.35

As *Table 10* shows, the goat population was quite variable – in fact, the morphological plasticity of the European goats has also been emphasized by Vigne (1988, 10). It was probably due to the expressed sexual dimorphism. In all likelihood, both goats with a withers height above 68 cm were males, while the other ones were females.

Pig and wild swine

Among the occurring domestic species in the site, the pig is the poorest from a qualitative point of wiew. Because it was a typical meat producing species its individuals were mostly killed young. In addition, their bones were very thoroughly chopped up and therefore both its characteristic and measurable bones are extremely rare.

The most usable skull fragments were a left and right cranium fragment, both subadult. These indicate an elongated, narrow skull with a straight profile line, that is, primitive skulls that are very close to the skull of the wild swine. This skull form is quite common in the Early Neolithic both in Central and Southeastern Europe.

Sexing was only possible in three cases: a right adult mandible fragment and a subadult mandible's oral fragment came from sows while another right mandible fragment was from a boar. This was a rather large individual, probably transitional between the domestic and wild population. Unfortunately it cannot be measured. In fact, this is the only proof of local pig domestication which means that the latter could not have been particularly common, although it was certainly practised.

The wild swine bones do not come from particularly large animals. An unmeasurable ramus and a distal humerus fragment with a 59 mm distal width represent really large individuals. All others are medium-sized and small. This is a bit surprising because wild swine had access to a favourable environment in the dense forest and the gallery forest. Also, the sexing of the upper and lower jaws showed that great majority of the wild swine killed were males.

The size distribution of lower M_3 is as follows:

size	41	42	43	44	45	46	47	48	49
n	1	1	-	2	3	-	-	1	1

Dog and wolf

As mentioned above, among the bone assemblage were parts of a dog sacrifice whose skull and two mandible halves were in quite a good state of preservation. The skull is medium-sized, belonging to a primitive type that is close to the wolf. Its brain-case is comparatively narrow and less arched. The frontal region is a little flat and the nasofacial part, particularly the praemolar region, is strongly shortened. The teeth are large, with a gap between the left P_3 and P_4. The P_2-P_3-P_4 are crowded on the right side and the P_3 stands nearly crosswise (*Fig. 16*). In the mandible halves, the P_1-P_2 are slightly and the P_4-M_1 are particularly crowded.

Of the mandibles, one left mandible is large, heavy and has gaps between its P_2-P_3-P_4. In another left mandible (middle-sized) the P_1-P_2 are a little crowded while a middle-sized right mandible half has a normal premolar row of teeth. The same is true for an upper premolar row, while a right premolar row (again middle-sized) shows a gap between its P_1 and P_2.

Using Koudelka's method (Koudelka 1884), two complete left radii and another left radius whose proximal epiphysis is missing, point to small dogs with a withers height of 41.86 and 39.61 cm, corresponding to the size of foxterriers. A ca. 150 mm long tibia comes from a somewhat larger individual of 43.80 cm, that is, the size of a Mittelschnauzer. Nevertheless, the majority of dogs belong to medium-sized animals showing that they were still very close to the wild form. The small ones, whose complete long bones have been described above, may perhaps represent the "turbary" dog which is not an independent type or breed but a group which simply shows a higher level of domestication.

The wolves killed by the inhabitants of the settlement, obviously when protecting their domestic stock, are small to medium-sized animals, although a large individual is also present among them (e.g. two calcanei with 60 and 64 mm greatest lengths).

Due to the small sample sizes for both groups, it cannot be determined whether any kind of connection existed between the two – domestic and wild – populations.

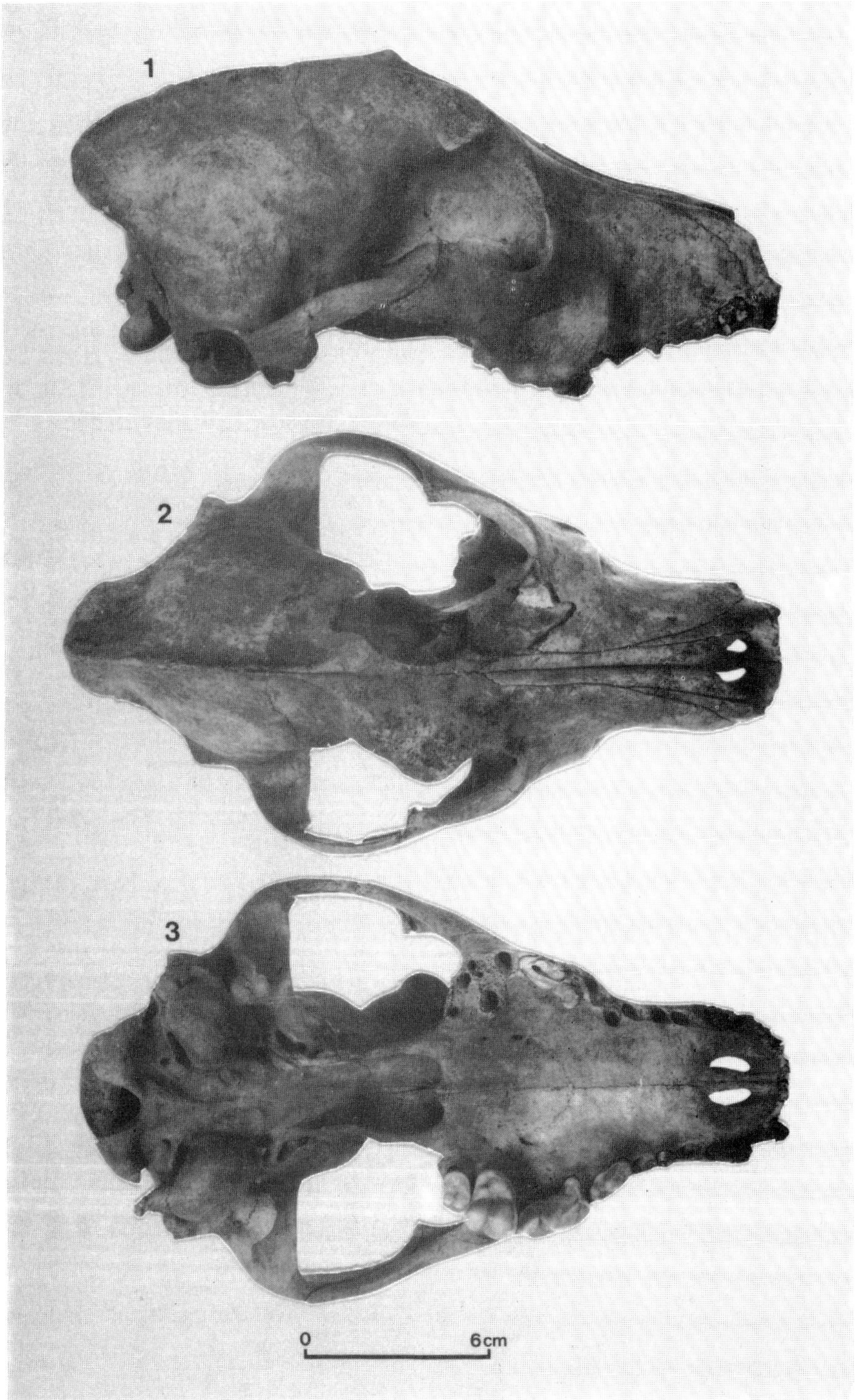

Fig. 16. Skull of sacrificed dog.

Wild species

Red deer

The total of 138 red deer remains from the site represent 16 individuals. The kill-off pattern of the species is a little unusual:

juvenile	6
subadult	2
adult	5
mature	1
undetermined	2

This means that the larger part of the red deer were killed as juveniles and subadults which points to a hunting tradition that did not care about the reproduction of the herd. This hypothesis was also strengthened by two other facts: 1. Antler fragments are very rare among the red deer remains (it is true that antler served as the most important raw material of animal origin for tool making, nevertheless, even in the presence of large-scale tool-making the number of waste fragments should be high). 2. Among the measurable red deer bones there is one single piece, a proximal radius fragment with a 67 mm proximal width, that definetely comes from a male. There are three medium to large bones that might be from male individuals but the majority of the bones (9 specimens) represent small and medium sized individuals, thus hinds.

The antler fragments come from small antlers, only one of them represents a large male; thus the red deer population from this site is comprised of smaller individuals than the typically large red deer of the Hungarian Neolithic. This could be the result of environmental factors. The preferred habitat of red deer is dense forest with thick undergrowth; the less favourable environment around the site comprised mainly grassland with small wooded areas or forest along the margins of rivers or lakes.

Roe deer

The roe deer was the most commonly hunted species at Endrőd 119. This was not the result of specialized hunting but rather reflected of the abundance of an ideal habitat (grassland or agricultural land with forested areas) for the species.

In the entire wild bone sample, that of the roe deer is the best preserved: several larger cranium fragments, 34 whole antlers and antler fragments, whole long bones (not counting the subadult and juvenile bones) and many measurable fragments have been found.

The kill-off pattern of roe deer is essentially better than that of the red deer because 63.63 % of the bones of determinable age are from adult and mature individuals and only 15.15 % are from juveniles. The age structure is as follows:

juvenile	5
subadult	7
adult	18
mature	3
undetermined	4

This picture reflects a more balanced hunting pattern than that for the red deer.

When such a large roe deer population lives in a favourable habitat it is not surprising that the size of the individual deer is large. When compared to the considerable data for the roe deer of Hungary (*Table 11*), one quarter of the roe deer bones in Endrőd 119 comprise large and medium to large specimens, another 31 % are medium sized, and less than 44 % are small and small-to-medium sized.

Table 11. Roe deer size distribution

	small	small-to-medium	medium	medium-to-large	large	total
n	49	19	48	7	32	155
%	31.6	12.3	31.0	4.5	20.6	100

In fact, the roe deer of Endrőd 119 were among the largest from prehistoric Europe; the 16.44 mm average length of the lower M3 is probably the largest found in Europe. These roe deer most resemble other Hungarian and Yugoslavian prehistoric individuals and are certainly larger than some Greek ones (Bökönyi 1976b, 147ff.; 1988b, 444; Boessneck 1962, 36f.; Amberger 1979, 133f.; Becker 1986, 135f.), though other Neolithic sites in Greece (Sitagroi and Achilleion) certainly yielded larger roe deer remains (Bökönyi 1986, *Table 5,33*; 1989, 322). In this way, the size decrease of roe deer that went from north to south in the Balkan Peninsula does not seem to have been a strict rule (Becker 1986, 137) but rather a general trend.

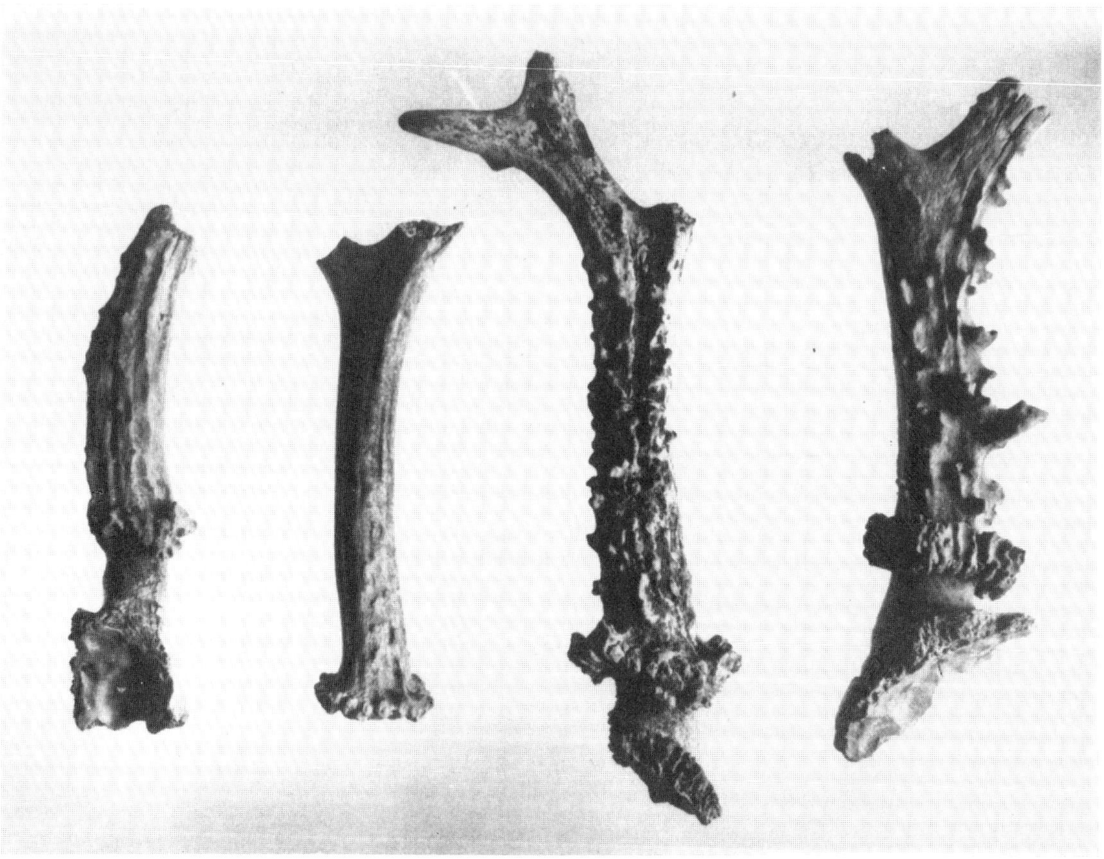

Fig. 17. Roe deer antler fragments.

The antlers and antler fragments reflect a similar picture. Of these, (see four nearly complete antlers on *Fig. 17*) one is very large, two are large, one is medium-to-large, 12 are medium, one is small-to-medium, and six are small. In four cases, the circumference of the burr can also be determined; these measure 132, 116, 110 and 96 mm respectively. The first is notably large, the second and third are medium in size and the fourth is small.

The size variation of the extremity bones is shown in *Figs. 18* and *19*. The size difference between bucks and does shows clearly on the scatter diagrams. The former are found in the upper part and the latter in the lower part of the range of variation.

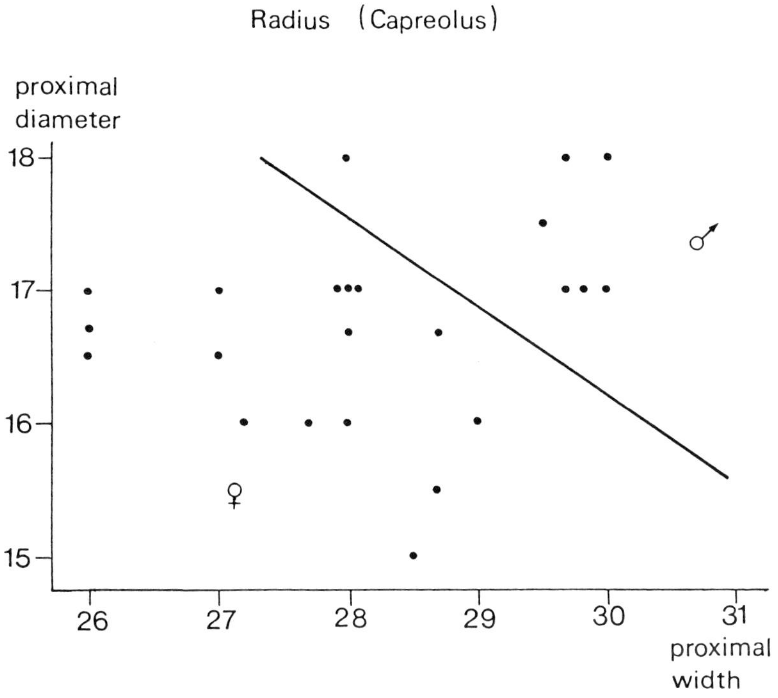

Fig. 18. Scatter diagram of proximal ends of radii. Capreolus.

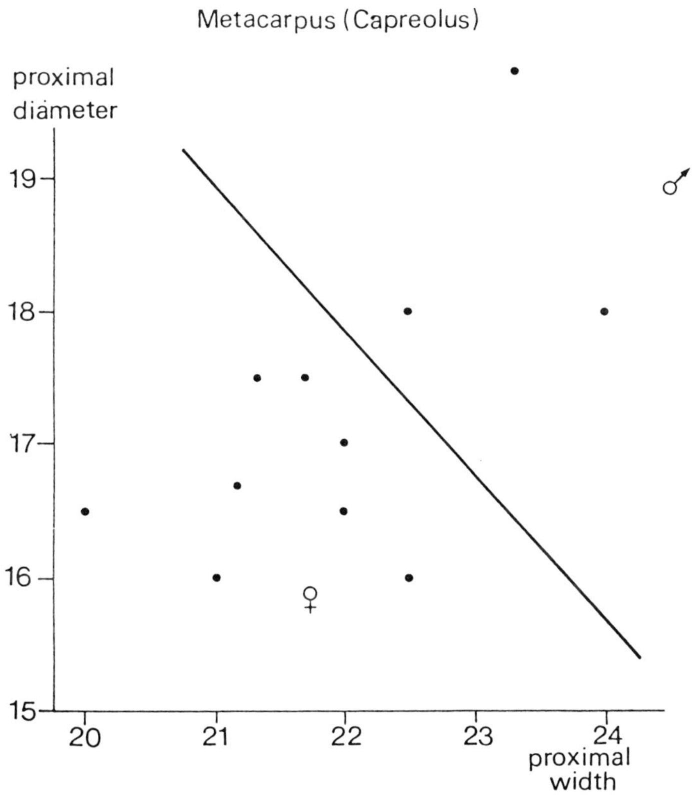

Fig. 19. Scatter diagram of proximal ends of metacarpals. Capreolus.

Wild ass

The European wild ass (Asinus hydruntinus Reg.) was represented by 16 items in the bone sample from Endrőd 119. They are as follows: a right maxilla fragment with the M_2 and M_3, two upper cheekteeth, an upper cheektooth fragment, a right lower M_3, a right lower P_3 or P_4, a right lower M fragment, a thoracic vertebra fragment, four rib fragments, two right proximal radius fragments, one right anterior phalanx, and one posterior phalanx I.

The 16 remains belong at least to four individuals; the maxilla fragment represents a senile animal: Its teeth are worn down to their roots. Most of the enamel folds and parts of the dentin are missing on the chewing surface (*Fig. 20*). The right lower M fragment is also heavily worn and its crown is fragmented pointing to a mature animal. All other teeth and bones are probably from two adult individuals.

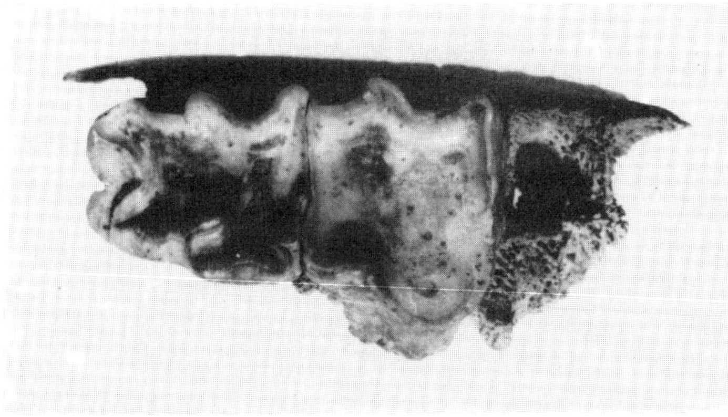

Fig. 20. Maxilla fragment with M_2 and M_3 of old Hydruntinus ass.

The identification of the teeth and the bones was based on the following points:

1. The small size, absence of pli caballin and the angular lateral (buccal) surfaces immediately exclude the caballin origin of these remains.

2. They are undoubtedly from a small equid although not from the smallest type (the length/width measurements of the two adult cheekteeth are 22.5x24 and 23x24 respectively. The same measurements for the M_2 of the maxilla fragment are 23x24 while those of the M_3 are 29x22. The length of the lower M_3 is 26.7, and that of the lower P is 24 mm. The sagittal length of the anterior phalanx I is 76 mm, and that of the posterior one is ca. 68 mm). Thus, considering the fact that half-asses diminish in size in an east to west direction, the finds could not be those of a hemione.

3. The short protoconus, particularly that of the right upper molar with its almost total absence of a mesial lobe (Bökönyi 1954, 14; 1972, 15; 1986b, 307; Dietrich 1959, 17) is very typical for asses in general and for Asinus hydruntinus in particular (Stehlin and Graziosi 1935, 6). On the lower M_3 one can clearly see the typical ectoflexid (lateral valley) of Asinus hydruntinus with is blunt, angular end (Bökönyi 1986b, 307).

4. The phalanges have a more gracile form than the horse toe bones and they are comparatively much longer than the first phalanges of half asses (Eisenmann 1986, 77) nicely falling into the range of the variation of Asinus hydruntinus.

Thus, these equid remains can be identified as teeth and bones of Asinus hydruntinus Reg., that enigmatic wild ass species whose survival into the Holocene could first be shown in bone material from a Hungarian Körös culture site, Hódmezővásárhely-Bodzáspart (Bökönyi 1954, 12ff.). Later, it could be demonstrated that it occurred in practically every Early Neolithic site in the Carpathian Basin and has also been found in Rumania (Necrasov 1961, 270ff; 1964, 141ff.) as well as in Yugoslavia (Bökönyi 1970, 1703; 1974, 435f.). There is even one site in Northeast Yugoslavia (Nosa-Gyöngypart) where Asinus hydruntinus was the most frequently encountered of the animal species (Bökönyi 1984, 30). This wild ass is present in the faunal assemblages of some Early Neolithic Italian sites too (Borzatti von Löwenstern 1969, 91ff.; Bökönyi 1977-82, 345f.; 1985, *Table 1*).

At first it was thought that this Pleistocene remnant species had become extinct at the end of the Early Neolithic, but now it appears not to be the case: Kratochvil (1972, 26f.; 1973, 195ff.) described remains of this species from a settlement of the Linear Pottery culture. Since then, its teeth and bones have also been found in bone assemblages from Middle and possibly also Late Neolithic sites in Hungary.

The Hydruntinus wild asses were hunted for their meat and felts. There is no proof of their having been domesticated. Their favourite habitat was the open grassland with small forested spots. In the Endrőd region good habitats abounded and their only serious predator must have been man. Nevertheless they could easily outrun hunters, and could only have been killed by organized communal hunting groups.

Wild cat

The wild cat is represented by a left femur and a metatarsal respectively. They probably come from one single adult individual.

The greatest length of the femur is 138 mm pointing to a very large individual. In comparison, the size variation of recent female and male wild cats from the northern Carpathians in Czechoslovakia is 120.3 to 144.4 mm (Kratochvil 1976, 38). Thus, this individual is seemingly a male wild cat from the upper range of the size variation.

Polecat?

A cranium fragment (*Fig. 21*) and a metacarpal of a large mustelid probably come from an adult polecat. The polecat is not a common prey animal at prehistoric sites. It was probably killed more for its precious fur since it represents almost no danger to livestock.

Fig. 21. Polecat cranium fragment.

Marten

A metacarpal comes from a marten. Unfortunately it cannot be determined whether this specimen comes from a pine or a beech marten. In general, marten is the most common mustelid found at archaeological sites.

Badger

Thirteen badger bones occur in the bone sample of Endrőd 119. Nevertheless, some of them derive from intrusive animals that died in their underground burrows. The certainly subfossil speciments represent five individuals - one mature, one adult, two subadult and one juvenile. The flesh of badger was certainly eaten. Some people in forest areas still consume it.

Fox

The 14 fox remains are from four individuals: three adult and a subadult. The two measurable mandibles with their 15 mm long lower M_1 point to rather large individuals.

Carnivore

A damaged left calcaneus probably comes from a small mustelid, although, it is impossible to identify it to species.

Brown hare

The brown hare is the most frequently found small mammal species, due to the favourable natural environment of grassland, "Parklandschaft" and agricultural land. Its meat was probably counted as a delicacy. The measurable bones point to medium and large individuals, e.g. a right mandible fragment with a 20 mm long cheeckteeth row.

Wild birds

The identification of the bird bones has not yet been concluded. Prof.Dr. D. Jánossy who is doing this work has nevertheless already identified as many as 20 different species which is the largest number of wild bird species ever found at a subfossil site in Hungary. He will discuss the bird species in detail. It is necessary, however, to mention, in respect to habitat requirements and the role of wild birds in human consumption, that out of the 20 bird species, 16 are waterfowl and only 4 are land species. This accords well with earlier statements concerning the Neolithic and Bronze Age avifauna of Hungary (Bökönyi and Jánossy 1965, 95 f f.; Jánossy 1985, 98).

Pond tortoise

There is one pond tortoise remain, a carapax fragment, in this large sample and even this may be secondary because it displays neither cut or burn marks. The absence of this species is certainly strange and not immediately understandable because it is certain that pond tortoises must have lived in great numbers in the swamps in the vicinity of the settlement.

Fishes

Interestingly enough, Mr. I. Takács could identify only four species : catfish, carp, pike and pikeperch out of the 671 fish remains collected during the course of the excavation. It is true that the fish bones were in a very bad state of preservation but the presence of these four species does not indicate the presence of a developed fishing industry in the settlement in spite of the occurrence of harpoons, hooks and net weights but rather points to a collecting of fish remaining in ponds after the water receded following extensive flooding.

Abnormalities and pathological changes found in the animal bones sample

Quite a few bones in the animal bone sample of Endrőd 119 show abnormalities and pathological changes particularly of a traumatic and infectious/inflammatory nature. Interestingly enough, all but one are from domestic animals. This is not surprising, because the domestic animals of the Early Neolithic were in a somewhat incipient stage of domestication when this new and previously unknown stress caused essential disorders in the metabolism, hormonal balance and other life-functions on the one hand while on the other hand, changes in diet, physical environment and mismanagement by inexperienced breeders also affected them badly.

These pathological cases will be discussed here according to Wäsle's grouping (1976, 51ff.).

1/ Ontogenic and dental abnormalities

a/ Ontogenic abnormalities

A domestic cattle rib belongs to this group: A bone bridge runs in dorsoventral direction on its medial side (*Fig. 22, 1*). Since there is no sign of inflammation, it must be considered an ontogenic abnormality that caused no trouble to the animal.

b/ Oligodontia

The P_2 is missing, in a left cattle mandible (*Fig. 22, 2-3*). In fact, it has not been knocked out. In early domestic animals this is unusual although it is a common phenomenon at a higher level of animal husbandry.

c/ Crowded teeth

Crowded teeth could be observed in five cases: the P_4 and M_1 were crowded, in the maxilla of a sheep or goat, the P_4 and M_1 are again crowded, in another right sheep/goat mandible (*Fig. 22, 4*) and the P_3 and P_4 are crowded (*Fig. 22, 5*) in a left mandible. In the sacrificial dog skull (*Fig. 16*) the left P_2-P_3 and the right P_2-P_3-P_4 are crowded (the right P_3 so strongly that it stands crosswise in the tooth row). In its mandible, the P_1-P_2 are lightly, and the P_4-M_1 strongly crowded. Such crowded teeth are quite common in pigs and dogs in the early stages of domestication. It is, however, not so common in cattle and caprovines because of the quadrangular form of their teeth.

d/ Interdental spaces

A small space could be observed between the P_1 and P_2 of a left mandible of a medium-sized dog resembling the tooth row of a wolf, thus an animal in an incipient stage of domestication.

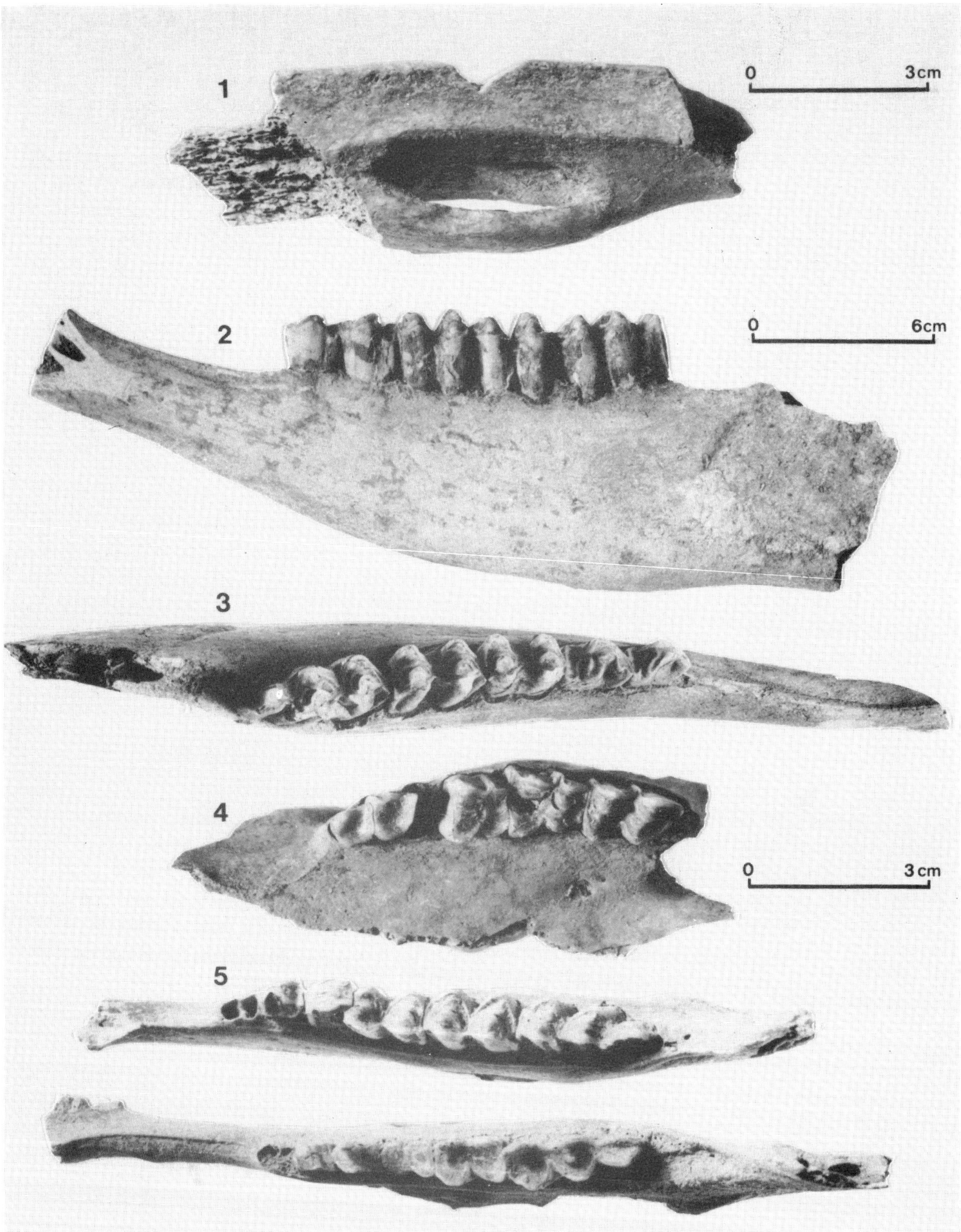

Fig. 22. Pathological bones. 1. Cattle rib with a bone bridge on its medial site. 2–3. Cattle mandible with missing P_2. 4. Sheep/goat maxilla with caries on its M_2. 5. Sheep/goat mandibles with crowded and deeply worn teeth.

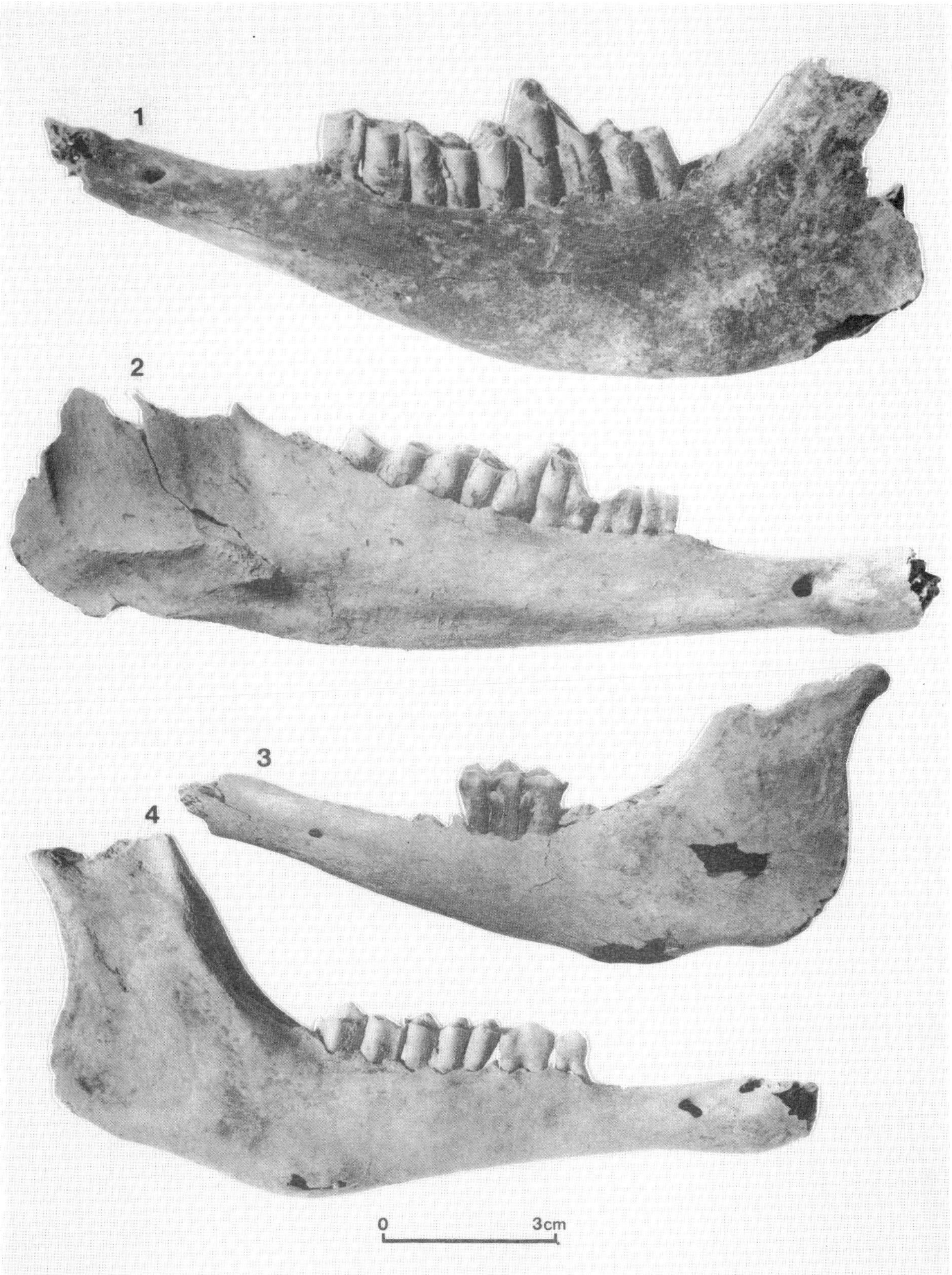

Fig. 23. 1–2. Sheep/goat mandibles with overgrown teeth. 3–4. Left cattle and right sheep/goat mandible with large size M_1 breaking through the ventral wall of the jaw.

e/ Irregular tooth wear.

Each left and right caprovine mandible show irregular tooth wear: in the left, the aboral part of M2 and the oral column of M3 and in the right most of M1 and the oral part of M2 are overgrown (Fig. 23, 1-2). In these cases the upper counterparts of the overgrown teeth are lacking, allowing the lower teeth to fill in the gaps.

f/ Abnormalities in tooth size

In a left cattle and a right sheep/goat mandible (both are juvenile) the large M1 is pushing out from the ventral side of the mandible. This phenomenon is not rare in newly domesticated cattle and caprovines: the teeth following the size decrease of the jaw with some delay do not have enough room and cause bumps on the lower surface of the mandible or even break through it (Fig. 23, 3-4).

2/ Effects of traumatic damage and infected wounds

a/ Fractures, particularly those of ribs, are quite frequently encountered in the sample. There is a fracture healed with a pseudarthrosis on a cattle rib. On another (juvenile) rib there is a not completely healed callus (Fig. 24, 1). There is a well-healed fracture with a flattened callus on one of the sheep/goat ribs. Four other rib fractures are not completely healed. A pig rib's fracture also healed with a callus and some dislocation (Fig. 24, 2). Finally a sheep/goat tibia fragment also healed with some distortion.

The cause of most of the above fractures was human maltreatment.

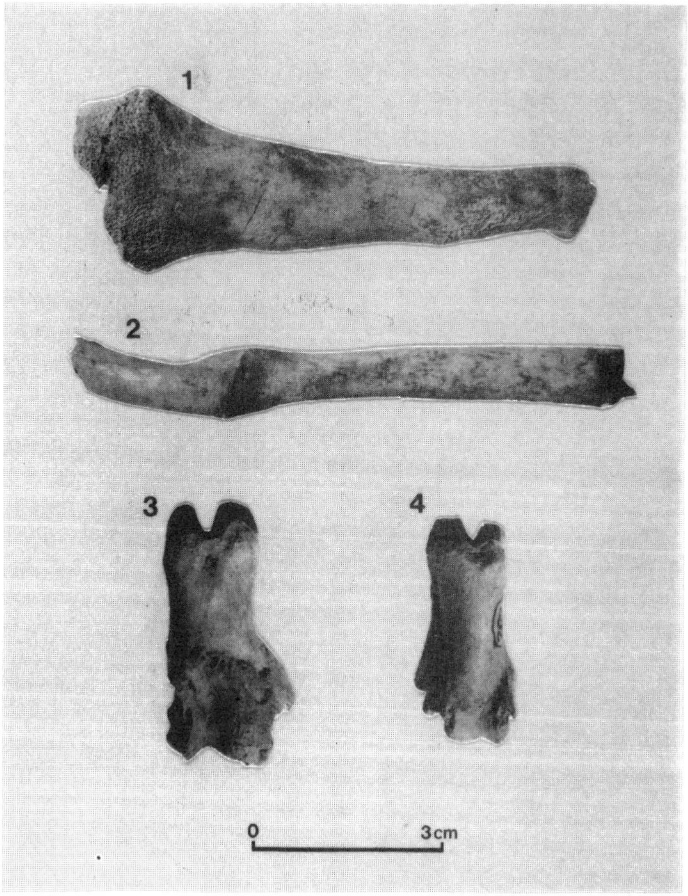

Fig. 24. 1. Cattle rib with incompletely healed fracture. 2. Pig rib fragment with callus of a healed fracture. 3–4. Sheep/goat first phalanges with large exostoses.

b/ Periostitis

Periostitis could be detected only on phalanges. A first phalanx of cattle and two of sheep/goat (*Fig. 24, 3-4*) show very strongly developed exostoses.

c/ Arthritis

All arthritic cases were found on caprovine bones. They can be divided into two groups: 1. In the first group there are four goat scapulae (three left and a right one) with a well developed chronic inflammation of the cavitas articularis (*Fig. 25, 1-4*). While *Fig. 25, 1* only gives a general idea of the four specimens, *Figs. 25, 1,2* and *4* clearly show the difference between the original size of the articular surface and the newly developed tissue that caused extension in the size of the surface which developed into a kind of capsula (in German "Schale") that certainly hindered the free motion of the shoulder joint. The other group consists of two sheep and two sheep/goat proximal radii fragments showing exostoses of different kinds as remmants of earlier chronic arthritis.

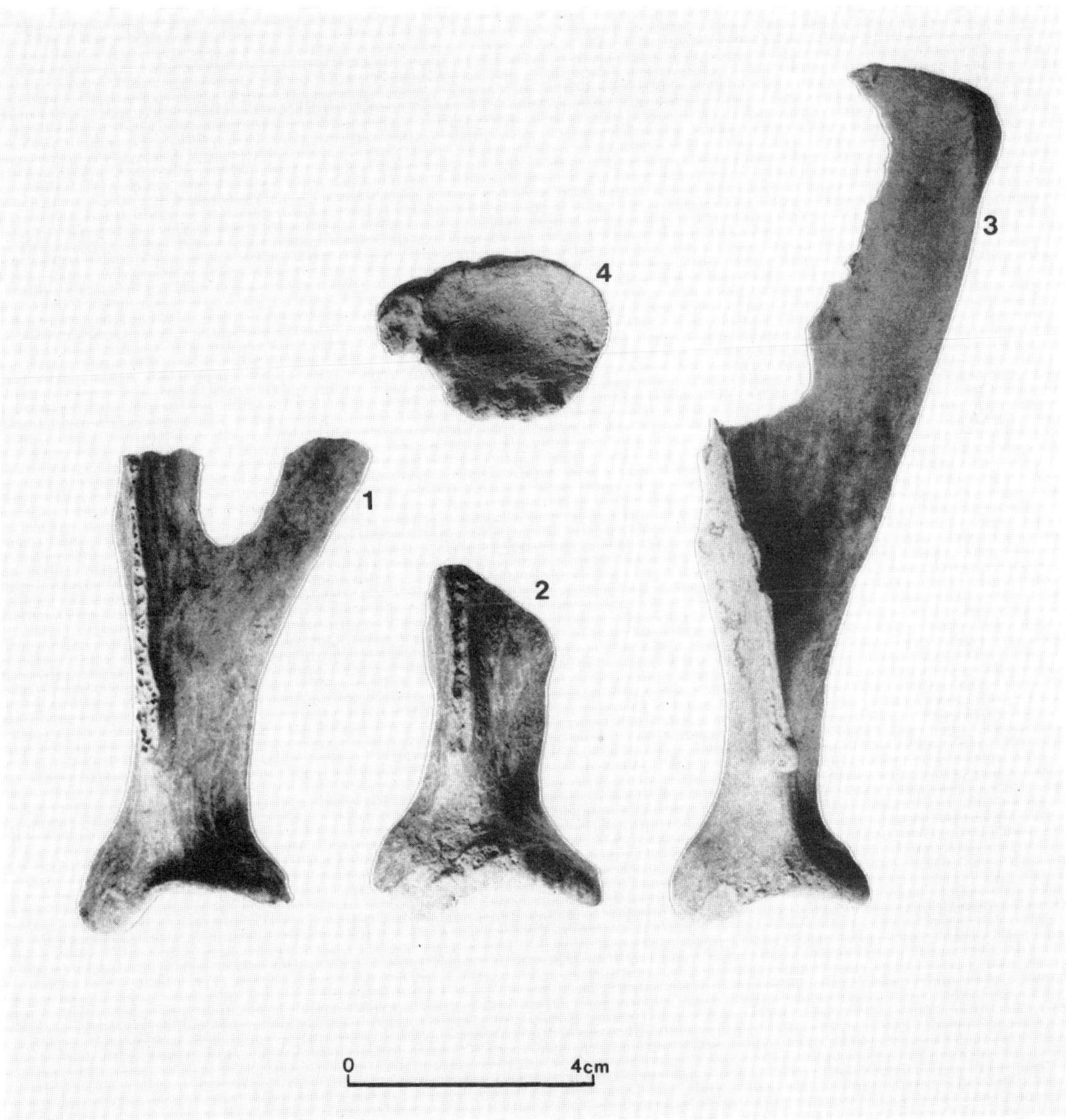

Fig. 25. Four goat scapulae with chronic arthritis.

d/ Periodontitis

The number of cases of periodontitis is unusually low. One can see a heavy periodontitis in a left sheep/goat mandible: the horizontal part of the mandible is swollen to twice its size and there is a large fistula with a sequester in the bottom of an excess. (*Fig. 26, 1*). This is a clear example of chronic periodontitis. The sequester consists of larger necrotic tissue from the spongious part of the mandible. The other case also concerns a chronic periodontitis in the lower jaw of a mature sheep/goat. Spongius thickening of the jaw and the loss of M_1 can be observed.

f/ Caries

Caries are extremely rare in subfossil animals. One appears here on a M from a sheep/goat's upper jaw. It seems to be an old cavity because it destroyed a large part of the tooth.

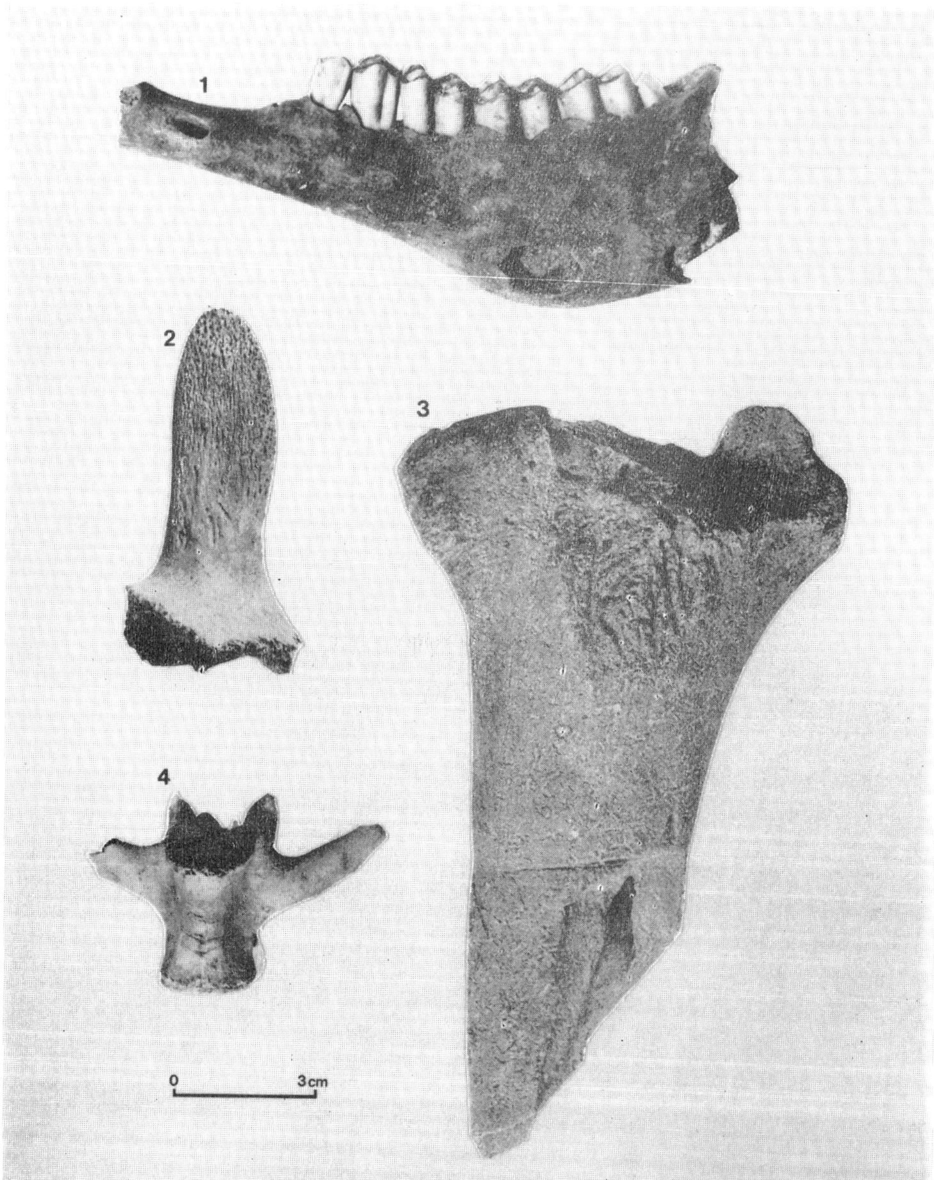

Fig. 26. 1. Sheep/goat mandible with chronic periodontitis, fistula and sequester. 2. Female sheep horn core with fingertip like impressions. 2. Circular cut marks on the diaphysis of a cattle radius. 4. Transversal cut marks on the lower surface of a sheep/goat lumbar vertebra.

3/ Chronic deformation

a/ Deformation of the spinal column
This is the only pathological case in the sample where a wild species is involved: large exostoses can be seen on the ventral surface of the fossa caudalis of a roe deer's cervical vertebra. They are probably connected with the age of the animal, although it does not appear to have been very old.

b/ Metabolical diseases
A left horn core of an ewe (*Fig. 26, 2*) may be placed within this group, having fingertip like impressions on its medial side. These impressions point to serious starvation probably during the winter period (Hatting 1975, 346). This is not surprising considering that in primitive animal husbandry herders had probably not collected enough hay for the winter months, and shepherds had to find their food in scrubland along rivers where long plants emerged from the snow that covered the open landscape.

Cut marks on the bones

In comparison to the large size of the bone sample, cut marks derived from chopping up the bones into smaller parts either in order to get pieces for cooking, removing the flesh from the bones, cutting through the articulations and/or removing the brain or marrow from them; or as the first step in the process of tool making, were not very numerous. Most of the bones showing cut marks are cattle bones which is understandable because of their large size. In fact, most of the cattle bones were broken up with a blunt, heavy object (stones, wooden clubs, etc), the heavy stone axes were probably too precious and were therefore used only in special cases. The essentially weaker bones of small mammals or birds were broken by hand and their carcasses were probably torn apart and cooked in larger chunks.

Of the larger cattle skull fragments five (*Fig. 1, 2-6*) show cut marks resulting from either the opening of the brain-case (*Fig. 1, 2-5*) or from the severing of the horns. In two cases (*Fig. 1, 3, 5*) cut and saw marks can also be seen on the basal part of the horn cores themselves. These marks were probably connected with the removal of the horn sheath which was a precious raw material.

One can also see cut marks on three cattle ribs, always on the lateral side (see *Fig. 24, 1*). There is a trace of sawing in a mediolateral direction on the diaphysis of a left distal humerus fragment. Connecting circular cuts forming an almost complete circle (*Fig. 26, 3*) can be seen on the anterior and posterior side of a right proximal radius fragment. These latter can be considered a first step in the preparation a bone tool. Finally, a sheep/goat lumbar vertebra shows several transversal cut marks on its ventral side (*Fig. 26, 4*).

Among the caprovine bones a distal humerus fragment has a deep circular cut on its diaphysis in order to break it into two parts for tool making while on the ventral surface of a lumbar vertebra (*Fig. 24, 5*) five transversal cuts are displayed.

Kill-off pattern and exploitation of domestic animals

The kill-off pattern (or to use a more exact expression – the age group ratios – because the bone sample of a settlement contains the bones of animals which were killed and which died as the result of an illness or from some other cause) gives important clues about the age structure of the different domestic species populations.

In Endrőd 119, only the bone samples of cattle and caprovines can be used for such studies because they are numerous enough to make up a statistically representative sample; the numbers of pig and dog individuals are so low that they can only suggest basic trends. At the same time, the kill-off patterns provide very useful information about the exploitation of the different species. It is quite obvious that domestic animals kept for their meat were slaughtered young not only to produce better quality meat, but also to avoid wasting forage during critical periods such as winter or during droughts. On the other hand, animals which only produced a "secondary" product as adults were kept for a longer time. It is not very difficult to understand that dairy cows or goats and wool producing sheep were of real value only when kept to adult or mature ages and were killed when their productivity declined. This is the way to judge the kill-off patterns given in *Table 12.*

Table 12. Kill-off pattern of domestic species

	neonate		juvenile		subadult		adult		mature		uniden-tified	total
species	ind.	%	ind.	%	ind.	%	ind.	%	ind.	%	ind.	ind.
cattle	1	0.70	60	41.96	28	19.58	47	32.87	7	4.89	40	183
sheep/goat	26	3.82	160	23.49	270	39.65	165	24.23	60	8.81	107	788
pig	-	-	6	40.00	3	20.00	5	33.33	1	6.67	5	20
dog	-	-	4	33.33	1	8.33	7	58.33	-	-	-	12

Based on the ratios of age groups, *Table 12* reveals that cattle, sheep, goat and pig were chiefly exploited for their meat. It can be seen that 62.24 per cent of cattle, 66.96 per cent of caprovines and 60.00 per cent of pigs were killed before having reached adult age. (Strangely enough, pig, a species which was certainly kept exclusively for its meat, was represented by the lowest ratio of immature individuals. This was probably only due to the small pig sample, hovewer.)

The immature (neonate+juvenile+subadult) samples of the above three groups show some differences. First, there is no newborn individual among pigs, only one among cattle and 26 (nearly four per cent) among caprovines (the cattle sample contains remains of an embryo as well, although it has not been included in *Table 12*). Thus juvenile individuals are most common among cattle and pigs while subadult animals are most frequently encountered among the caprovines.

The adult and mature individuals plus some of the subadults should have comprised the breeding stock. Nevertheless, 35 to 40 per cent is too large for a breeding stock. It is particularly true for the multiparous pig: According to the experiences at Sitagroi-Photolivos, Greece, a breeding stock of 10 per cent can keep a pig population on a steady level or even increase it (Bökönyi 1986a, 85). In the uniparous cattle and sheep certainly somewhat larger breeding stock is needed, though during the breeding period of a cow (between its 5th and 22nd year) she and her female descendants produce at least 60 calves.

Thus, the breeding stock, in addition to keeping the population at least on an even level, had some rudimentary "secondary" use. It was not a true "secondary" use since it could be carried out without killing the animals. Here animals skins were used because, although the early domesticated sheep already had some wool (Bökönyi 1974, 159; 1977, Fig. 17) these early sheep mainly had kemps rather than wool in their fleece (Ryder 1969, 495 ff). These skins would have functioned well as clothes. The role of cattle skins would have been similar.

A real secondary use may have been milk. In fact, the earliest evidence of cow milking comes from the 4th millenium B.C. from the Middle East and South Russia where cows appear with conspicuously large udders (Brentjes 1965, 38). The first pictoral evidence of cow milking is known from the temple of Nin-Hursag, Ur, after 2400 B.C. (Zeuner 1963, 219). On that frieze, the cows are milked from behind like sheep or goats which supposes that cattle milking derived from sheep or goat milking. This is supported by the fact that caprovines were domesticated about one thousand years earlier than cattle. Finally, there is linquistic evidence for the early exploitation of milk: according to information kindly supplied by Prof. J. Harmatta, the western Indo-European languages have a common word for milking: **mlg-/*xmelg-* and for milk: **mlag-ti*, actually "milking, milked" which originate from the time prior to the separation of the western Indo-European languages, at least, at the beginning of the Neolithic (Bökönyi 1974, 109). In fact, however, Dennel (1974, 34) suggested, without any solid proof, that cows were probably milked in the Neolithic of Bulgaria.

Returning to the kill-off patterns, that of the dog was quite different from the other species: most of the dogs were killed as adults indicating that the dog, in spite of the fact that its flesh was undoubtedly eaten, was exploited in other ways too: as a hunting companion, herd and watch dog. These uses were probably more important than its meat.

Meat quantity, consumption, habitation

It is rather difficult to determine the quantity of meat consumed by the inhabitants of a prehistoric settlement. One cannot start using the weight of the bones found there as Kubasiewicz did (Kubasiewicz 1956, 243f.) because this method has too many limitations. In cases where only parts of settlements were excavated the determination of the absolute quantity of consumed meat is practically impossible. It is enough to determine the ratios of meat quantities of the different species. Within the domestic species one counts in caprovine units (a cow is equivalent to 7, a pig to 1.5, and a dog to 0.5 caprovine). Wild animals are counted in roe deer units (which are about the same as caprovine units since early sheep/goats are about the same size as roe deers). Thus an aurochs is equivalent to 30, a red deer to 10, a wild swine to 6 and a wild ass to 9 roe deers etc. (In this way the ratio of domestic meat to that of wild meat can also be roughly compared.)

Table 13. Numbers of individuals and meat quantities in caprovines units for domestic and roe deer units for wild animal species

a/ domestic	number of individuals	%	caprovine unit	%
cattle	183	18.25	1281	60.86
sheep/goat	788	78.56	788	37.43
pig	20	1.99	30	1.43
dog	12	1.20	6	0.28
total	1003	100.00	2105	100.00

b/ wild	number of individuals	%	roe deer unit	%
aurochs	23	17.56	690	62.17
red deer	16	12.21	160	14.42
roe deer	37	28.24	37	3.33
wild swine	29	22.14	174	15.68
wild ass	4	3.05	36	3.24
badger	5	3.82	2.5	0.23
fox	4	3.05	1.2	0.11
wolf	5	3.82	7.5	0.68
hare	8	6.11	1.6	0.14
total	131	100.00	1109.8	100.00

(Among wild animals only the ungulates, the certainly consumed carnivores and the brown hare were counted.)

Table 13 clearly shows that there was an essential difference between the frequency and the economic importance of a species depending on its size. The Early Neolithic faunas of Southwest Asia, Southeast and South Europe were clearly caprovine-dominated, however, the cattle's importance in food production was much higher in most cases. After the switch to cattle keeping, its importance became even more marked: over 90 per cent of the consumed domestic meat was beef.

The case was similar in the wild sample too: aurochs, only fourth in the frequency list of wild animals, yielded more than sixty per cent of all the wild meat. In fact, venison comprised more than one third of the total quantity of meat consumed by the inhabitants. If one also adds the meat from birds and fishes to it, it becomes close to the quantity of domestic meat. (One must consider that an essential part of the fish bones would have been eaten by dogs while another part consisted of so small pieces that they were simply not collected.)

As regards the absolute quantity of meat, uncertainty will surely increase. While in the former approach the only risk factor was the determination of the number of individuals, in this, three factors play essential roles: the determination of the number of individuals, the determination of the different meat yields in the different species, and the effect of the different age groups as well. As a result, it is very hazardous to attempt to determine the absolute meat quantity. Nevertheless, since it is rare that one has the opportunity to calculate it in a sample from a totally excavated settlement, it is well worth trying.

It does not make much sense to try to determine the meat quantity of rare small mammals and of birds because they have minimal importance in human consumption. Close attention must be paid to the domestic animals and the five wild ungulate species because these groups provided the main part of the meat consumed by the inhabitants.

Table 14 shows the number of individuals, the weights of meat (including fat and bones) (in kg), the percentages of meat quantities, the kg calories of the different meat sorts, and the total calorie quantities as well as their ratios in five domestic species and the five wild ungulates.

Table 14. Numbers of individuals, meat and calorie quantities

	number of individuals	meat quantity	%	number of calories in 1 kg meat	the whole calorie quantity of the species	%
cattle	183	27.450	64.41	1500	41.175.000	56.91
sheep/goat	788	14.184	33.28	2000	28.368.000	39.21
pig	20	840	1.97	3000	2.520.000	3.48
dog	12	144	0.34	2000	288.000	0.40
domestic	1003	42.618	100.00	-	72.351.000	100.00
aurochs	23	9.660	65.57	1500	14.490.000	55.24
red deer	16	1.600	10.86	1500	2.400.000	9.15
roe deer	37	555	3.77	1500	832.500	3.17
wild swine	29	2.436	16.54	3000	7.308.000	27.86
wild ass	4	480	3.26	2500	1.200.000	4.58
wild	109	14.731	100.00	-	26.230.500	100.00
Total	1112	57.349			98.581.500	

Thus, the whole quantity of meat from the main meat producing animals was 57.349 kg or as many as 98.581.500 calories. And if one also adds meat quantities from the other mammals, the birds and the huge number of fish (which can only be guessed at), the calorie level would be around 150.000.000.

If one supposes that an average person needs ca. 3000 calories per day, this meat quantity would be enough for 50.000 days for one person and consequently for 5.000 days or 13.7 years for two nuclear families each consisting of 5 persons living in the two houses of the hamlet. Nevertheless, since the inhabitants were also consuming cereals, wild fruits, berries, mushrooms, honey, eggs from wild birds, shellfish, snails etc., one may rightly suppose that they only partly supplied their food demands with meat. This means that the above meat quantity could have been consumed over 30 to 40 years. Even this would not represent a particulary long stay and would certainly contradict the suggestion that this was a permanent occupation.

According to J. Makkay, the excavator of the site, the settlement was inhabited throughout the whole span of the Körös culture (Makkay, pers. comm.). This would mean ca. 500 years (from about 5000 to 4500 years B.C.; uncalibrated C_{14} date).

The small, satellite type settlement was more a hamlet than a village in the same way that Hammond (1981, 215 ff.) and Dennel (1983, 171) found with Linear Pottery culture settlements. This hamlet consisted of two or possibly three houses housing two or three nuclear families of ca. 5 members each. Glancing both at this picture, on the one hand, and at the numbers of individuals of domestic and wild animals on the other, one can immediately see that the life span of the settlement, the number of its inhabitants, and the numbers of domestic and wild animals kept and killed in the settlement or hunted around it are not in accordance with each other.

It is certain that the settlement was inhabited in the early phase of the Körös culture. Earlier (Bökönyi 1989a, 14f.), the site was placed in the middle phase of the Körös culture (based on a smaller bone sample from early excavations). Nevertheless, the identification of the enormously large, complete bone assemblage from the site has revealed that a great part of the finds belonged to the early phase of the culture. This is chiefly shown by the following: 1/ the large numbers of wild mammals, birds (mainly waterfowl) and fishes, 2/ the low frequency of cattle. In this respect, the site or at least its early phase is similar to Deszk-Olajkút, Ludas-Budzák, Röszke-Lúdvár, and Endrőd 39 (Bökönyi 1989a, 15). Unfortunately, there is neither a vertical nor a horizontal stratigraphy at the site, and although two houses certainly belong to a later phase, the bone sample from them cannot be distinguished from the rest because the outlines of the houses can not always be determined and the houses themselves were disturbed by later pits.

As Kosse has pointed out (Kosse 1979, 35), all Körös settlements were situated in the forest steppe belt. In fact, 96 per cent of them lay in the forest steppe itself and the rest on its borderline. This type of environment is ideal for animal husbandry, especially cattle and then for caprovine breeding as well as for the hunting of steppe species (aurochs, roe deer and wild ass). Since the Körös sites are strongly connected to gallery forests – according to Kosse, 59 per cent of them lay in forests along streams and watercourses (Kosse 1979, 62) – it was very suitable for pig keeping and especially good for agriculture and the hunting of forest animals. Another reason why pig keeping may not have attained greater importance at Körös sites is because old caprovine-keeping traditions coming from the Near East and the southern half of the Balkans.

These factors were all valid at Endrőd 119 too. It was situated in an ideal environment, in the vicinity of the forest steppe and close to the gallery forest and water. Therefore, it is not surprising that it was permanently occupied. This may be seen in the occurrence of remains of – chiefly – caprovines killed in all seasons of the year. Earlier, Kayser and Voytek suggested a semi-sedentary way of life (with some kind of transhumance) in the Körös-Starčevo group based on the high frequency of caprovines (Kayser and Voytek 1983, 330f.). However, it is very difficult to find direct or even indirect evidence for the existence of transhumance in prehistoric settlements. Recently, it has become a fashion to suppose transhumance or other forms of semi-sedentary ways of life if the ratio of caprovines is high in the bone sample at a settlement. But given the low population density and sufficient grazing land around the settlements it would not be necessary to wander seasonally to remote regions (Bökönyi 1989a, 14).

Nevertheless, there remains a discordance between the life span of the site (ca. 500 years) and the number of inhabitants (10 to 15 persons) at the settlement on the one hand, and the numbers of individuals in the domestic and wild species found there, on the other. In fact, these latter figures are too low in spite of the fact that they do not represent the minimum but rather the approximate numbers of individuals. This was calculated in such a way that, in addition to the number of most common bones types, the age and size categories were also taken into consideration (Bökönyi 1970a, 291f.) resulting in larger figures and in this way getting closer to the real numbers of individuals.

Taking cattle as an example, the occurrence of 183 individuals does not appear too low because assuming the average lifetime of a cow to be 20 years, the 47 adult and 7 mature cows plus half of the 40 individuals of undetermined age (see *Table 12*) lived in total about 1480 years, while 209 immature (neonate+juvenile+subadult) individuals plus the remaining 20 individuals of undetermined age lived on average 3 years, or a total of 327 years. This works out as an average of roughly 3 adult and 1 immature cattle living in the settlement at a given time. Based on similar considerations (but on a shorter lifetime) it may be supposed that ca. 5 adult and 2 immature caprovines were kept in the settlement at any given time. The numbers of pigs and dogs were insignificant.

These calculations suggest that a family owned one or two cows and two to four sheep in the Early Neolithic of Endrőd 119. These are surprisingly low figures when compared to the stocks of recent nomads (the Basseri tribe) in South Persia where a household has about 6-12 donkeys and somewhat less than 100 sheep and goats (Barth 1961, 13). It is true, that these nomads do not intensively hunt, fowl or fish and thus, have secondary sources to complement their meat diet. They do not carry out serious agriculture either as did the inhabitants of Endrőd 119. Although the essential role of hunting, fishing and fowling has been seen earlier, there convincing proof of agricultural activity from Körös sites mainly situated in gallery forests (see above). Their average distance from the nearest watercourse or lake was 0.83 km (Kosse 1979, 74). Such areas provide good soils and favourable water conditions for agricultural production. Added to this is the fact that in a 2.5 km radius around Körös sites, 70 percent of the soil lies in a flood zone, and even in the remaining 30 per cent, water plays an essential role in the formation of soil and vegetation (Kosse 1979, 125). The Körös culture was considered a hunter, gatherer and animal keeping culture (Banner 1937; Kutzián 1944), however, it has recently been discovered that plant cultivation was important too. Cultigens including wheat, einkorn, barley and also possibly flax have been found at Körös sites (Trogmayer 1968; Kosse 1979, 128). These plants provided a large amount of human food, animal fodder and raw material of vegetable origin. The low figures of occurring animals seem even more conspicuous if one calculates the numbers of different domestic and wild animals killed during the lifetime of the settlement. For example, taking cattle, it can be calculated that in 500 years only 183 cattle were killed. This means that a cow was slaughtered only every third year. The numbers are somewhat better with the caprovines (they were more common anyway): 1.5 caprovines were killed each year. There is not much sense in looking at the pigs either because calculations yield a total of only one individual killed every 25th year, while one dog died or was killed ca. every 40th year. The situation is no better with the hunted animals either: an aurochs was killed every 20th year, a red deer every 30th year, a roe deer every 13.5 years, a wild swine every 17.2 years, and a wild ass every 0.25 year. There is no point in discussing the rarer wild species.

This somewhat strange situation can be explained in different ways. First a considerable number of bones could have been carried away by dogs and by hunters, going out on long trips. Thus, the number of individuals determined in the bone sample would be essentially smaller than it had been in realty.

Secondly, it can also be suggested that the site was not continuously inhabited during the whole period of the Körös culture. It is not necessary to hypothesize a seasonal habitation of the site (this has already been discussed and rejected in an earlier part of this paper). Instead, one can reckon with longer or shorter habitation periods alternating with times when the site was deserted. Since the population was very small, it did not leave clear habitation levels just as there were no sterile levels to mark periods of abandonment either. In this way, if one supposes that the site was only inhabited over half the time of the Körös culture, the number of animals kept and hunted in and around the site would be closer to that amount which one would expect to find in an Early Neolithic animal keeping plus hunter-fisher-gatherer site. This would mean that about 8 (adult and young) cattle and 14 caprovines lived at one time in the settlement. Their products had to be complemented by hunting, fishing and gathering.

There could also be a third possibility. This was that the large central site and the small "satellite" sites were not independent economic units but rather the small sites were loosely connected outposts of the central settlement. They were created in order more successfully to exploit the environment. In this way, it could easily be that the subsistence of these "satellite" sites was not completely based on local resources but came from a central stock. It is highly possible that e.g. when butchering a cow or killing an aurochs or a large red deer the meat quantity could not have been consumed by the two or three nuclear families living in the small site alone but some parts might have gone to the central settlement too. It is even more probable that the slaughter and butchering of large-size animal species took place at the central settlement as did the distribution of the meat. In this way, the low number of animals both domestic and wild, killed and butchered in this "satellite" site is more understandable. Nevertheless, a conclusive demonstration of this would necessitate full excavation of the central settlement and evaluation of its complete animal bone sample.

At any rate, an exact determination of the full calorie intake in the form of animal protien and fat could even then be impossible because the quantity of animal bones carried away or gnawed by dogs and destroyed by chemical and physical agents in the soil cannot be determined. In the same way, the amount of consumed fish that comprised a considerable part of the diet would remain difficult to determine precisely. Nevertheless, this opportunity to study the complete animal bone assemblage from a "sattelite" site, clearly shows that the consumption of meat was a highly significant factor.

References

Amberger, K.-P., 1979. *Neue Tierknochenfunde aus der Magula Pevkakia in Thessalien*, II. Die Wiederkäuer. Diss. München.

Barth, F., 1961. *Nomads of South Persia. The Basseri tribe of the Khamseh confederacy.* London.

Banner, J., 1931. A kökénydombi neolithkori telep. (The Neolithic settlement at Kökénydomb.) *Dolgozatok*, 6 (1930) 49-159. Szeged.

Becker, C., 1986. *Kastanas. Die Knochenfunde.* Prähistorische Archäologie in Südosteuropa, 5, Berlin.

Bökönyi, S., 1952. Die Wirbeltierfauna der Ausgrabungen in Tószeg vom Jahre 1948. *ActaArchHung*, 2, 71-113.

Bökönyi, S., 1954. Eine pleistozäne Eselart im Neolithikum der Ungarischen Tiefebene. *ActaArchHung*, 4, 9-24.

Bökönyi, S., 1964. A Maroslele-panai neolitikus telep gerinces faunája - The vertebrate fauna of the Neolithic settlement at Maroslele-Pana. *ArchÉrt*, 91, 87-93.

Bökönyi, S., 1969. Archaeological problems and methods of recognizing animal domestication. In: Ucko, P. J. and Dimbleby, G. W. (eds.) *The domestication and exploitation of plants and animals.* London, 219-229.

Bökönyi, S., 1970a. A new method for the determination of the number of individuals in animal bone material. *AJA*, 74, 291-292.

Bökönyi, S., 1970b. Animal remains from Lepenski Vir. *Science*, 167, 3926, 1702-1704.

Bökönyi, S., 1971. The development and history of domestic animals in Hungary. *Amer. Anthrop.*, 73:3, 640-674.

Bökönyi, S., 1972. Once more on the osteological differences of the horse, the half-ass and the ass. In: Firouz, L., *The Caspian miniature horse in Iran.* Field Research Projects, 64, Miami, 12-23.

Bökönyi, S., 1973a. Some problems of animal domestication in the Middle East. In: Matolcsi, J. (ed.) *Domestikationsforschung und Geschichte der Haustiere.* International Symposium in Budapest 1971. Budapest, 69-75.

Bökönyi, S., 1973b. Stock breeding. In: Theocharis, D.R., *Neolithic Greece.* Athens, 165-178.

Bökönyi, S., 1974. *History of domestic mammals in Central and Eastern Europe.* Budapest.

Bökönyi, S., 1976a. The development of early stock rearing in the Near East. *Nature* 264, 19-25.

Bökönyi, S., 1976b. The vertebrate fauna of Obre. *Wiss.Mitt.d.Bosn.-herzeg. Landesmus.*, IVA. Sarajevo, 55-154.

Bökönyi, S., 1977. *Animal remains from the Kermanshah Valley, Iran*, BAR Suppl. Ser., 34.

Bökönyi, S., 1977-82. The Early Neolithic fauna of Rendina. *Origini*, XI, Roma, 345-354.

Bökönyi, S., 1978a. The introduction of sheep breeding to Europe. *Ethnozootechnie*, 21, 65-70.

Bökönyi, S., 1983. Animal bones from test excavations of Early Neolithic ditched villages on the Tavolière, South Italy. In: Cassano, S. M. and Manfredini, A., *Studi sul neolitico del Tavolière della Puglia. Indagine territoriale in un' area-campione.* BAR Internat. Ser., 160, 237-249.

Bökönyi, S., 1984a. Die frühneolitische Wirbeltierfauna von Nosa. *ActaArchHung*, 36, 29-41.

Bökönyi, S., 1984b. Die Herkunft bzw. Herausbildung der Haustierfauna Südosteuropas und ihre Beziehungen mit Südwestasien. In: Schwabedissen, H. (ed.) *Die Anfänge des Neolithikums vom Orient bis Nordeuropa.* IX. Der Beginn der Haustierhaltung in der "Alten Welt". Köln-Wien, 24-43.

Bökönyi, S., 1985. A comparison of the Early Neolithic domestic and wild faunas of the Balkans, Italy and South France. *Cahiers Ligur de Prehist. et Protohist.*, N.S., 2, 182 -192.

Bökönyi, S., 1986a. Faunal remains. In: Renfrew, C., Gimbutas, M. and Elster, E. S. (eds.) *Excavations at Sitagroi. A prehistoric village in Northeast Greece,* I. Monumenta Archaeologica, 13, Los Angeles, 63-132.

Bökönyi, S., 1986b. The equids of Umm Dabaghiyah, Iran. In: Meadow, R.H. and Uerpmann, H.-P. (eds.) *Equids in the Ancient World.* Beihefte zum Tübinger Atlas des Vorderen Orients, Reihe A (Naturwissenschaften), Nr. 19/1, Wiesbaden, 302-317.

Bökönyi, S., 1988a. *Környezeti és kulturális hatások későneolitikus Kárpát-medencei és balkáni lelőhelyek csontanyagán.* Értekezések, emlékezések. Budapest.

Bökönyi, S., 1988b. The Neolithic fauna of Divostin. In: McPherron, A. and Srejović, D. (eds.) *Divostin and the Neolithic of Central Serbia.* Ethnology Monographs, No. 10, Department of Anthropology, University of Pittsburgh. Pittsburgh, 419-445.

Bökönyi, S., 1989a. Animal husbandry of the Körös-Starčevo complex: its origin and development. In: Bökönyi, S., (ed.) *Neolithic of Southeastern Europe and its Near Eastern connections.* VAH, 2. Budapest, 13-17.

Bökönyi, S., 1989b. Animal remains. In: Gimbutas, M., Winn, Sh. and Shimabuku, D., *Achilleion. A Neolithic settlement in Thessaly, Greece, 6400-5600 BC.*, Monumenta Archaeologica 14, Los Angeles, 315-332.

Bökönyi, S. and Jánossy, D., 1965. Szubfosszilis vadmadárleletek Magyarországon - Subfossile Wildvogelfunde aus Ungarn. *VertebrHung.* VII, 1-2, 85-99.

Bökönyi, S. and Kretzoi, M., 1983. La faune. In: Arnal, G. B., *La Grotte IV de St. Pierre de la Fage (Hérault) et le Néolithique ancien du Languedoc.* Montpellier, 128-147.

Borzatti von Löwenstern, E., 1969. Industrie romanelligue e neolitiche nella Grotta delle Prazziche (Novaglie-Lecce). *RivSciPr*, 24, 91-143.

Boessneck, J., 1962. Die Tierreste aus der Argissa-Magula vom präkeramischen Neolithikum bis zur mittleren Bronzezeit. In: Milojčić, V., Boessneck, J. and Hopf, M., *Die deutschen Ausgrabungen auf der Argissa-Magula in Thessalien,* I. Bonn, 27-99.

Boessneck, J. , Müller, H.-H. and Teichert, M., 1964. Osteologische Unterscheidungsmerkmale zwischen Schaf (Ovis aries Linné) und Ziege (Capra hircus L.). *Kühn-Archiv*, 78, 1/2, Berlin.

Brentjes, N., 1965. *Die Haustierwerdung im Orient.* Die Neue Brehm Bücherei, 344, Wittenberg.

Cornevin, Ch. and Lesbre, F., 1891. Caractères ostéologiques differentiels de la chèvre et du mouton. *Bull.de la Soc. d'Anthrop. de Lyon*, 10. 47-72.

Dennel, R. W., 1972. Stone age farming in Bulgaria. *The Ill. London News*, Sept., 61-62.

Dennel, R. W., 1974. The economic development of Bulgaria from the Neolithic to the Early Bronze Age. In: *Thracia, Primus Congressus Studiorum Thracicorum*, III, Sofia, 33-37.

Dennel, R. W., 1983. *European economic prehistory. A new approach.* London.

Dietrich, W. O., 1959. Hemionus Pallas im Pleistozän von Berlin, *VertebrPalaeas.*, 3, 13-22.

Ducos, P., 1976. Quelques documents sur les débuts de la domestication en France. *La Préhistoire de France*, II 165-167.

Eisenmann, V., 1986. Comparative osteology of modern and fossil horses, half-asses, and asses. In: Meadow, R. H. and Uerpmann, H. P. (eds.) *Equids in the Ancient World.* Beihefte zum Tübinger Atlas d.Vord. Orient, A. Nr. 19/1. Wiesbaden, 67-116.

Geddes, D., 1980. *De la chasse au troupeau en Mediterranée occidientale. Les débuts de l'élevage dans le bassin de l'Aude.* Toulouse.

Geddes, D., 1981. Les débuts de l'élevage dans la vallée de l'Aude. *Bull. de la Soc. Préhist. Franc.*, 78, 10-12, 270-378.

Gejvall, M.-G., 1969. *Lerna I. The fauna.* Princeton.

Gromova, V. I., 1953. Osteologičeskie otlicia rodov Capra (kozly) i Ovis (barany), *Trudy komissii po izučeniu četvertičnogo perioda*, X,1. Moscow.

Hammond, F., 1981. The colonisation of Europe: the analysis of settlement process. In: Hodder, I., Isaac, G. and Hammond, N. (eds.) *Pattern of the past: studies in honour of David Clarke*. Cambridge, 211-248.
Hatting, T., 1975. The influence of castration on sheep horns. In: Clason, A. T. (ed.) *Archaeozoological Studies*. Amsterdam - Oxford - New York, 245-351.
Herre, W., 1955. *Das Ren als Haustier*. Leipzig.
Higgs, E .S., 1962. The fauna of the Early Neolithic site at Nea Nikomedeia, Greek Macedonia. *PPS*. XXVIII. 271-274.
Hole, F. and Flannery, K. V, 1967. The prehistory of Southwestern Iran: a preliminary report. *PPS*. XXXIII., 147-206.
Jánossy, D., 1985. Wildvogelreste aus archäologischen Grabungen in Ungarn (Neolithikum bis Mittelalter.). *Fragm. Mineral. et Palaeont.*, 12, 67-103.
Jarman, M. R. and Jarman, H. N., 1968. The fauna and economy of Early Neolithic Knossos. *BSA*, 63, 241-264.
Kayser, T. and Voytek. B., 1983. Sedentism and economic change in the Balkan Neolithic. *Journ. of Anthrop. Arch.* 2, 323-353.
Kosse, K., 1979. *Settlement ecology of the Körös and Linear Pottery cultures in Hungary*. BAR Internat. Ser., 64.
Kouldelka, F., 1884. Das Verhältnis der Ossa longa zur Skeletthöhe bei Säugetieren. *Verh. d. Naturforsch. Ver. Brünn*, 63. Brünn.
Kratochvil, Z., 1969. Species criteria on the distal section of the tibia in Ovis ammon f. aries L. and Capra aegagrus f. hircus L. *Acta Veter.*, 38, 483-490.
Kratochvil, Z., 1972. Knochenüberreste von der neolithischen Siedlung "Jeleni louka" bei Mikulov (Bez. Breclav). *Prehled vyzkumu* 1971, AU ČSAV Brno 24-27.
Kratochvil, Z., 1973. Der Fund von Equus (Hydruntinus) hydruntinus (Regalia, 1907) und andere Säuger aus dem südmährischen Neolithikum. *SlA*, XXI-1, 195-210.
Kratochvil, Z., 1976. Das postkraniale Skelett der Wild- und Hauskatze (Felis silvestris und F. lybica f. catus). *Acta Sci. Nat. Acad. Bohemoslov.*, Brno, N.S. X, 6, 1-43.
Kratochvil, Z., 1988. Telesna velikost obecneho (Capreolus capreolus L.) v coucasnosti a munulosti - Body size of roe deer (Capreolus capreolus L.) in the present and the past. *Folia Venatoria* 18, 115-133.
Kubasiewicz, M., 1956. O metodyce badan wykopaliskowych szczatkow kostnych zwierzecych - Über die Methodik der Forschungen bei Tierausgrabungsknochen. *Mat. Zachod. – Pomorsk.*, II, 235-244.
Kutzián, I., 1944. *A Körös kultúra - The Körös culture*. Budapest.
Makkay, J., 1982. *A magyarországi neolitikum kutatásának új eredményei*. Budapest.
Matolcsi, J., 1970. Historische Erforschung der Körpergrösse des Rindes auf Grund von ungarischem Knochenmaterial. *Zeitschr. f. Tierzüchtg. u. Züchtgsbiol.*, 87, 2, 89-137.
Necrasov, O., 1961. K izučeniu domaşnix i dikix životnix ranne neolitičeskoi kultury Kris. *Anal. Stiint. al. Univ. d. Iasi* 7., 2, 265-272.
Necrasov, O., 1964. Sur la significance de certains caractères morphologiques de Equus (Asinus) hydruntinus Reg. *Rev. Roum. de Biol., Ser. Zool.*, 9, 3, 141-149.
Payne, S., 1969. A metrical distinction between sheep and goat metacarpals. In: Ucko, P. J. and Dimbleby, G. W. (eds.), *The domestication and exploitation of plants and animals*. London, 295-305.
Payne, S., 1973. Animal bones (of the Frachthi Cave). *Hesperia*. XLII, 1, 59-66.
Perkins, D., 1969. Fauna of Çatal Hüyük: evidence for early cattle domestication in Anatolia. *Science*, 164, 177-179.
Poulain-Josien, Th., 1975. Les aminaux domestiques en France à l'époque néolithique. *L'homme et l'animal*, 1er Coll. d'Ethnozool., Juin 1975, Inst. Int. Ethnosci., Paris 409-415.

Reitsma, G. G., 1932. *Zoologisch Onderzook der Nederlandsche Terpen*, 2. Het Schaap. Wageningen.
Rütimeyer, L., 1961. Die Fauna der Pfahlbauten der Schweiz. *Neue Denkschr. d. Allg. Schweiz. Ges. d. ges. Naturwiss.*, 19, Basel.
Ryder, M. L., 1969. Changes in the fleece of sheep following domestication. In: Ucko, P. J. and Dimbleby, G. W., (eds.) *The domestication and exploitation of plants and animals*. London, 495-521.
Schramm, Z., 1967. Kosci dlugie a wysokosc w klebie u kozy - Skeletal materal of early medieval sheep and goat from Northwest Poland excavations. *Roczn. Wyzsz. Szkoly Roln.w Poznan.*, XXXVI, 89-174.
Schwartz, C. A., 1976. The vertebrate fauna from Rug Bair. In: Gimbutas, M. (ed.) *Neolithic Macedonia*. Monumenta Archaeologica I, Los Angeles, 364-374.
Schwartz, C. A., 1981. *The animal remains from Sesklo*. Manuscript. Los Angeles.
Stehlin, H. G. and Graziosi, O., 1935. Ricerche sugli asinidi fossili d'Europa. *Mem. de la Soc. Paleont. Suisse*, 56, 1-68.
Teichert, M., 1975. Osteometrische Untersuchungen zur Berechnung der Wiederristhöhe bei Schafen. In: Clason, A. T. (ed.), *Archaeozoological Studies*. Amsterdam - Oxford - New York, 51-69.
Trogmayer, O., 1968. A Körös-csoport barbotin kerámiájáról. (The barbotine pottery of the Körös group.) *ArchÉrt*, 95, 6-11.
Vigne, J.-D., 1988. Les grandes étapes de la domestication de la chèvre: une proposition d'explication de son statut en Europe occidentale. *Ethnozootechnie*, No. 41, Paris, 1-13.
Vörös, I., 1980. Zoological and palaeoeconomical investigations on the archaeozoological material of the Early Neolithic Körös culture. *FA*, XXXI, 35-64.
Wäsle, R, 1976. *Gebissanomalien und pathologisch-anatomische Veränderungen an Knochenfunden aus archäologischen Ausgrabungen*. Diss. München.
Whitehouse, R., 1971. The last hunter-gatherers in Italy. *WorldArch* 2, 239-254.
Zeuner, F.E ., 1963. *A history of domesticated animals*. London.

Measurement tables

SKULL	dog	fox
basal length	158x	130
overall length	185x	142
I_1-M_1	71x	62
M_1- basion	89	68
I_1- aboral and of palate (length of palate)	88	73
aboral end of palate - basion	70	57
I_1- aboral end of nasal bones	90x	71
aboral end of nasal bones - opisthion	104	75
I_1- the middle of the line connecting the width of frontal bones	101x	86
this latter point - opisthion	95	60
length of row of teeth (I_1-M_2)	87x	74
length of row of incisors	10x	5
length of diastema	16	15
P_1- P_4	48 /s/ 45 /d/	42.5
M_1- M_2	19 /s/ 20 /d/	12.5
length of P_4	18	13.5
greatest width of the skull	103	76x
greatest frontal width	52x	35
greatest width of the brain case	61	49
distance between medial canthuses	34x	29
width at foramina infraorbitalia	40.5	27
width of row of incisors	30	15
width at the canines	38x	25
P_1 - P_1	36	20x
M_1 - M_1	65	40
width at the mandibular joints	95	62
width at the external auditory meatuses	63	48

width at the condylus occipitales	39.5	26
length of foramen magnum	17	12
width of foramen magnum	20	14
occipital height (basion - opisthion)	49	32

HORN CORE

Measurements: 1. greatest length
2. greatest diameter
3. smallest diameter
4. circumference of basis

Bos

210ˣ	55	45	167ˣ
320ˣ	60	54	190
377	91ˣ	71ˣ	275 w
-	55	48	170
-	58	48	177
-	59	50	170
-	60	47	178
-	60	48	174
-	60	52	178
-	61	48	180
-	63.5	53	187
-	66	53	187
-	66	57	-
-	87	71	268 w

Ovis

25	22	16	62
30ˣ	36	24	97
35	29	21	83
38	21	16	60
40	25	18	71

45	33.5	22	90
47	27	15.5	72
50^x	30^x	-	-
55	28	14	64
55	31	15	135
55^x	-	21^x	-
57	26	47	73
60^x	28	16	78
60	28.5	16	79
60	31	18	82
60	31.5	20^x	87^x
65^x	30	17.5	80
68	28.5	19	82
70	29	18	86
75	32	17	88
95^x	44	29	118
105^x	52	33	143
115^x	44	33	122
115^x	65^x	37	180^x
120^x	57	37.5	150
-	20	15	60
-	23	16	62
-	29	18	85
-	29.5	20	82
-	29.5	21	83
-	30	19	83
-	30	22	88
-	30x	33	-
-	31	18	82
-	31.5	19	90
-	32	18	88
-	33	20	94
-	34	20	88
-	35	23	90^x

-	37.5	22.5	100
-	46[x]	31	120
-	46	35	-
-	49	33	134
-	50	34	135
-	51	34	142
-	54	36	142
-	59	40.5	158
-	60	41	173

Capra

230	59	46	175
250	40[x]	28[x]	-
-	35	25.5	98
-	38	25	106
-	40	31	120
-	68	49.5	182
-	70	51[x]	-

UPPER ROW OF TEETH

Measurements: 1. length of premolar row
2. length of molar row
3. length of P_4 (M_3)

Bos

-	84	-
-	90.5	-
-	91	-
-	92[x]	- w

Sus

-	76	36
-	79	36

-	83	43 w
-	-	44 w

Canis

-	20.5	19.5
-	-	19

Capreolus

30	33	-
30	35	-
27	34	-

LOWER ROW OF TEETH

Measurements: 1. length of premolar row
2. length of molar row
3. length of M_1 (M_3)

Bos

48	91	41
50	91	39
51	92.5	40
52	91.5	39
53	-	-
54	91	42 w
54	95	40
54.5	95	40
55	-	-
57	101	45 w
58	100	43 w
58	-	-
61	104	44 w
-	91	42 w

-	92	40
-	96	41.5 w
-	-	37
-	-	37
-	-	38
-	-	38
-	-	39
-	-	39
-	-	39
-	-	39
-	-	39
-	-	40
-	-	40
-	-	40
-	-	41^x
-	-	41^x
-	-	41^x
-	-	42^x
-	-	42^x
-	-	43^x
-	-	42^x
-	-	43^x
-	-	45^x

Sus

44	-	-
64	-	- w
-	81	41.5 w
-	82	45 w
-	83	45 w
-	92	49 w
-	-	42 w
-	-	44 w

-	-	44 w
-	-	45 w
-	-	48 w
Canis		
36	35	21
37	32	20
39	33.5	20.5
-	35	21
-	-	21
38	35	21
Cervus		
52	-	-
Capreolus		
26	36.5	15
26	38	16
26	39.5	16.5
26.5	-	-
26.5	-	-
27	39	16
27	-	-
27	-	-
28	39	16
28	39.5	16.5
28	40	16.5
28	40	17.5
28	-	-
29	38.5	16
29	40.5	16
29	41	18
29	42	17

30.5	39.5	16.5
-	39	16.5
-	41	16
-	41	17.5
-	-	16

Meles

19	22.5	16
20	22	16

Vulpes

34.5	25.5	15
37	28	15

ATLAS

Measurements: 1. length of arcus ventralis
2. length of arcus dorsalis
3. width of cranial articular surface
4. width of caudal articular surface
5. greatest width
6. greatest height

Bos

42	-	100	91	145	-
51	57[x]	106[x]	107	-	94 w
51	57	113	108	186	84 w
53	57	117	113	212	98 w

Ovis

19.5	22.5	45	42	57	32
24	21.5	43	42.5	-	-

24	22	45.5	41	62	34
24	-	45.5	44	63	33
25	22	45	41	60[x]	31.5
25	27	39	36	57	32
25	-	47	45	-	34
25.5	25	44	42	62	36.5
26	25	65	62	-	-
27	25	50	50	-	40
28	27.5	49.5	48	-	-
28	-	44	43	-	-
Capra					
25.5	28	43	39	-	35
Canis f.					
10	14.5	41[x]	35	78[x]	27
Capreolus					
26	25	38	37	58[x]	33
27	28	38.5	36.5	55[x]	34

AXIS

Measurements: 1. length of body
2. length of arch
3. length of dens
4. width of dens
5. width of cranial articular surface
6. width of fossa caudalis
7. greatest width
8. height of cranial articular surface
9. height of fossa caudalis
10. greatest height

Bos									
118	55	22	41	92	45.5	94^x	54	41	100
Ovis									
47.5	29	9	18	39	20	-	22.5	16	-
50	27.5	10	19	40	33.5	40^x	26	17	43
50	31	9	18	37.5	19	-	24	16	44
50.5	28	10	19	40	23	-	27	-	-
51.5	32	9	19	41	22	30^x	28	15.5	43
52	30	9	21	42.5	21	-	26	15	-
52	31	8	19	38.5	22	40^x	16	28	43
52	-	11.5	18	41	-	-	27	18	-
53.5	31	9	20	39	21	39	25	16	-
53.5	-	10	-	33^x	21^x	-	25	17	-
54	29	9	20	42	21^x	-	24	17	-
54.5	-	29	19.5	39	22	-	26	17.5	-
55	-	10	20	41	22.5	-	25	17	-
Capra									
60	33	10	22	46	26	49	33	21	58
Sus									
55	21	18	13	63	41	-	29	23	- w
55	23	12	14	67	42.5	-	35	28	93 w
56	27.5	20	13.5	63.5	41	-	28	28	- w
Canis f.									
47.5	21	14.5	7.5	31	19	30	13	12	38
Capreolus									
60	34	9	20	40	22	-	27	16	46

58	33	10	18.5	40	21	40	27	15	-
60	40.5	8.5	18	36	18^x	-	24	15	-
61	39.5	9.5	17	36	18	40^x	27	17	-
61	-	9	19	35	17	-	25	15	-
65	40	11	21	40	21	-	26	18	-

SCAPULA

Measurements: 1. greatest length
2. greatest width
3. smallest width of collum scapulae
4. width of angulus articularis
5. diameter of facies articularis

Bos

-	-	49	69	47
-	-	49.5	72^x	51
-	-	51	66	50
-	-	51	67	50
-	-	51	69	48.5
-	-	53	71	52
-	-	55	75	51
-	-	56	76	57w
-	-	62	77^x	63^{x}w
-	-	-	65	49
-	-	-	67	-
-	-	-	69	47^x
-	-	-	73	51
-	-	-	74	52^x
-	-	-	77	58 w
-	-	-	90	62^x

Ovis

122	80[x]	17.5	32	20
-	-	15	28	17.5
-	-	15	30	18[x]
-	-	16	31.3	17
-	-	16	32	21
-	-	17	28	18.5
-	-	17	29	18
-	-	17	30.5	19.5
-	-	17	32.5	19[x]
-	-	17.3	29	18.5
-	-	17.3	31	20
-	-	17.5	28	19
-	-	17.5	29	20.7
-	-	17.5	30	19
-	-	17.5	30.5	18
-	-	18	29	18
-	-	18	29	18.7
-	-	18	29.5	19
-	-	18	31.5	20.5
-	-	18	32	20
-	-	18	32	20
-	-	18	32.5	20.5
-	-	18.3	30	19.5
-	-	19	29	19.5
-	-	19	31	19
-	-	19	31.5	19
-	-	19	32.5	21.5
-	-	20	31	20
-	-	-	30	19

Capra

110^x	-	15	29^x	18
120	78	17.5	-	19
135^x	-	17	29^x	20
136	92	18.5	30.2	20
-	-	13.5	28	18.5
-	-	16	30.5	19
-	-	16.5	28	18
-	-	16.5	28	20
-	-	17	30	20^x
-	-	18	30	21
-	-	18	30	22
-	-	18	30	22
-	-	18	30	23
-	-	18	30	23
-	-	18	30	23.5
-	-	18	30.5	22
-	-	18	30.5	23
-	-	18	31	22.5
-	-	18.5	30	22^x
-	-	18.5	30.5	21
-	-	18.7	30.5	22.8
-	-	19	29	21
-	-	19	29	24^x
-	-	19	33^x	24.5
-	-	19.2	30	23
-	-	19.5	31	22.5
-	-	19.7	34	23.7
-	-	-	31	22

Sus

-	-	35	50	35 w
-	-	36	47	34.5 w
-	-	-	52	33^xw
-	-	-	54	37^xw

Canis

-	-	22.5	25	15

Capreolus

-	-	16	27.5	22
-	-	16.5	28	20.3
-	-	17	29.7	22.5
-	-	17.5	28.5	21
-	-	18	29	22
-	-	18	29.8	20.7
-	-	17	32	19
-	-	17.5	29	19
-	-	17.5	30	18.5
-	-	17.7	30	19
-	-	17.5	31	19.5
-	-	18	28.8	18.5
-	-	18	29	17.7
-	-	18	30.5	19.5
-	-	18	31	19.7
-	-	18	31.7	-
-	-	18	32.5	20
-	-	18.3	29	18
-	-	19	29^x	21
-	-	19	32	20
-	-	20	32	19

-	-	20.5	30	18.2
-	-	21.3	35	24
-	-	22.3	35	22
-	-	-	32.5	21.3

Vulpes

86	-	16	19	11

Lupus

-	-	8	16	12
-	-	8.5	15	12
-	-	8.5	16	12
-	-	8.8	17.5	12

HUMERUS

Measurements: 1. greatest length
2. proximal width
3. smallest width
4. distal width
5. proximal diameter
6. smallest diameter
7. distal diameter

Bos

284	93	37	83	101	39	75
-	-	-	84	-	-	80
-	-	-	84[x]	-	-	82
-	-	-	84	-	-	83
-	-	-	85	-	-	80

-	-	-	85	-	-	82
-	-	-	85	-	-	84
-	-	-	86	-	-	81
-	-	-	86	-	-	82
-	-	-	91	-	-	88

Ovis

127[x]	-	13.5	26	37	13	22
128[x]	-	13	28	-	12	23.5
131[x]	36[x]	13.5	39	15	23.5	-
133	-	13	27	-	14	23.5
133	35	13.5	28[x]	38	15	-
143[x]	-	16	30	44	16	27
-	35	-	-	47.5	-	-
-	37.5	-	-	48	-	-
-	-	11	29	-	13	25
-	-	11.3	25	-	13	23
-	-	11.5	26	-	14	24
-	-	12	26	-	13.2	23
-	-	12	26.5	-	11.7	23
-	-	12	27	-	12	24
-	-	12	27	-	13	23.5
-	-	12	27.5	-	12.5	23
-	-	12	27.7	-	14	24
-	-	12.5	25.5	-	13	21.5
-	-	12.5	26	-	13	22
-	-	12.5	26	-	14	23
-	-	12.5	27	-	14	25
-	-	12.5	28	-	13.5	23[x]
-	-	13	26	-	13	24
-	-	13	27	-	13	23
-	-	13	27	-	13	23.2
-	-	13	27	-	14	24

-	-	13	27	-	14	23
-	-	13	27.2	-	12.8	23.7
-	-	13	27.5	-	13	24.5
-	-	13	28	-	12.8	23
-	-	13	28	-	13	24
-	-	13	28	-	13.5	24
-	-	13	28	-	14	23.3
-	-	13.3	28.5	-	14.7	25[x]
-	-	13.5	26	-	14	22.3
-	-	13.5	28	-	14	25
-	-	13.5	28.5	-	13	24
-	-	13.5	28	-	15	23
-	-	13.5	28	-	15.5	25
-	-	13.7	27.7	-	14	23
-	-	13.5	29	-	13	25
-	-	13.7	29	-	14.5	26
-	-	13.8	28	-	14.8	24.8
-	-	14	25.5	-	14.7	23
-	-	14	26	-	15	23
-	-	14	27	-	14.5	23
-	-	14	27.5	-	14.5	23
-	-	14	28	-	13	21
-	-	14	28	-	14	28.5
-	-	14	28	-	14.7	24.7
-	-	14	28.5	-	14	25
-	-	14	28.5	-	14	25
-	-	14	28	-	14	23
-	-	14	29	-	14.5	24
-	-	14.3	30	-	16	26.5
-	-	14.5	27	-	13	23
-	-	14.5	29	-	15.5	28
-	-	14.7	28.5	-	14	24.7
-	-	14.7	28.5	-	16.2	25
-	-	14.7	29	-	15	26

-	-	15	27.5	-	15	24
-	-	15	27	-	14	24
-	-	15	29	-	14.5	22.5
-	-	15	29	-	15	24
-	-	15	29.5	-	16	26
-	-	15	31	-	15	27
-	-	16	27	-	14	23
-	-	16	30.5	-	15	26
-	-	-	25	-	-	22.7
-	-	-	25	-	-	23
-	-	-	25.5	-	-	23
-	-	-	25	-	-	22.7
-	-	-	25	-	-	23
-	-	-	26	-	-	22.5
-	-	-	26	-	-	24
-	-	-	25.5	-	-	23
-	-	-	26.3	-	-	23
-	-	-	26.3	-	-	20
-	-	-	26.5	-	-	24
-	-	-	26.5	-	-	23
-	-	-	26.5		-	23.7
-	-	-	26.7	-	-	20
-	-	-	26.7	-	-	22
-	-	-	26.8	-	-	23.5
-	-	-	27	-	-	22.7
-	-	-	27	-	-	23
-	-	-	27	-	-	23
-	-	-	27	-	-	23
-	-	-	27	-	-	22.5
-	-	-	27	-	-	24
-	-	-	27	-	-	23.5
-	-	-	27	-	-	24
-	-	-	27	-	-	24.3
-	-	-	27	-	-	25

-	-	-	27	-	-	24.7
-	-	-	27	-	-	24
-	-	-	27	-	-	25
-	-	-	27	-	-	25
-	-	-	27.2	-	-	23
-	-	-	27.5	-	-	24.5
-	-	-	27.5	-	-	25.5
-	-	-	27.5	-	-	26
-	-	-	27.7	-	-	24
-	-	-	27.7	-	-	25.5
-	-	-	28	-	-	23
-	-	-	28	-	-	23.5
-	-	-	28	-	-	24.8
-	-	-	28	-	-	24.7
-	-	-	28	-	-	24
-	-	-	28	-	-	24
-	-	-	28	-	-	24
-	-	-	28	-	-	24
-	-	-	28	-	-	24
-	-	-	28	-	-	24.5
-	-	-	28	-	-	26
-	-	-	28	-	-	25
-	-	-	28	-	-	25
-	-	-	28	-	-	25
-	-	-	28	-	-	26
-	-	-	28.2	-	-	25
-	-	-	28.5	-	-	24.5
-	-	-	28.5	-	-	25.5
-	-	-	28.5	-	-	25
-	-	-	28.5	-	-	24.5
-	-	-	28.5	-	-	24.8
-	-	-	28.8	-	-	24.5
-	-	-	29	-	-	24.8
-	-	-	29	-	-	25

-	-	-	29	-	-	27
-	-	-	29	-	-	25.5
-	-	-	29	-	-	27
-	-	-	29.5	-	-	24
-	-	-	29.5	-	-	26
-	-	-	29.5	-	-	27
-	-	-	29	-	-	27
-	-	-	30	-	-	24.5
-	-	-	30	-	-	26
-	-	-	30	-	-	26
-	-	-	31	-	-	27

Capra

-	-	14	31	-	17	26
-	-	15.5	30	-	16	26
-	-	-	32.5	-	-	26
-	-	-	32.5	-	-	26
-	-	-	32.5	-	-	26.2
-	-	-	33.5	-	-	27
-	-	-	33.5	-	-	28
-	-	-	33.8	-	-	30.8

Sus

-	-	-	51	-	-	50 w
-	-	-	51	-	-	51 w
-	-	-	51.5	-	-	52 w
-	-	-	53	-	-	51 w
-	-	-	55	-	-	56 w
-	-	-	59	-	-	53 w
-	-	-	57	-	-	54 w

Canis

-	40x	-	-	56	-	- w
-	-	8	22.5	-	9	17x
-	-	11.7	27	-	12	21

Cervus

-	-	-	62	-	-	66

Capreolus

-	-	13.7	30.5	-	16	29
-	-	15	31	-	16	28.8
-	-	-	29	-	-	27
-	-	-	30	-	-	28
-	-	-	30	-	-	28
-	-	-	30	-	-	28.5
-	-	-	30.5	-	-	30
-	-	-	31x	-	-	29
-	-	-	31x	-	-	29
-	-	-	31	-	-	30
-	-	-	31	-	-	30
-	-	-	31	-	-	30x
-	-	-	31.3	-	-	29.7
-	-	-	31.3	-	-	29.8
-	-	-	32	-	-	29.5
-	-	-	33.5	-	-	31

Meles

-	-	9	29	-	12.5	27

Lepus

110	20	6	14	21.3	8	10.2
114	-	7	14	-	6.5	11
-	18^x	-	-	22	-	-
-	-	7	14	-	6	10
-	-	7	14	-	6	11
-	-	-	14	-	-	10.5
-	-	-	14	-	-	10.5
-	-	6.8	14.2	-	7	10.8
-	-	-	15	-	-	10

RADIUS

Measurements: the same as those of humerus

Bos

-	81	-	-	41	-	-
-	82	-	-	44	-	-
-	83	-	-	40	23.5	-
-	83	-	-	45^x	-	-
-	85	-	-	42^x	25	-
-	86	-	-	43	-	-
-	86	-	-	44^x	-	-
-	88	-	-	45^x	-	-
-	89	-	-	46	-	-
-	90	-	-	48	-	-
-	91	-	-	44.5	-	-
-	91	-	-	47	29	-
-	92.5	-	-	44	26	-
-	96	-	-	49.5	-	-
-	103	-	-	53	-	- w

-	-	-	66	-	-	43
-	-	-	79	-	-	46
-	-	-	72.5	-	-	43
-	-	-	74	-	-	45
-	-	-	74.5	-	-	43
-	-	-	76	-	-	46
-	-	-	76	-	-	45
-	-	-	75	-	-	46
-	-	-	90	-	-	57

Ovis

128	26.5	15	25	14	8	16.5
130	28	14	24	14	9	18
130	28	16	26.5	13.5	8	17
130	30	16.5	27	14	9	18
132	28.5	15	25	14	7.7	17.7
135	27.7	15	24.5	14	8	17.5
135	28[x]	16	25	15	7.5	18
135.5	29	15	26	15	8	18
136	26.5	13.5	24	13.5	7	16.7
137	26.5	14.5	26.5	15	8	16.5
137	27	14	26	15	7	17
137	27	15	26	13	8	17.5
137	27	16	26	15	7.5	17.5
137	28	15	26	14.5	8	18
137	29.7	15	16	15	8	17.5
138	25	15	25	15	8	17
138	27.5	14	28	14.5	7	20
138	28	16	27	15	8	18
139	30	15.7	26.5	15	9	17.5
138	-	14.3	25	15	8	18.5
139	-	15	25.5	-	7	17
140	27	15	25	14	8	17.5

140	27[x]	15	26	15.5	8.5	17
138[x]	30	17	-	-	9	-
140[x]	28	15	26	14	7.5	18
140	28.5	15	26	14.5	8	18
141	28.5	15	25.5	13	8	18
141	30	15.5	27	15	8	19.5
141.5	28.5	16	25	15	8	18.5
142	25.7	15.3	25	13	7.5	17.3
142	27.3	15.3	25	14.5	7.7	16.5
142	27.7	15	25	15	7.7	17
142	29	15.5	24.7	14.5	8.5	18
14.2	30	16.5	27	15	8	18
142	-	15	26	-	7.7	18
142.7	28	15	26.3	15.5	-	18
143	28.7	15	26.5	14.7	7.5	18
143.7	29	15.5	27	14.5	9	19
144	30	15.5	28	15.7	8	18.7
144	30	16	27	16.2	7.8	18
144	30	16	28	16	9	18.5
145	28	15	26	14	8	-
145	30[x]	16	27	15	8.8	19
145	30	16.5	27	15	9	19
145	30.5	15	28	15	8.2	18
146	30	16	27	16	8.3	18
147	30	16	28	13	8	19
148	31	16	27.5	15.5	8.3	18
148.5	29	15	25	15	8	18
149[x]	28.5	16	27	15	8	18
149	30	15.5	28	15	8	19.5
149	30	16	17.5	16	8.5	19
150	29	15.5	26.3	14.5	8.5	19
150	29	16	27.5	15	8.5	19
151.5	29.5	15	27.5	14.5	8.5	18.5
153	29.5	16	26	14.5	8.8	19

142	31	17	28	16	8.5	18.5
153	31	15.2	27	15	10	19
-	23.5	-	-	13	-	-
-	24	-	-	14	7	-
-	24.5	-	-	14.8	8	-
-	24.7	-	-	14.7	8.8	-
-	24.8	-	-	14	7	-
-	25	-	-	13	7.5	-
-	26	-	-	13.5	7.5	-
-	26	-	-	14	7	-
-	26	-	-	14	7	-
-	26	-	-	14	7	-
-	26	-	-	14	8	-
-	26	-	-	15	7.5	-
-	26.5	-	-	13.3	8	-
-	26.5	-	-	13.5	8.5	-
-	26.5	-	-	14	7.7	-
-	26.7	-	-	14	8	-
-	27	-	-	13	7	-
-	27	-	-	14	7	-
-	27	-	-	13.7	7.5	-
-	27	14	-	14	8.2	-
-	27	-	-	15	8	-
-	27	-	-	15	8	-
-	27	-	-	13	-	-
-	27.2	-	-	14	8.3	-
-	27	-	-	14	8	-
-	27.5	-	-	14	7	-
-	27.5	-	-	14	7.8	-
-	27.5	-	-	15	8	-
-	27.5	-	-	13	-	-
-	27.5	-	-	13	7	-
-	27.7	-	-	16	8	-
-	27	-	-	14	7	-

-	27.5	-	-	14.7	-	-
-	28	-	-	14	-	-
-	28	-	-	14	8	-
-	28[x]	-	-	13	-	-
-	28	-	-	14	8	-
-	28	-	-	14	-	-
-	28	-	-	14.3	-	-
-	28	-	-	14	8	-
-	26.8	-	-	14	7	-
-	27	-	-	13.5	7	-
-	27.8	-	-	14.8	7	-
-	28	-	-	14	7	-
-	27.2	-	-	14	-	-
-	28	-	-	14.5	8.2	-
-	28	-	-	15.2	-	-
-	28	-	-	14.5	8	-
-	28	-	-	15	9	-
-	28	-	-	15.5	7.5	-
-	28	-	-	15	8	-
-	28	-	-	14	8	-
-	28	-	-	14.5	8	-
-	28	-	-	15	8	-
-	28	-	-	15[x]	8	-
-	28	-	-	15	-	-
-	28	15.5	-	15	8.5	-
-	28.5	-	-	14.2	7.7	-
-	28.5	-	-	14	-	-
-	28.5	-	-	14.5	9	-
-	28.5	15	-	15	7.5	-
-	28.5	-	-	15	7	-
-	28.5	-	-	15	8.5	-
-	28.7	-	-	14.8	-	-
-	29	14	-	14.5	8	-
-	29	-	-	14.5	9	-

-	29	-	-	14.5	9	-
-	39	-	-	14	7	-
-	29	-	-	14	7	-
-	29	-	-	15	7.5	-
-	29	-	-	15	8	-
-	29	-	-	15	-	-
-	29	-	-	15	7.5	-
-	29	-	-	16	7.5	-
-	29	-	-	15	8.5	-
-	29	-	-	15	-	-
-	29	-	-	15	7	-
-	29	-	-	15	8	-
-	29	-	-	15	8	-
-	29	-	-	15	8	-
-	29	-	-	15	8	-
-	29	-	-	15	-	-
-	29	-	-	15	8	-
-	29	-	-	15	-	-
-	29	-	-	18.8	-	-
-	29.2	-	-	16	-	-
-	29.3	14.5	-	5.5	7.7	-
-	29.5	-	-	15	9	-
-	29.5	-	-	14.5	-	-
-	29.7	-	-	15.2	9	-
-	29.7	-	-	15	-	-
-	29.7	-	-	14.8	-	-
-	29.7	-	-	15	8.7	-
-	30	26.5	-	15.5	8	-
-	30	15	-	15.5	8	-
-	30[x]	-	-	14	8	-
-	30	-	-	14	8	-
-	30	-	-	15	9	-
-	30	-	-	15	7	-
-	30	-	-	15	8.5	-

-	30	-	-	15	8	-
-	30	-	-	14.7	8	-
-	30	-	-	15	8	-
-	30^x	-	-	15	9	-
-	30	-	-	15.5	8.5	-
-	30	-	-	16^x	8.5	-
-	30	-	-	15.5	-	-
-	30	-	-	15.5	-	-
-	30	-	-	16	8	-
-	30.5	-	-	15	7.5	-
-	30	-	-	16	-	-
-	31	-	-	15.2	9	-
-	31	7	-	17^x	8	-
-	31	17	-	17^x	8	-
-	32	-	-	15	8.5	-
-	30	-	-	15	8	-
-	30	-	-	14.7	8	-
-	30	-	-	15	8	-
-	30^x	-	-	15	9	-
-	30	-	-	15.5	8.5	-
-	30	-	-	16^x	8.5	-
-	30	-	-	15.5	-	-
-	30	-	-	15.5	-	-
-	30	-	-	16	8	-
-	30.5	-	-	15	7.5	-
-	30	-	-	16	-	-
-	31	-	-	15.2	9	-
-	31	17	-	17^x	8	-
-	31	17	-	17^x	8	-
-	32	-	-	15	8.5	-
-	-	14.2	26	-	-	19
-	-	16	27	-	-	20
-	-	15.7	27.5	-	-	19.5
-	-	-	26	-	-	19.3

-	-	-	26	-	-	17
-	-	-	27	-	-	18

Capra

131	20	30.5	17	10	20	-
143	30	16	28	15	9	19
145	34	20	-	17	9	21
151	30	15.5	27	14	8	19
-	25	-	-	13	7.7	-
-	26	-	-	16	9	-
-	28	-	-	15	-	-
-	28	-	-	15	8	-
-	28	-	-	16	-	-
-	28.5	-	-	14.5	7.5	-
-	29.5	-	-	16	8	-
-	30	-	-	15.5	-	-
-	32	-	-	18	9	-
-	33	-	-	16	-	-
-	-	-	28	-	-	19

Sus

-	36	21	-	23	15	- w
-	37	25	-	27	19	- w
-	39	-	-	29.5	-	- w
-	40	23	-	29	18	- w
-	40	-	-	38.5	-	- w
-	41.5	-	-	20.5	-	- w
-	-	-	41	-	-	31 w

Canis

123^{x}	24.5	9.5	-	-	5	-
130	12.3	9	17	8	4	12
213	24	18.5	32^{x}	18	8	-
-	24	16	-	17	7.5	-
-	25	18.5	-	18	9	-
-	-	10	8.5	-	6	10.5

Cervus

-	62.5	-	-	34	-	-
-	67^{x}	-	-	38	-	-
-	-	-	56	-	-	44
-	-	-	61	-	-	45

Capreolus

130	26	16	24	15	8.8	18
177	26	15	25	15	8.7	19
183	26	15	26.5	17	8	20
185	26	15	26	16.5	9	19.7
-	26	-	-	16.7	10.8	-
-	27	-	-	16.5	-	-
-	27	-	-	17	-	-
-	27.2	-	-	16	-	-
-	28	-	-	16.7	9	-
-	28	-	-	17	8.5	-
-	28	-	-	16	-	-
-	28	-	-	17	-	-
-	28	-	-	17	10	-
-	28.7	-	-	15.5	-	-
-	28.7	16	-	16.7	10.7	-
-	29	-	-	16	-	-

-	29.5	17	-	17.5	10.2	-
-	29.7	19	-	18	11	-
-	29.7	-	-	17	-	-
-	29.8	19	-	17	11	-
-	30	-	-	17	10	-
-	30	-	-	18	-	-
-	-	14	25	-	-	18
-	-	-	25	-	-	19
-	-	-	25.5	-	-	19
-	-	-	26	-	-	20
-	-	-	26	-	-	20
-	-	-	27	-	-	20
-	-	-	26	-	-	22

Lepus

120	10	5.2	11	7.2	4.8	7
125	10.8	6	10.8	6.8	4	7
-	10	5	-	6.5	4	-
-	11	6	-	7	45	-
-	-	-	11	-	-	7

METACARPUS

Measurements: the same as those of humerus

Bos

196	57	29.5	58[x]	37	23	33
201	58	32	52	37.5	24	34
202	62	31	62	40	24	34.5
205	60	31	63.5	39	24	35
205.5	60.5	31	62	39	24	34

213	60	35	62	39	24	34
218	72	39	74	44	26	38
-	57	-	-	40	-	-
-	59	-	-	36	-	-
-	60	-	-	35	-	-
-	60	-	-	39	-	-
-	61	-	-	33	-	-
-	61.5	34	-	39.5	-	-
-	63.5	-	-	38	-	-
-	-	-	62.5	-	-	34.5
-	-	-	63	-	-	34.5

Ovis

109	20	11.3	22.5	15.8	8	14.8
111	19	12	22	14	7	14
111.5	20	11.5	22.7	14.7	8.7	15
113	20	11	22	16	9	15
113[x]	-	12	22.5	-	8.5	14
114	21	12.5	23	15	9	15
114.5	21.7	23	13	8.8	16	-
115	20	11.5	22.7	15.5	8.5	15
115.3	20.2	12	22.2	16	8	15
116[x]	20	13	-	15	8	14[x]
116	20	13	24	15.5	8	16
116	21	12	23.2	15	9	15
116[x]	-	12.5	24	-	8	16
116.5	21	11	22.3	10.5	8	14.5
117	20	-	22.7	-	9	14.8
111.7	21	12.5	22.7	15.5	7	15
117.5	20	12	22.8	15	8.8	14.8
118	19.5	11	22	15.8	8	15
118	20.5	13	24	16	8	15
118	21	12.5	22	15	8.5	15

118ˣ	-	12	23ˣ	-	8	-
119.5	19	12	21.8	14.2	8.5	13.5
120	20ˣ	11.8	23	15ˣ	8	14.5
120	20	13	22.2	15	8	15
120	20.5	12	23	14	8	14.8
120ˣ	-	13	22.8	-	8.3	14.5
121	19	12	22	14	8	14
121	20.7	13	23	15.2	8.7	15
122	20	11.5	22	-	-	-
123.7	21	12	23.8	16	8.5	15.8
124ˣ	-	12	-	-	8	15
124.5	21	-	23	15	9.5	16
-	18	-	-	13	-	-
-	19	11	-	15	-	-
-	19	11.7	-	15	-	-
-	19	12	-	15	-	-
-	19	-	-	13	-	-
-	19	-	-	13.5	-	-
-	19	-	-	15	-	-
-	19.3	12	-	14.8	-	-
-	19.3	12	-	14.5	-	-
-	19.5	12	-	14.5	-	-
-	19.5	19.7	-	14	-	-
-	19.5	21.5	-	14	-	-
-	19.5	-	-	14.5	-	-
-	19.8	11	-	14.7	-	-
-	20	11.8	-	15ˣ	-	-
-	20	12	-	14	-	-
-	20	12	-	15	-	-
-	20	12	-	15.5	-	-
-	20	12	-	15.8	-	-
-	20	13	-	14.8	-	-
-	20	13	-	15	-	-
-	20	17.3	-	15	-	-

-	20	-	23.5	16.3	9	15
-	20	-	-	15	-	-
-	20	-	-	15	-	-
-	20	-	-	15	-	-
-	20.2	12.2	-	14.7	-	-
-	20.2	-	-	14.8	-	-
-	20.2	-	-	15	-	-
-	20.5	11.5	-	15.5	-	-
-	20.5	-	-	11	-	-
-	20.8	12.5	-	15	-	-
-	20.8	12	-	15.3	-	-
-	21	12	-	15	-	-
-	21^{x}	12	-	16	-	-
-	21	12	-	16.5	-	-
-	21	12.5	-	15	-	-
-	21	13	-	16	-	-
-	21	-	-	14.8	-	-
-	21	-	-	15.5	-	-
-	21	-	-	16	-	-
-	21	-	-	16	-	-
-	21.2	12.5	-	16	-	-
-	21.3	-	-	15^{x}	-	-
-	21.5	12.7	-	15	-	-
-	21.5	14	-	16	-	-
-	21.5	-	-	15.5	-	-
-	21.5	-	-	16	-	-
-	21.7	12	-	16	-	-
-	22	12.5	-	15.5	-	-
-	22	12.8	-	15.7	-	-
-	22	-	-	15.5	-	-
-	-	20.8	-	-	9.7	14.5
-	-	11.5	22.5	-	8	14.8
-	-	12.3	23.5	-	8	14.8
-	-	-	21.5	-	7.7	14

-	-	-	22	-	7.5	13.5
-	-	-	22	-	7.5	14.5
-	-	-	22	-	7.8	15
-	-	-	22	-	8	14.5
-	-	-	22	-	8	14.7
-	-	-	22	-	8.5	15
-	-	-	22	-	-	13
-	-	-	22	-	-	14
-	-	-	22	-	-	14.2
-	-	-	22.2	-	7.5	14.7
-	-	-	22.2	-	8	15.5
-	-	-	22.3	-	7.8	14.5
-	-	-	22.3	-	8	15
-	-	-	22.3	-	8	15
-	-	-	22.5	-	8	15
-	-	-	22.5	-	9.2	14.2
-	-	-	22.5	-	-	14
-	-	-	22.5	-	9	14.5
-	-	-	22.8	-	9	14
-	-	-	22.7	-	8	14.5
-	-	-	23	-	8	15
-	-	-	23	-	9	15.5
-	-	-	23	-	8	15
-	-	-	23	-	-	15
-	-	-	23.3	-	8.5	15
-	-	-	23.3	-	8.5	15.3
-	-	-	23.2	-	9	15
-	-	-	23.5	-	-	14
-	-	-	23.5	-	9.3	15.5
-	-	-	23.7	-	8.7	15.8
-	-	-	23.7	-	8	15
-	-	-	24.5	-	9.8	16
-	-	-	23.8	-	8	15

Capra

119.5	23	15.5	26^{x}	15	10	15
-	20	-	-	15	-	-
-	22	15	-	16	-	-
-	23.3	14	-	16	-	-
-	24	15.8	-	17	-	-
-	26	16	-	18	-	-
-	-	-	24	-	-	15
-	-	-	28^{x}	-	9	16
-	-	-	28.8	-	10	17.5
-	-	-	33.5	-	12.5	20

Cervus

-	47.5	-	-	34	-	-

Capreolus

169.5	21.3	12.7	22	17.5	10	15
183	21	12.7	23	16	10	15
-	22	12	-	17	-	-
-	22	-	-	16.5	-	-
-	22.5	-	-	18	-	-
-	20^{x}	12	-	16.5	-	-
-	21.7	13	-	17.5	-	-
-	23	-	-	16.7	-	-
-	23.3	13.3	-	19.8	-	-
-	-	-	21	-	8.5	14^{x}
-	-	-	21.5	-	8	14.5
-	-	-	21.5	-	9	14
-	-	-	22	-	-	15
-	-	-	22	-	-	15
-	-	-	22.5	-	-	15

-	-	-	23	-	10.2	15.7
-	-	-	23.5	-	10.3	16

FEMUR

Measurements: 1. length to the trochanter major
2. length to the caput
3. proximal width
4. smallest width
5. distal width
6. proximal diameter
7. smallest diameter
8. distal diameter

Bos

-	-	118	-	-	60^x	-	-
-	-	-	39	102	-	43	131

Ovis

160^x	157	40^x	14.5	33.5	21	13.5	41
166	162	41.5	15	35^x	20	14	43
-	160	-	15	35.7	-	15	41^x
-	163.5	-	14.7	36	-	15	43.5
-	164.5	-	15	-	-	14.5	-
-	-	21.5	-	-	16	-	-
-	-	39	-	-	18	-	-
-	-	39	-	-	20	-	-
-	-	40	-	-	20	-	-
-	-	40	-	-	23	-	-
-	-	41	15	-	22	15	-
-	-	41	-	-	20	-	-

-	-	41	-	-	21^x	-	
-	-	41	-	-	21	-	-
-	-	41.3	-	-	20^x	-	-
-	-	42	16	-	24	17	-
-	-	42.5	-	-	21	-	-
-	-	43	16	-	35	14	-
-	-	45	-	-	22	-	-
-	-	46	-	-	21	-	-
-	-	49.5	-	-	22	-	-
-	-	-	16	35	-	15.5	34
-	-	-	-	33	-	-	44
-	-	-	-	34	-	-	41
-	-	-	-	35	-	-	43
-	-	-	-	35	-	-	41
-	-	-	-	35	-	-	43

Capra

-	-	-	33	-	-	41	
-	-	-	-	36.5	-	41.5	
-	-	-	-	39	-	47	

Sus

-	-	85	-	-	51^x	-	- w
-	-	-	-	63	-	-	70w

Capreolus

-	-	45	-	-	25	-	-
-	-	46.5	15	-	34	-	-
-	-	47	-	-	23	-	-
-	-	48	-	-	24	-	-
-	-	-	16	40	-	17	52

-	-	-	17	42	-	20	52
-	-	-	-	38	-	-	49
-	-	-	-	39.5	-	-	49.5
-	-	-	-	40	-	-	47
-	-	-	-	40	-	-	48
-	-	-	-	42	-	-	48

Felis silvestris

138	137	25.5	9.7	23^x	15	9	-

TIBIA

Measurements: the same as those of humerus

Bos

-	-	37	61.5	-	27	46
-	-	38.5	66	-	28	49
-	-	-	62^x	-	-	49
-	-	39	64.5	-	30	50
-	-	39.5	67^x	-	37	50
-	-	-	64^x	-	-	47^x
-	-	-	64	-	-	48
-	-	-	64	-	-	50
-	-	-	64	-	-	51
-	-	-	65	-	-	48
-	-	-	65	-	-	49
-	-	-	65	-	-	50
-	-	-	66	-	-	50^x
-	-	-	66^x	-	-	50
-	-	-	66.5	-	-	48.5
-	-	-	67	-	-	48.5

-	-	-	69	-	-	53
-	-	-	68	-	-	53^x
-	-	-	77	-	-	59 w
Ovis						
189.5	35^x	14	22.5	-	11	17.5
194	37	13	24	40	11	18
194	-	13.2	24	38	11.8	18
198	38	14	25	38	11	20
199	-	14	24	41	11	19
204	40^x	14.7	25.8	43	11.5	21
-	35	-	-	39	-	-
-	36	-	-	39	-	-
-	37	-	-	38	-	-
-	37	-	-	39	-	-
-	37	-	-	39	-	-
-	37	-	-	40	-	-
-	37	-	-	40.5	-	-
-	37	-	-	41.5	-	-
-	38	-	-	39.5	-	-
-	38	-	-	40	-	-
-	38	-	-	41^x	-	-
-	39	-	-	39	-	-
-	39	-	-	39^x	-	-
Capra						
-	35	13	-	38	-	-
-	37	-	-	43	-	-

Sus

-	64	-	-	70	-	- w
-	65	-	-	78	-	- w
-	65.5	-	-	77	-	- w
-	-	17	29	-	13	24
-	-	-	36	-	-	32 w

Canis

150[x]	29	10.5	-	31.5	10	-
-	-	11	22	-	12	16
-	-	15	28	-	15	22 w
-	-	17.5	31	-	16	21.5 w

Cervus

-	-	-	57	-	-	43.5

Capreolus

209	42	16	27.5	45	13	21
-	36.5	-	-	41.5	-	-
-	37	-	-	39.5	-	-
-	40	-	-	44	-	-
-	43	-	-	45	-	
-	43.5	-	-	48	-	-
-	46	-	-	49	-	-
-	49	-	-	55	-	-
-	-	15	25	-	12	20
-	-	15	25	-	13	20
-	-	15	27	-	12.7	20
-	-	15	25	-	14	20

-	-	16	28	-	15	22
-	-	17.5	30x	-	14.8	22.5
-	-	18	28.5	-	????	22.5
-	-	-	26.5	-	-	21.5
-	-	-	27	-	-	21.7
-	-	-	27.5	-	-	21
-	-	-	28	-	-	20
-	-	-	28.3	-	-	22
-	-	-	29	-	-	22
-	-	-	29x	-	-	22
-	-	-	29	-	-	23

Vulpes

-	-	8	15	-	6.8	10.7

Lepus

157.5	22.5	8	15.7	23.5	7	11
158.5	22	8	15	25	7	11
-	21.5	-	-	24	-	-
-	22.3	-	-	25	-	-
-	23	8	16	25	8	11
-	23	-	-	25.5	-	-
-	23.8	-	-	25.5	-	-
-	-	7	15	-	6.5	11
-	-	8	15	-	7	11
-	-	9	16	-	8	12
-	-	-	15.2	-	-	11
-	-	-	15.5	-	-	11.3
-	-	-	16	-	-	11
-	-	-	16	-	-	11
-	-	9	16.5	-	6.5	11

ASTRAGALUS

Measurements: 1. greatest length
2. greatest width
3. greatest diameter

Bos

64.5	44	35.5
66	42	36
66	44	35.5
67ˣ	44	36.5
68	47	38
69	45	38
70	51	40
71	48	-
73	47.5	40
73	51	40
74	48.5	41
76ˣ	48	41
77	49	43.5
81ˣ	57	- w
85ˣ	-	43ˣw

Ovis

14	17	14
24	18	16
25	18	16
25.5	18	15.7
26	18	15.5
26	18.5	15
26	18.5	15.5

26	18.5	16
26	19	15
26	19	16.5
26	19	-
26.5	18	16
26.5	18.5	16.7
26.5	19	16.5
26.5	19	16.5
26.5	19.5	15.5
26.5	19.7	15
27	19	17
27	19.5	16.7
29	19	17.2
58	19.5	17
75	18	14

Capra

26	17.5	15
27	19	16

Sus

53	32	30 w
53	32	32 w
56	36.5	33 w

Canis

33	21	- w

Cervus

52	35	31
55	35	31
55	34	28.5
60	37^{x}	32

Capreolus

31	20	17
31	20	18
31.5	20	-
32	20.5	18
32.5	22	18.7
33	21.5	19.5
34.5	22	20

Lepus

19	10	10

CALCANEUS

Measurements: the same as those of astragalus

Bos

120^{x}	41	51
131	44	58
133	45.5	57
135	44	58
136	47	53
136	51	59

137	50	-
137.5	46	58
139	47	58
140	47	55
140	48	61
140	52	58
143	50.5	59
144.5	49	57.5
145	49	58
147	50	67

Ovis

33	18	16
47[x]	17	19
48.5	17	18.5
49	16	19
50	18	19
51	17.5	20
51	18.5	20.5
51	18	21
51	19	21
51	-	19.7
51.5	19	20
52	18	20.5
52	19	20.5
52	19	21
52	19	21
52	20	19
52.5	18.5	21.5
52.5	19	21
52.5	21	21
52.7	18	20
53	18	20

53.5	18	21
53.5	20	22
54	20	21
54	20	-
54.5	19	19
55.2	19	22
55.5	20	22
56	18.5	20
57	19	23
58.3	18	19.3

Capra

54	19	20
150	16.5	19

Sus

105^x	29	40 w

Canis

60^x	35	27 w
64	27.5	27.5 w

Capreolus

53	21	20
67	20	26
67	24	26

Lepus

33.7	13	13
39	23.5	13

METATARSUS

Measurements: the same as those of humerus

Bos

216x	46x	27	54	-	24	32
220	48	26	58x	47	25	32x
228	49	73.5	55x	46.5	26	33
245x	50	27	-	48	23.5	29x
246	53	27	58	50	28	36
-	48	-	-	48	-	-
-	49	26	-	48	-	-
-	50	26	-	48	-	-
-	53	-	-	51	-	-
-	54	-	-	52	-	-
-	-	-	59	-	-	33
-	-	-	54	-	24	31
-	-	-	56	-	25	33
-	-	-	58	-	27	35
-	-	-	58	-	25	32.5
-	-	-	54	-	-	31
-	-	-	61	-	27	35
-	-	-	64	-	-	35.5

Ovis

119	165	9.7	21	18	7	13
119[x]	-	10.7	-	-	8.7	-
120	17	10	20.5	18	8	14
120[x]	197	10	22.8	20	8	16
121[x]	-	9.7	21.3	-	6.5	14
122[x]	-	10	20.5	-	8	14
123	18	11	22	19	8	14
123	-	11	22.5	-	8	15
124	17.7	10	22	18	18.5	14.5
124	18	10	22.3	18	8	14.7
124[x]	18.8	11.5	23	19	8.8	15
124.8	17	10	20	18	9	14.5
125	17.5	10.7	22	18.5	8.2	15
125[x]	17.7	10.2	22	18.7	8	14.5
125	17.8	10	21	18	8	14.8
125	18	10.7	22	19.5	8.7	15
125[x]	17.7	10.2	-	18.5	8.7	14
125[x]	-	10.3	23	-	9.5	-
125.3	17	10	21.3	17	8	14.3
125.3	17.7	10	21.5	18.5	8.5	15
125.5	17.5	10	21	17.2	7.5	14.5
126	17	10	21	18	8	14
126.7	17	10.7	22	18.8	8	14.8
127	18	10	23[x]	19	8	13.5
127	18	11	21.7	18.5	9.5	15
127	19	10.5	22.3	19	8	15.5
127	19	10.7	22.3	19	8.2	15.5
128	19	11.8	22	19	7.8	14.7
129[x]	18	10	22	18.5	7.8	14.5
129	18	10.7	22.3	19	8.5	14.5
129	18	11	21.8	19	8.2	15
129	18.5	11	21.2	-	8	15

129	18.5	11	22	19	8	15
129.5	18	10	22	18.5	8	14.7
130	18	10	21.2	18.3	8	15
130[x]	-	11	22	-	-	-
131	17	10.7	21.5	18.5	10	14.8
131	18	12	21.7	18.5	9.5	15
132	19	10.2	22.7	18.3	8.5	15.3
132	19.2	11	23	20	9	15.5
132[x]	19.7	10	23[x]	19.5	8.7	15.5
133	17	11	21.3	18.5	7.8	17
133[x]	18	11.5	22	-	8.8	14
134	18.5	10.2	23	19	9	16
134	19	12	23	21	9.5	15
134.5	19	11.8	23	21	9	15
135	18.5	11	21.5	19	9	15
135	19	11	22.5	20	9	14.5
136	19	11.5	22.3	19.7	9	15
136.5	20	10.7	22	20	8.5	15.5
137	20	11	23	20.5	9	16
137.5	18.5	10.3	21.8	18	9	14.7
138	18.8	11	22.5	19	8.7	16
141	21	12	24.5	21.2	9.5	17
-	12	-	-	20	-	-
-	16.7	10	-	17.5	-	-
-	17	9	-	17	-	-
-	17	9.7	-	18	-	-
-	17	-	-	17.5	-	-
-	17	-	-	19	-	-
-	17.5	9	-	18	-	-
-	17.5	10	-	18	-	-
-	17.5	10	-	19	-	-
-	17.5	11	-	18.5	-	-
-	17.5	-	-	18	-	-
-	17.5	-	-	19	-	-

-	17.7	-	-	18.2	-	-
-	17.8	10.2	-	18	-	-
-	18	9.7	-	18.3	-	-
-	18	10	-	18[x]	-	-
-	18	10	-	18.5	-	-
-	18	10	-	19	-	-
-	18	102	-	18.3	-	-
-	18	10.3	-	18.3	-	-
-	18	10.5	-	19	-	-
-	18	10.5	-	19.7	-	-
-	18	10.5	-	20	-	-
-	18	11	-	19	-	-
-	18	-	-	18	-	-
-	18	-	-	18	-	-
-	18	-	-	18	-	-
-	18	-	-	18.2	-	-
-	18	-	-	18.2	-	-
-	18	-	-	18.3	-	-
-	18	-	-	18.5	-	-
-	18	-	-	18.5	-	-
-	18	-	-	18.5	-	-
-	18	-	-	18.7	-	-
-	18	-	-	19	-	-
-	18	-	-	19	-	-
-	18	-	-	19	-	-
-	18	-	-	19	-	-
-	18	-	-	19	-	-
-	18	-	-	19.2	-	-
-	18.2	11.2	-	19	-	-
-	18.2	-	-	19.2	-	-
-	18.3	11	-	19.2	-	-
-	18.3	11	-	19.5	-	-
-	18.5	-	-	18[x]	-	-
-	18.5	-	-	19	-	-

-	18.5	-	-	20	-	-
-	18.5	-	-	20	-	-
-	18.5	-	-	20.3	-	-
-	18.7	10	-	18.8	-	-
-	18.7	-	-	19.3	-	-
-	18.8	11	-	19	-	-
-	18.8	-	-	18	-	-
-	18.8	-	-	19.3	-	-
-	19	10.2	-	19.7	-	-
-	19	10.7	-	19	-	-
-	19	10.7	-	20	-	-
-	19	11	-	19	-	-
-	19	11	-	19	-	-
-	19	11	-	19.2	-	-
-	19	11.8	-	19.3	-	-
-	19	12	-	19	-	-
-	19	-	-	18	-	-
-	19	-	-	18.5	-	-
-	19	-	-	19	-	-
-	19	-	-	19	-	-
-	19	-	-	19	-	-
-	19	-	-	19.5	-	-
-	19	-	-	20	-	-
-	19	-	-	20	-	-
-	19	-	-	21.5	-	-
-	19.2	-	-	19.7	-	-
-	19.5	10.5	-	20	-	-
-	19.5	11.7	-	20.7	-	-
-	19.7	11	-	20	-	-
-	19.7	-	-	19	-	-
-	19.7	-	-	19.5	-	-
-	19.8	12	-	19	-	-
-	19.7	-	-	20	-	-
-	20	11	-	21.2	-	-

-	20	12	-	21	-	-
-	20	11.5	-	20	-	-
-	20	12.5	-	20	-	-
-	20	-	-	20.5	-	-
-	20	-	-	20.5	-	-
-	-	9.5	20.5	-	6.8	14
-	-	10	21	-	9	15
-	-	10	21.2	-	7	13.7
-	-	10	22	-	8	14.2
-	-	10	22	-	9	15
-	-	10	22	-	9	15.7
-	-	10.5	21.7	-	8.5	14.7
-	-	10.5	22.3	-	8	15
-	-	10.5	21.5	-	8	14.5
-	-	10	25	-	7	14.8
-	-	-	21	-	7	14.8
-	-	-	21	-	8	15
-	-	-	21	-	-	14.5
-	-	-	20.7	-	-	14.8
-	-	-	21.3	-	9	14
-	-	-	21.5	-	8.5	15
-	-	-	21	-	8	14
-	-	-	21.8	-	8	15.2
-	-	-	21.7	-	8.8	14.7
-	-	-	21.2	-	9	14.3
-	-	-	21	-	-	15
-	-	-	21.7	-	8	15
-	-	-	22	-	8	15
-	-	-	22	-	9	15
-	-	-	22	-	-	14.8
-	-	-	22	-	-	14
-	-	-	22	-	-	15
-	-	-	22	-	-	15
-	-	-	22	-	-	15.5

-	-	-	22	-	9	15.8
-	-	-	22	-	-	15
-	-	-	22	-	-	14.5
-	-	-	22	-	-	15
-	-	-	22	-	8	15.5
-	-	-	22.5	-	-	15
-	-	-	22.5	-	9	15
-	-	-	22.2	-	-	14.7
-	-	-	21.3	-	87.8	14.5
-	-	-	22	-	8.8	15
-	-	-	22.5	-	9	15.5
-	-	-	22.5	-	9	15
-	-	-	23	-	9	15.5
-	-	-	23	-	9	14
-	-	-	23	-	9	15.5
-	-	-	23.7	-	-	15
-	-	-	23	-	-	15

Capra

110.5	19	11.5	22.5	17	9	14.8
113	-	-	24.2	-	10.5	16
114.5	20.5	13.5	25.2	18.3	10	16
115	20	-	25.5	19	9	16
115	20	12	-	18.5	-	-
116^{x}	20	12.5	24.5	19.3	10.5	16.3
118.2	18.5	11.8	22.5	17	8.7	14.5
128^{x}	23^{x}	13.2	26.5	20	10.2	17
-	19	13.5	-	19	-	-
-	21	-	-	19	-	-
-	-	-	23	-	8.5	14.3
-	-	-	24	-	10	16
-	-	-	25	-	9	15.5
-	-	-	25^{x}	-	10	15.7

-	-	-	27.5	-	10	18
Cervus						
-	39.5	-	-	44	-	-
-	-	-	46.5	-	20	31
Capreolus						
205	22	12.5	25.5	24	11.5	17
209	21	12.5	25.7	23.7	12.5	17
211	21	11.7	24	22	11.2	17
-	20	11	25	21	-	17
-	21	12	-	23.5	-	-
-	21.5	12.5	-	23	-	-
-	21.5	13	-	22.5	-	-
-	22	13	-	23.5	-	-
-	22	-	-	24	-	-
-	22[x]	13	-	23[x]	-	-
-	22.5	-	-	23.8	-	-
-	24	-	-	25[x]	-	-
-	-	13	26	-	12.5	17
-	-	-	23.3	-	12	16
-	-	-	25	-	12	16
-	-	-	25	-	-	17
-	-	-	25.5	-	12.5	17
-	-	-	25.5	-	12	17.2
-	-	-	25.5	-	12	17.8
-	-	-	26	-	12.5	17.8
-	-	-	26	-	-	16.5
-	-	-	25	-	12.5	17
-	-	-	26	-	12	17
-	-	-	26.5	-	12	17.8

OS PHALANGIS

Measurements: 1. sagittal length
2. proximal width
3. smallest width
4. distal width
5. proximal diameter
6. smallest diameter
7. distal diameter

Cervus

56	22	17.5	21.5	28	16	18

Capreolus

35.5	12.8	8	10	16	8	9.8
38	12	8	9	16	8	10
38	13	9	11	16	8	10
38	17	9	11.5	17	8	10
38.5	12.5	8	10.7	16.5	8	10
40	13.5	9.5	12	37	8	11
41	123	9	11.5	17	8	10

Asinus

68^x	-	22.5	30.5	-	15	20^{x}p
76	42	25.5	37	32	16.5	21 a

Fish remains from the Early Neolithic site of Endrőd 119

ISTVÁN TAKÁCS

The Neolithic settlement of Endrőd 119 was excavated by János Makkay within the framework of the Microregion Research Project conducted on the Great Hungarian Plain by the Archaeological Institute of the Hungarian Academy of Sciences.[1] This is the first Körös culture settlement which has been completely excavated. It additionally includes are important zoological assemblages.

The zoological assemblage from this site yielded 592 fish bones. These were analyzed by the Hungarian Agricultural Museum. The remains were well preserved and uniformly brownish in colour. They are usually fragmented so that, with some exceptions, their measurements could not be taken (*Figs. 1-3*). The finds included bones from four species: catfish (Silurus glanis L.), carp (Cyprinus carpio L.), pike (Esox lucius L.) and pikeperch (Lucioperca lucioperca L.). Most of the bones (56.7 percent of the fragments) are from catfish. This species is followed by carp (37.3 percent) while pike makes up only 5.0 percent of the assemblage. Pikeperch is represented only by a single dentale (*Table 1*). When the number of individuals is estimated the contributions of catfish and carp are somewhat reduced (to 54.4 and 32.2 percent respectively), while that of pike significantly increases (12.2 percent; *Table* 2). Catfish similarly dominates within the species distribution of pits and layers treated as closed deposits as well, although in two cases the bones of carp are more numerous.

The distribution of skeletal elements

Twenty-four bones of the skeleton (mostly elements of the cranial region) were found, although their distribution varies across the species. Among the remains of catfish, cleithra (90 pieces) clearly dominate. This bone is followed by fragments of the first radiale (61 pieces) while vertebrae (49 pieces) form only the third largest group. Of the carp bones, on the other hand, vertebrae are most numerous (46 pieces) and cleithra in second place (35 pieces). It is interesting that the 30 bones of pike recovered are composed of only four types of skeletal elements. With the exception of four specimens, pike skeletons are represented by cleithra (14 pieces) and dentalia (12 pieces).

When the species is not taken into consideration, cleithra form the largest skeletal part group within the total number of bones (139 pieces, 23.4 percent). This is followed by vertebrae (95 pieces, 16.0 percent) with the first radiale taking only third place (65 pieces; 10.0 percent; *Table 3*).

Fish remains make up 2.7 percent of all zoological finds from this Endrőd site.[2]

The survival rate of fish bones is well illustrated when the estimated number of individuals identified at the site is multiplied by the number of vertebrae occurring in a complete specimen of each species (*Table 4*).[3]

---

[1] Animal bone finds of the settlement were analyzed by Sándor Bökönyi. His permission to identify the fish bones should be acknowledged here.

[2] S. Bökönyi: The Early Neolithic vertebrate fauna of Endrőd 119 (paper published in this volume).

[3] P. Széky: *A halak élete* (The life of fish). Budapest 1967, 11.

Table 1.

The number of bones found in features

(number of fragments)

Features	Catfish	Carp	Pike	Pikeperch	Varia
General, Square 55 west:					
Pit 1 (-120-130 cm)	17	1	-	-	4
Pit 1 (-90-120 cm)	18	3	-	-	-
Square 33 bottom:					
layer (105-125 cm)	10	4	-	-	-
Square 29:					
Ash pit (-90-120 cm)		4	37	-	-
Ash pit (-110-130 cm)		3	16	-	-
Square 33:					
Large pit (bottom)	57	33	3	-	-
Large pit (-150-200 cm)	53	4	8	-	-
Squares 37-38:					
Pit (bottom)	-	7	-	-	-
1986 season:	25	20	3	1	-
Total:	336	221	30	1	4
Percentage:	56.7	37.3	5.0	0.1	0.6

Table 2.

The number of individuals estimated in features

(number of fragments)

Features	Catfish	Carp	Pike	Pikeperch
General	21	9	5	-
Square 55 west: Pit 1	7	2	-	-
Square 33: bottom layer	3	1	-	-
Square 29: Ash pit bottom	2	5	-	-
Square 33: Large pit bottom	9	5	4	-
Squares 37-38: Pit bottom	-	2	-	-
1986 season:	7	5	2	1
Total: 90	49	29	11	1
Percentage:	54.0	32.2	12.2	0.02

Table 3.

The distribution of fish bone by body regions and species at Endrőd 119

Species	Catfish		Carp		Pike		Pikeparch		Total	
Bone	n	%	n	%	n	%	n	%	n	%
1. praeoperculum	1	8.3	11	91.6	-	-	-	-	12	2.0
2. operculum	2	8.3	22	91.6	-	-	-	-	24	4.0
3. suboperculum	-	-	2	100.0	-	-	-	-	2	0.3
4. maxillare	-	-	2	50.0	2	50.0	-	-	4	0.6
5. dentale	12	48.0	-	-	12	48.8	1	4.0	25	4.2
6. articulare	11	100.0	-	-	-	-	-	-	11	1.8
7. quadratum	9	81.8	2	18.1	-	-	-	-	11	1.8
8. sinplecticum	1	100.0	-	-	-	-	-	-	1	0.1
9. hyomandibulare	16	50.0	16	50.0	-	-	-	-	32	5.4
10. epihyale	3	100.0	-	-	-	-	-	-	3	0.5
11. ceratohyale	43	95.5	1	2.2	1	2.2	-	-	45	7.6
12. urohyale	-	-	1	100.0	-	-	-	-	1	0.1
13. branchialia	2	33.3	3	50.0	1	16.6	-	-	6	1.0
14. neurocranium	6	40.0	9	60.0	-	-	-	-	15	2.5
15. parasphenoideum	16	69.5	7	30.4	-	-	-	-	23	3.3
16. basioccipitale	12	50.0	12	50.0	-	-	-	-	24	4.0
17. pharyngeal tooth	-	-	15	100.0	-	-	-	-	15	2.5
18. cleithrum	90	64.7	35	29.1	14	10.0	-	-	139	23.4
19. coracoideum	4	100.0	-	-	-	-	-	-	4	0.6
20. basipterygium	-	-	4	100.0	-	-	-	-	4	0.6
21. I. radiale	61	93.8	4	6.1	-	-	-	-	65	10.1
22. vertebrae	49	51.5	46	48.4	-	-	-	-	95	10.1
23. costae	-	-	23	100.0	-	-	-	-	23	3.3
24. spina pinnae	-	-	8	100.0	-	-	-	-	8	1.3

Table 4.

The proportion of bone survival calculated from the estimated number of individuals

Species	Number of vertebrae per individual	Number of vertebrae found	Number of individuals estimated	Theoretical number of vertebrae
Catfish	71-75	49	49	3479-3675
Carp	36-37	46	29	1044-1073
Pike	61-64	-	11	671- 704
Total:		95	89	5194-5452

Of the three most characteristic fish species 95 vertebrae represent 89 individuals.[4]
Assuming a theoretical 100 percent recovery of all vertebrae from these individuals, between 5194 to 5452 vertebrae should have been found. This hypothetical situation is never encountered in reality. It is still worth considering, however, why only 1.7-1.8 percent of this ideal number was found.

This realitively small number of finds may be attributed to several factors.

1) Excavation loss:
Loss of artifacts may be significant when traditional methods of excavation are used. Using more advanced techniques, such as collection of micro finds wet by sieving (as was the case at the site of Endrőd 119), the number of fish bones shows a quantum leap. Thus, one may assume that in this case the deficit in fish bone numbers is not the result of recovery bias.

2) Effect of soil, corrosion:
As mentioned in the introductory paragraph of this paper, the fish bones from Endrőd 119 are well preserved. They are not porous and have a uniform color. Thus, it is unlikely that many of the missing bones would have been dissolved in the soil.

3) Prehistoric loss:
Since factors examined under the two previous headings did not significantly reduce the recovery of fish bone, the answer must be sought among the effects of prehistoric conditions.

[4] Pikeperch, represented by only one bone, was not taken into consideration in this comparison. Since the aim of calculations was no more than demonstrating the order of magnitude of such losses, only vertebrae were used.

First, the question must be raised whether the fish caught were brought to the settlement after primary butchering (that is the removal of the head, tail and other fins, intestines etc.) off site. The 24 types of skeletal elements observed in the Endrőd material represent all regions of the fishes' body, therefore, the animals must have been taken in one piece to the site. Any judgement about "valuable" and "non-valuable" body parts must be treated cautiously as well. In the case of smaller fish, the head may have no major alimentary value, but the head of large fish contains many edible parts. Among the finds, several articulated bones of a carp skull were identified. This also suggests that the reason behind the reduced number of fish bone finds must be sought within the settlement (*Fig. 4.*).

Dogs and pigs living at a prehistoric settlement (or sometimes even wild animals) may have played a very important role in cleaning up kitchen refuse scattered around such sites. This activity is unquestionably evidenced by marks of gnawing on larger bones.[5] Smaller bones, such as remains of birds and fish, however, can be completely annihilated by even a medium size dog. It is important to note, that none of the fish bones from Endrőd showed marks of dog gnawing, although dog bones were identified in the animal bone assemblage from this settlement.[6] It seems likely, however, that most of the fish bones missing were destroyed by animals with the exception of those pieces which were thrown into especially deep pits or immediately enterred, thus, preventing access by scavengers. This theory would simultaneously explain the relatively small number of fish bones recovered and the lack of gnawing marks on the pieces available for study.

The quantity of bones in the kitchen refuse may also be reduced by secondary use such as tool making. As far as is known today, such exploitation of fish bone did not exist or was only very sporadic during the Neolithic, so that it may be discounted as a significant effect reducing the number of fish remains recovered.[7]

Exploitation, fish processing

Fishing with nets is evidenced by a number of net weights in Neolithic artifactual assemblages. Methods of catching individual fish are well illustrated by the presence of harpoons and hooks from the same period.[8]

The dominance of catfish in the species inventory of Endrőd 119 (and other Neolithic settlements), however, may be indicative of another technique as well. Fish could be gathered in pits and other depressions on the floodplain once water withdrew from these areas following periodic floodings. Catfish may have been trapped in such residual waters in relatively greater numbers due to their behavioral characteristics. While carp, pike and pikeperch live in the middle and surface layers of water and move over longer distances, catfish prefer the deep water and often rest in depressions on the bottom. Thus, when seasonal flooding is over, carp, pike and pikepirch move with the receding body of water while part of the catfish population may be trapped in the floodplain's relief. Naturally, some of the other fish remain in these backwaters as well, but they may be relatively underrepresented.

---

5 S. Bökönyi, personal communication.

6 S. Bökönyi, *op. cit.* in footnote 2.

7 L. Bartosiewicz - I. Takács: Pre-depositional modification on fish bone from Hungarian excavations. *Proceedings of the the Fifth Meeting of the ICAZ Fish Working Group. Stora Kornö 1989* (in press).

8 N. Kalicz: *Agyag istenek* (Clay gods). Budapest 1970, 17-23.

None of the fish caught by the inhabitants of this settlement were particularly big.[9] The catfish ranged in length between70 and 120 cm. Carp remains originated from quite small individuals although a couple of bones came from larger fish. The remains of pike and pikeperch are also represented by medium size specimens.

Traces of fish processing are rare at the Endrőd settlement. Cutmarks or contemporary fragmentation could not be identified. Therefore, butchering or lengthwise splitting of the body for the purposes of preservation could not be studied at all. Evidence of fish food preparation is scarce. Only a few burnt bones, possibly indicative of grilling on open fire, were found. However, no porous bones were encountered which could have been identified as remains of cooked fish.

Summary

Among the zoological finds at the Early Neolithic site at Endrőd 119, 595 fish bones were identified. These included (in decreasing order) catfish, carp, pike and one pikeperch.

It is remarkable that, in comparison to the estimated number of individuals, only very few of the fish bones survived. This calculation was based on the comparison between the number of vertebrae in one individual of each species and the number of individuals estimated on the basis of the find material. When these two values were multiplied, the 95 vertebrae recovered represent only 1.7-1.8 percent of the theoretical number of vertebrae (5194-5452).

The reason for this loss should be sought within the prehistoric settlement itself since all parts of the body were represented in the bone assemblage. This shows that the fish caught were brought to the site in one piece.

It is likely that kitchen refuse discarded at the site was scavenged by dogs and pigs. The few bones that have survived must have been buried or swept into deep pits inaccessible to these animals.

Bones of catfish usually dominate among the ichthyological finds of Neolithic sites. A possible explanation is that, in addition to fishing with special equipment, fish naturally trapped in pits of the floodplain could be regularly gathered following flooding. Characteristic behavioral traits influence the species distribution of fish trapped in such depressions (deep water is the preferred habitat of catfish).

Of the fish identified at Endrőd 119 were of average size and remains of especially large individuals were not found.

There is only sporadic evidence of fish processing in the form to modifications of the bone material.

---

9 I. Takács: Segédlet szubfosszilis harcsák testnagyságának kiszámításához a fő vázalkotók méretei alapján (An aid to the body size calculation of subfossil catfish using measurements of the main skeletal elements). *Magyar Mezőgazdasági Múzeum Közleményei* 1986-87, 105-126.

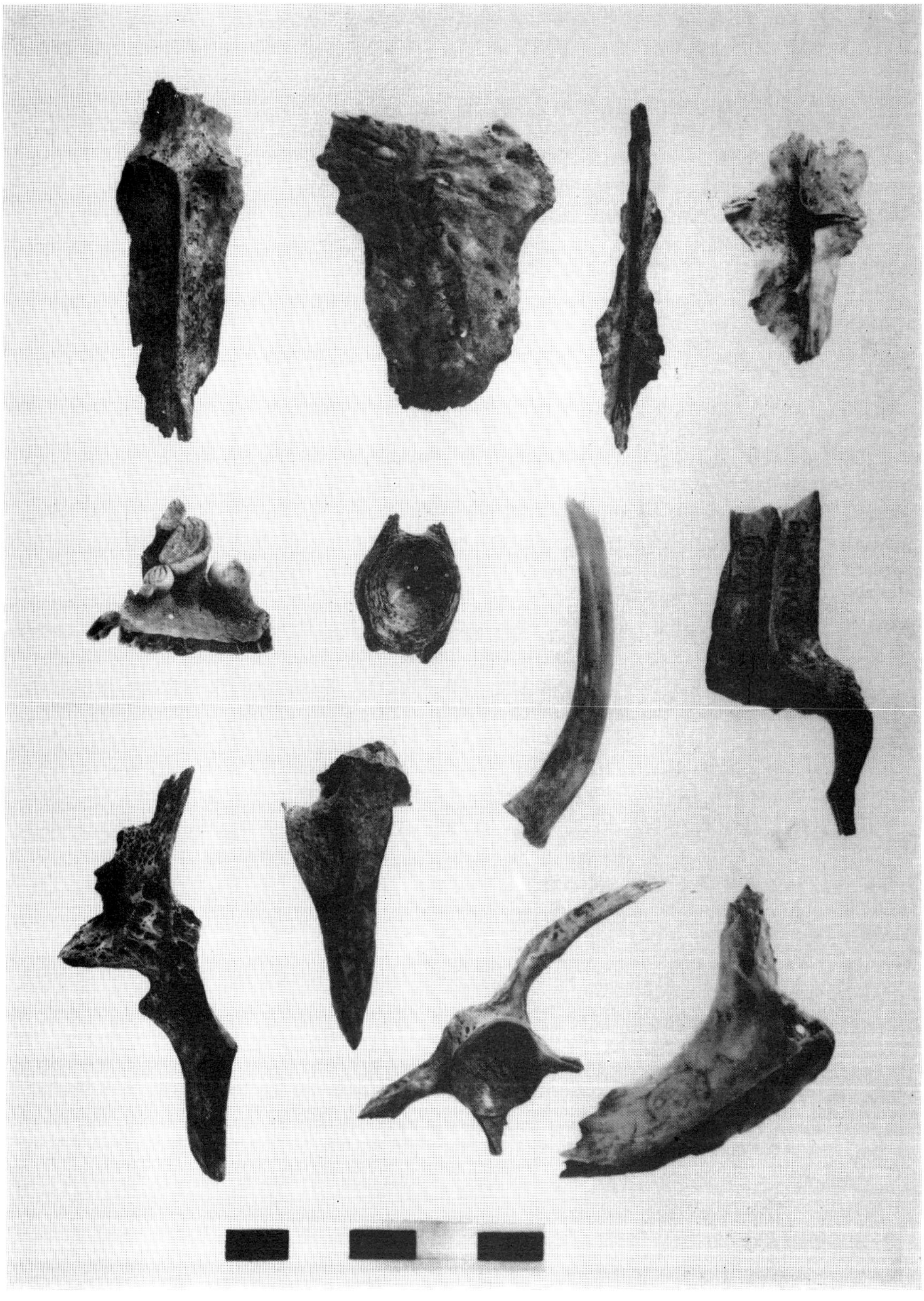

Fig. 1. Typical carp finds from Endrőd 119.

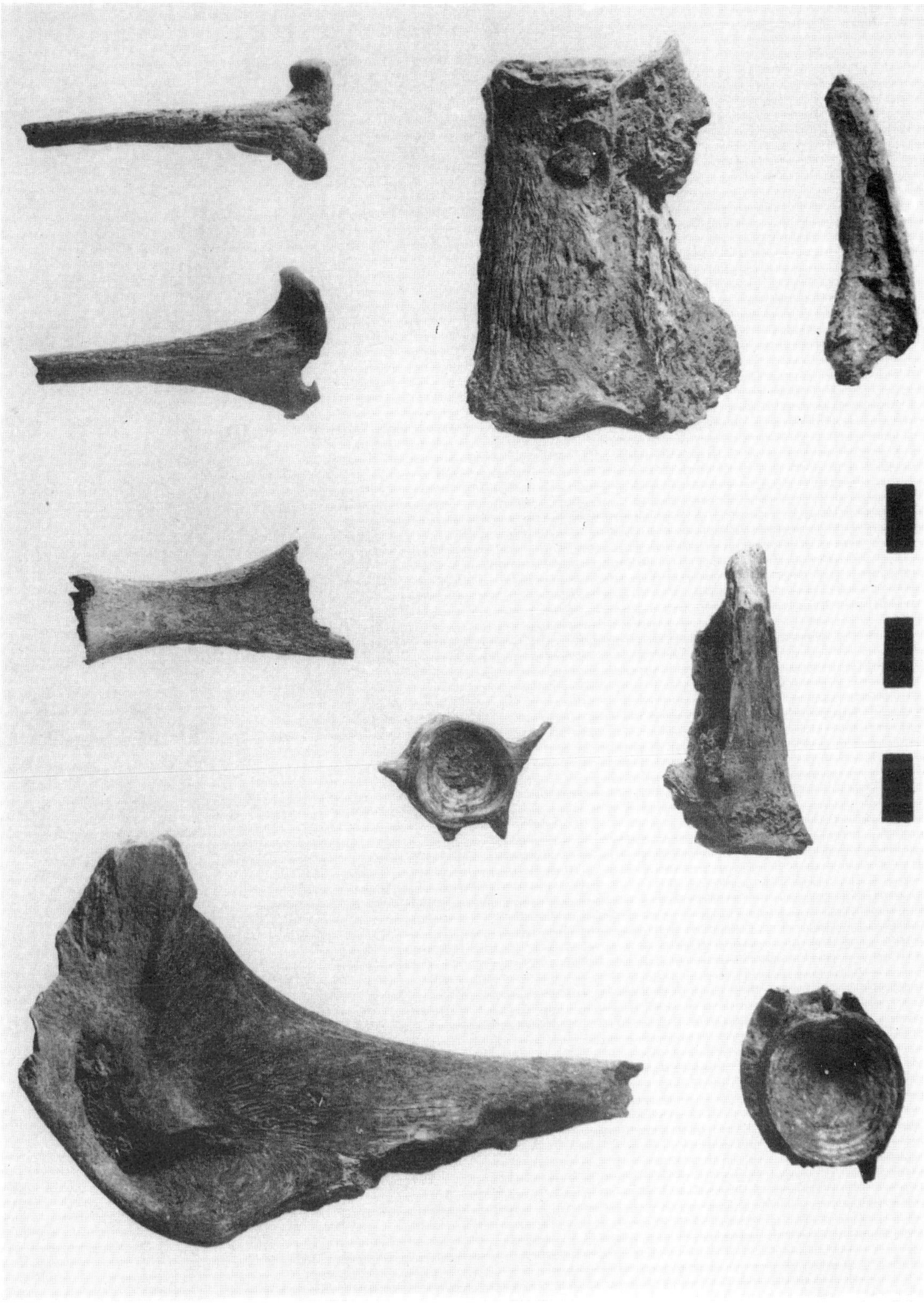

Fig. 2. Typical catfish finds from Endrőd 119.

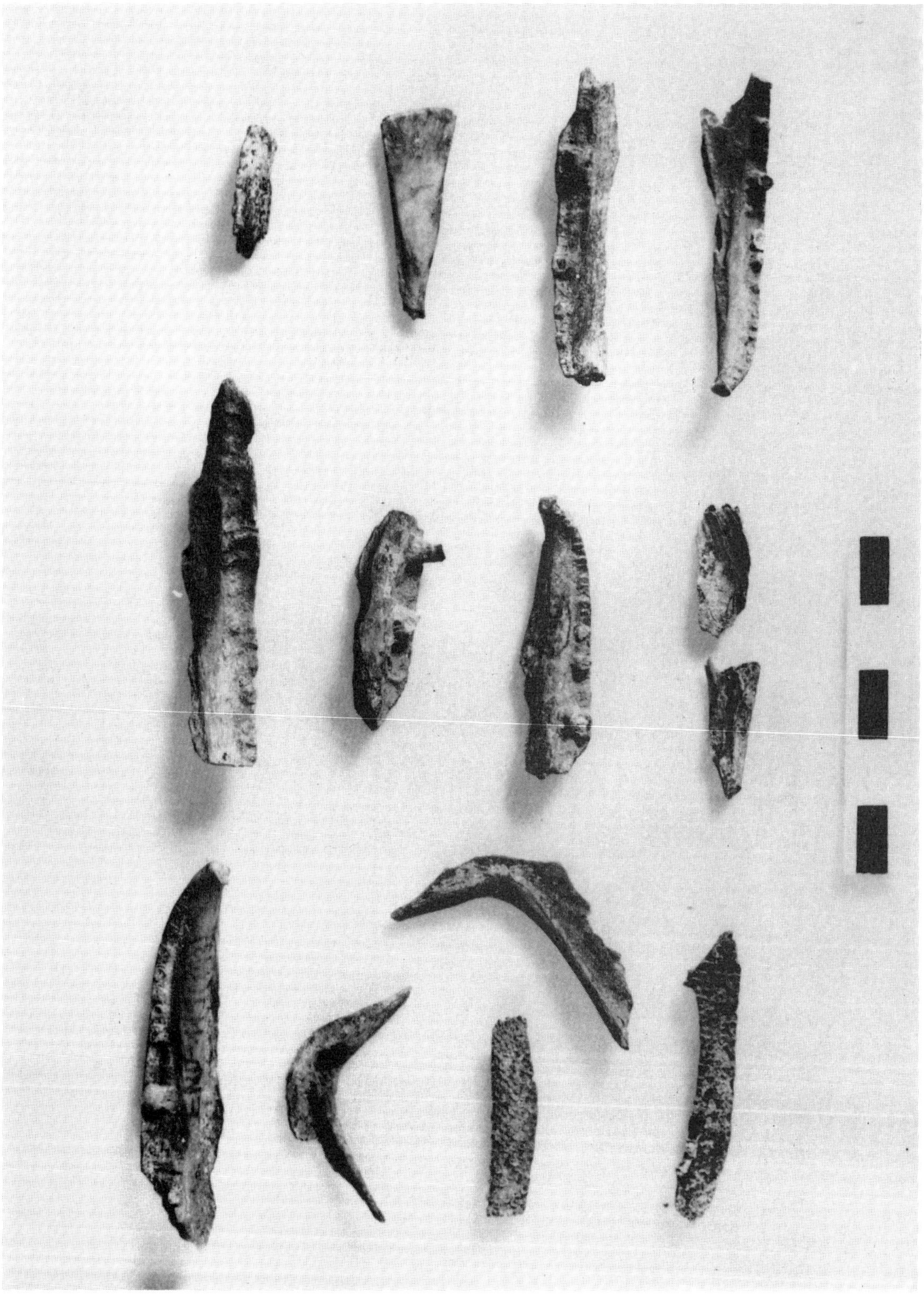

Fig. 3. Typical pike finds from Endrőd 119.

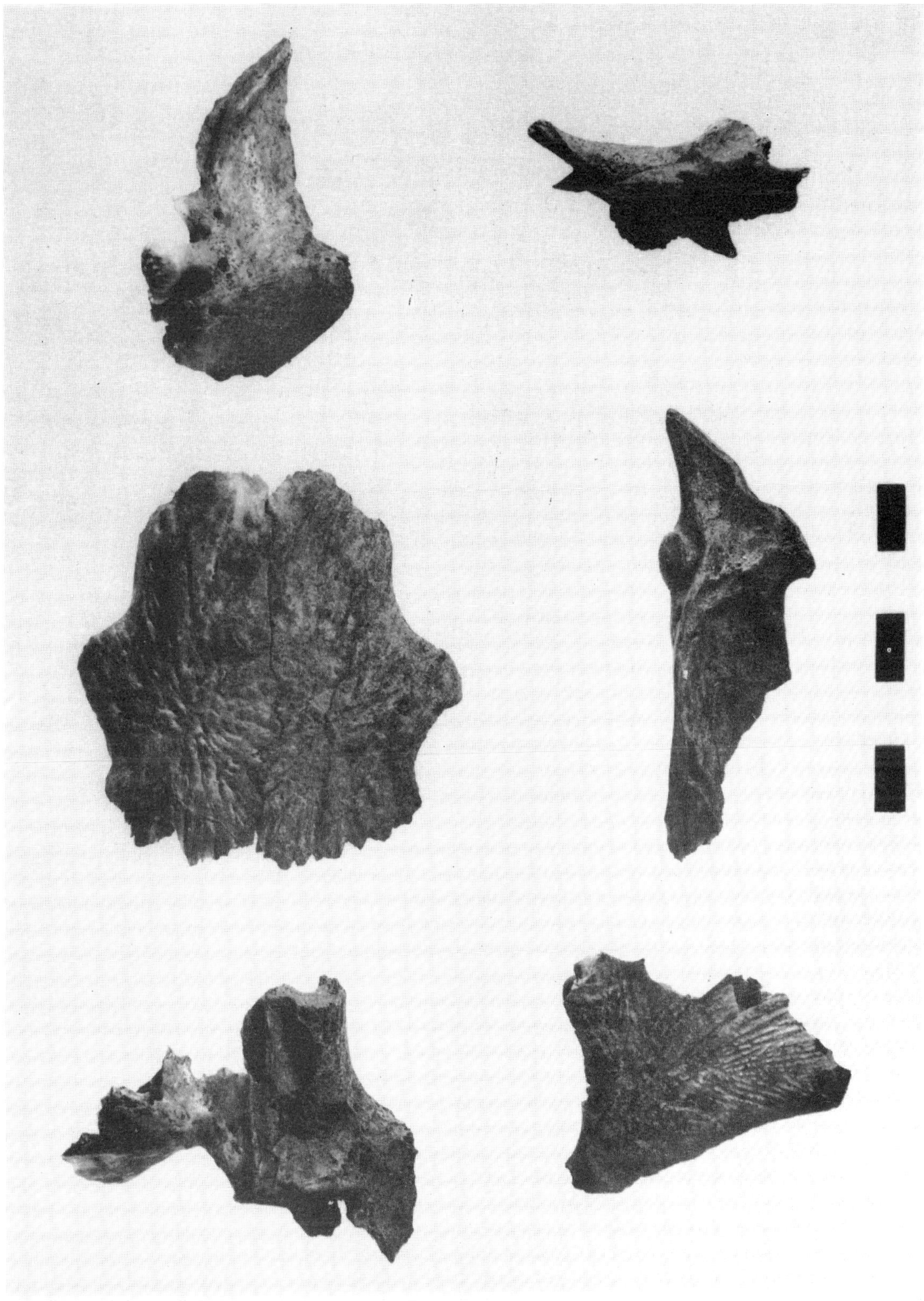

Fig. 4. Cranial bones of a large carp.

A Linear Pottery culture refuse pit and two contracted burials from the site of Endrőd - Öregszőlők XI.

(The Archaeological Topography of Hungary. [Vol 8. Békés county] IV.2. Site 3/42)

JÁNOS MAKKAY

The small settlement probably consisting of only a few features (refuse and dwelling pits) was first noted at this location during field surveys of the Hungarian Archaeological Topography project in 1974. This was possible partly because the area was not covered by modern buildings or disturbed by other recent features. On the other hand, surface collection resulted in a rich material exclusively showing the characteristics of the Alföld Linear Pottery culture (ALP). The good quality of these finds is indicated by the high proportion of decorated sherds.[1] Although ALP finds were brought to light at a variety of sites in the outskirts of Endrőd between 1975 and 1979 during excavations associated with the topographic surveys (such as the sites of Endrőd 3/6, 3/35 and 3/158[2]), the site under discussion here was chosen to study the ALP culture within the Microregion Program of the Archaeological Institute of the Hungarian Academy of Sciences. This small scale excavation took place between September 22 and 25, 1985.

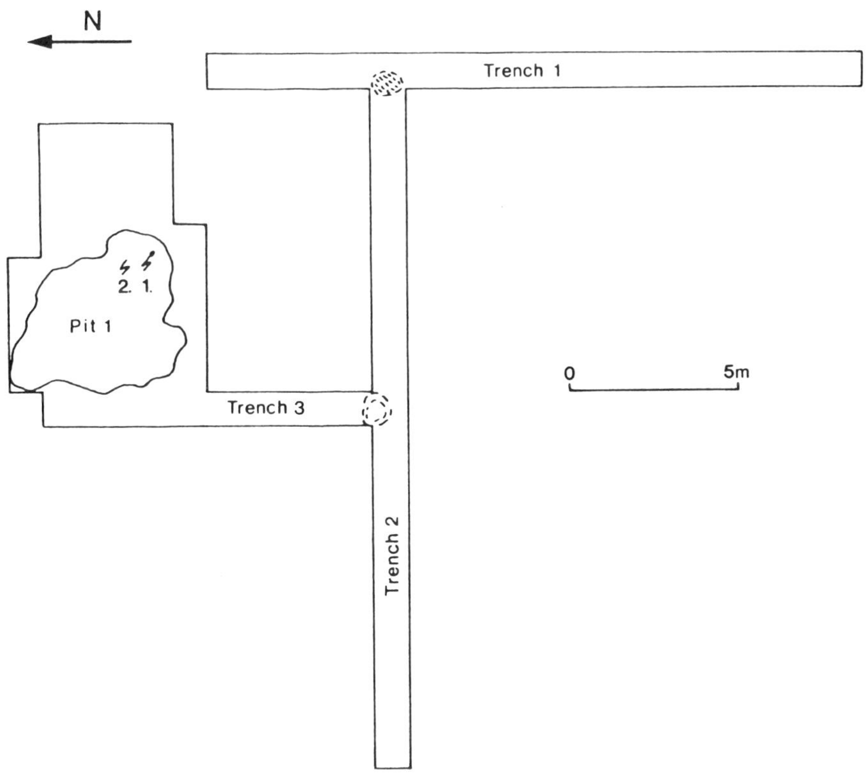

Fig. 1. Excavated areas of site Endrőd 3/42

[1] *MRT 8: Békés megye régészeti topográfiája.* A szarvasi járás. (The Archaeological Topography of Békés county. The Szarvas district. IV/2). Ed. by J. Makkay. Budapest, 1989, 147.

[2] Ibid., Sites 3.6, 3.7, 3.158.

Trench 1 was opened at the spot where the greatest number of sherds was located on the surface during previous surveys. Almost nothing was discovered here down to the bottom soil at 60 cm, except for a few ALP sherds between 30 and 60 cm. At this point, Trench 2 was started (*Fig. 1*). This attempt was similarly unsuccessful. The refuse pit sought after was found in Trench 3. The edge of this pit's dark soil spot was identified in this trench. Square I opened at this point revealed a relatively small and shallow refuse pit (*Fig. 2,1*). The yellow sterile soil was usually found at a depth of 60 cm and this is the only place where the pit's outline could be precisely observed. As is usual with the pits of this period, finds dominated by sherds are spread over a greater area above sterile soil and ancient humus covering it than in the sterile soil itself. This shows that these pits were larger in the upper strata than their lower portion dug into the sterile soil. This phenomenon was observed with the pit discussed here as well. The finds were particularly densely deposited near the pit's edge at the intersection of Trench 3 and Square I. Here a flat oval bowl with a handle was found at a depth of approximately 50 cm. This artifact was filled with mussel shells, both open and closed (*Plate 2,5*).

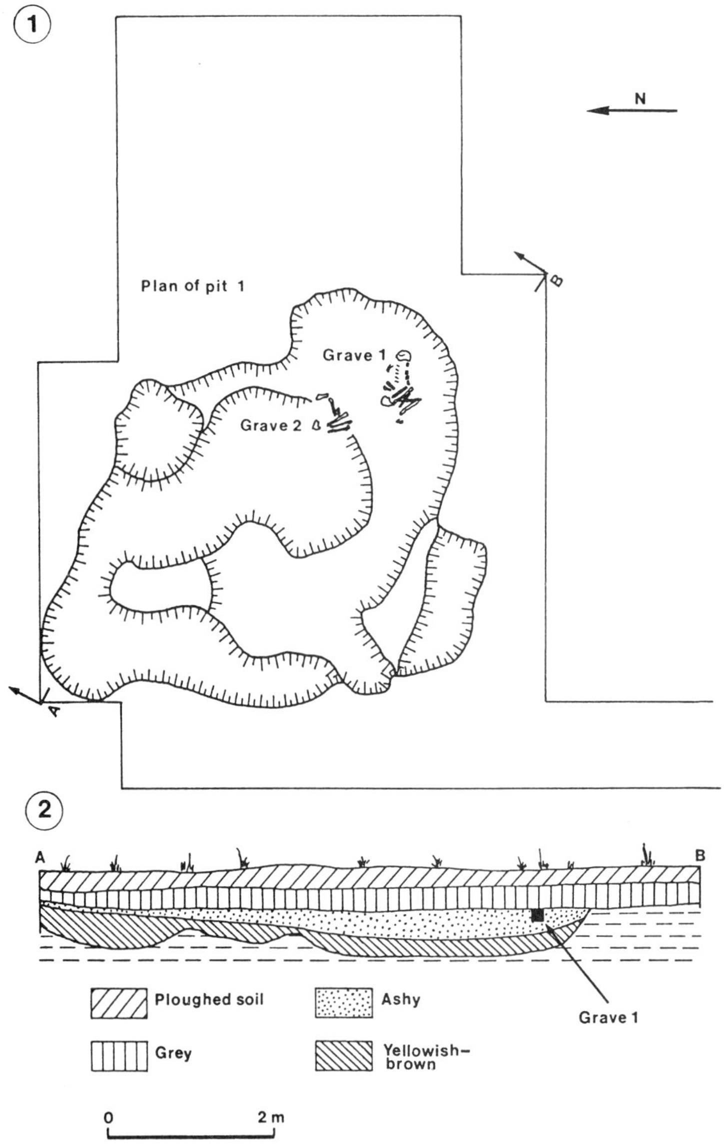

Fig. 2. Plan and section of pit 1.

Square I revealed generous amounts of fragmented Neolithic pottery by a depth of 30 cm, that is, the plough zone. This shows that land cultivation had reached and disturbed the uppermost part of the pit's fill. Below a depth of 30 cm, the material was in primary position. There was a loose, ashy layer between 25/30 and 50/55 cm which contained a plethora of pot sherds. Under this fill came an almost pure ashy layer which was thickest in the pit's southeastern corner, thinning towards the opposite direction. Below this ashy fill a dense, yellowish-brown clay and silt layer was recognized with only a few finds as is usual with ALP refuse pits (*Fig. 2,2*). Actually, the pit is quite shallow, no more than 1-1.1 m deep. It is remarkable that a black, almost tar-like hard fill, that is very typical of ALP refuse and dwelling pits was missing here.[3] A possible explanation for this phenomenon is that the function of this shallow pit was different from those with black fill which – at least in the Körös river valley – evidently served as pit houses.

Two contracted inhumations (Graves 1 and 2 in *Figs. 2, 1-2* and *Fig. 3*) were discovered in the southeastern part of this pit exactly on top of the ashy layer. Both skeletons lay on their left side in a contracted position. The skull, vertebral column and left humerus are missing from Grave 2. This may be explained by disturbances caused by modern activity (a few years ago buildings of a farmstead were still standing some 25 m southeast of the excavated area). The right humerus is missing from Grave 1 as well, which may be similarly the result of recent activities. Although spots indicative of these burials could not be observed either in the upper top layer or in the ashy deposit, the two bodies were probably enterred at the same time and the refuse pit was then abandoned. No grave goods were found in association with these two skeletons and neither could the use of ochre be observed. Their orientation was almost exactly east-west. At the time of excavation, the bones in Grave 1 were especially well preserved.

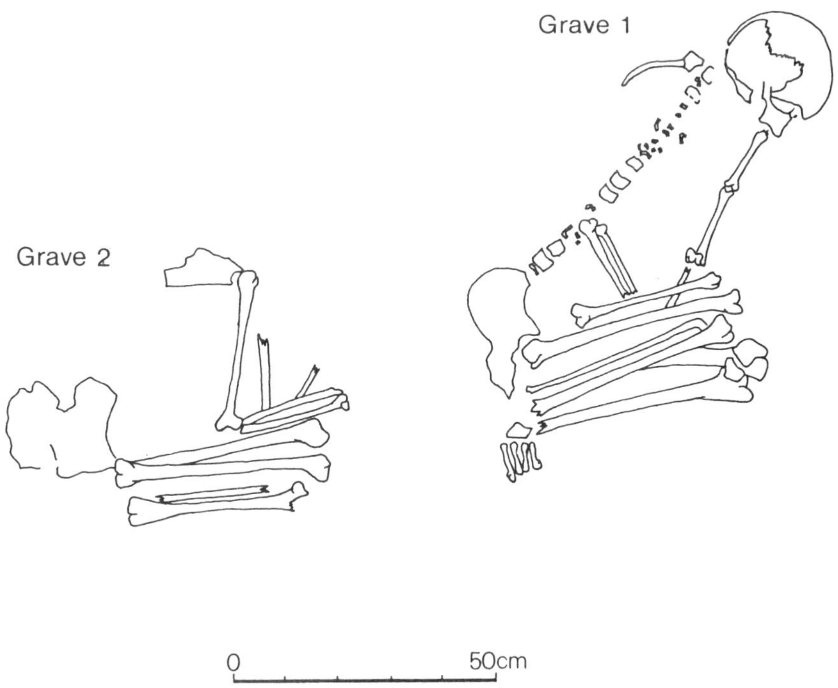

Fig. 3. Graves 1 and 2 from pit 1.

[3] J. Makkay: Some comments on the settlement patterns of the Alföld Linear Pottery. In: *Siedlungen der Kultur mit Linearkeramik in Europa.* Internationales Kolloquium Nové Vozokany, 17-20. November 1981. Nitra 1982, 160-161.

Eighteen pots could be restored from the pit's rich ceramic material (*Plates 1-3*). These plates show above all the fine decorated ceramics. This group is, in general, characterized by good execution: fine temper, a good polish, a surface covered by a brown slip and thin walls. The incised line decoration is dominated by a single line pattern. It may be composed of straight lines, meanders, wavy or even spiral patterns. The base of pedestalled bowls is medium high while the bowl part itself is not particularly wide (*Plate 1,1-3, 5-6, Plate 4,1-21, Plate 7,4-22, Plate 8,1-16*). Another typical form is a high cup decorated with a pattern divided into two or four sections (*Plate 2,2-3, Plate 6,1-14*). It often also contains an arrow pattern (*Plate 6,7,9,15*). There are smaller, undecorated cups as well with a pronounced belly line (*Plate 2,7*) or even with handles on the belly's edge (*Plate 2,6*). Simple bowls of an upside down truncated cone shape are usually grayish with a slightly arched wall (*Plate 3,3-5*). The broad bellied bottle with a slightly arched neck and elaborate surface is an especially unusual form (*Plate 2,1*) although it has many variants in this find material (*Plate 5,2, 5, 8, 12, 13*). The common tall bottle form with a cylindrical neck is represented mostly by miniature vessels, one of which could even be completed (*Plate 3,2*). Other miniature vessels belong to the bowl type with a truncated cone form.

Among the thick walled coarse forms occur bottles with a tall neck (*Plate 9,5*) and bomb shape (*Plate 16,6, 7, 10*), a wide bowl (*Plate 9,4*) and another special form. This latter is characterized by a slightly broadening rim which is smoothed flat on top (*Plate 10,3*). These forms are usually decorated by a shallow but broad depression running obliquely on the outside. The surface is often made rough (*Plate 9,4,7, Plate 11,4,7*), by brushing or combing (*Plate 10,7, Plate 12,16*). Some linear patterns are sharply incised (*Plate 12,5, 7, 9, 13, 15*) or are reminescent to the classical Linearband motifs (*Plate 12,17*). Plastic design is rare (*Plate 9,5, Plate 12,1, 4, Plate 13,6*) as well as pinched decoration. This latter may only be sporadically observed in the material. Only a few pieces show a technique that was customary with the rough types of ALP ceramics: some pots already in use, were thickly plastered with a mixture of chaff and clay which sometimes was even burnt through (*Plate 13,12*). This method, which often covers the patterns as well, is only characteristic of the ALP of all Neolithic cultures present in Hungary.[4] It may be observed on numerous vessels and sherds recovered from the rich artifactual assemblage of a house at the site of Szarvas 102 (unpublished finds). According to ethnographic parallels (for example ceramics used in the village of Csákvár at the end of the last and at the beginning of this century) this plastering was necessary in the case of large but relatively thin walled vessels which were used inside ovens.[5]

A semi-spherical cup with a flat base and thin wall is also noteworthy (*Plate 1,4*). Its unpolished surface without a slip is decorated by three incised spiral motifs running around the cup parallel to each other. Although this line-technique is typical of the ALP, the line itself is not. Parallels may only be found in the Transdanubian (Central European) LP. In that case, however, this motif is naturally combined with patterns typical of the early note decorated ('Notenkopf') or the following Zseliz culture.[6] Three other fragments were found in the pit's assemblage which did not fit into the ALP material on the basis of their decoration. One fragment originates from a vessel with a spherical belly which was black on the outside with a

---

[4] N. Kalicz and J. Makkay: *Die Linienbandkeramik in der Grossen Ungarischen Tiefebene*. StudArch 7. Budapest, 1977, Plate 50,2.

[5] Author's observation in 1968.

[6] J. Pavúk and S. Šiška: Neolitické a eneolitické osidlenie Slovenska. *SlA* 19:2 (1971), 21-22, Figure 3; H. Quitta: Zur Frage der ältesten Bandkeramik in Mitteleuropa. *PZ* 38:3-4 (1960), Abb. 4, a, g, m etc.

delicate burnish. The fragment has a small knob on it. Its whole surface is decorated by dense, fine, shallow parallel lines (incised and smoothed) which form interrelated but non parallel sections (*Plate 15,9*; the decoration is not shown on this plate). The other piece has a fine, shiny black polish on the inside and is decorated by patterns similar to the one described previously. The only difference is that this surface has undecorated bands as well, while the lines are fine and delicately incised (*Plate 15,10*). Although both specimens resemble Bükk culture sherds made using a similar technique, it is unlikely that we are dealing with an import originating from that area. Finally, a third piece (*Plate 15,13*) is smoothed on the outside and light brown in color. The two parallel incised lines on its surface are broken at a sharp angle. Although similar sherds (as well as the previously described type) occur in the early Zseliz culture as well[7], it may be suggested here that this vessel is coeval with finds from the later period of the Notenkopf style. Answering this question would be greatly facilitated by the few finds which were brought to light in Squares I and III at the site No. 35 of Endrőd. At that site, a small ALP pit was dug into the western edge of a major Körös culture pit (Square III, Pit 3). Both the western part of Pit 3 and the ALP pit were covered by a thin layer of Szakálhát ceramics at a depth between 30 to 60 cm. Grave 1 of the Szakálhát period was deepened from the top or the uppermost layer of this deposit. The burial was in part, located above the Körös culture pit and partly overlapped the ALP pit as well (its vessel is shown in *Plate 16,8*). Due to this complicated configuration it is not easy to take a stand concerning the three sherds that obviously represented the Transdanubian Linear Band Ceramic in Squares I and III of Endrőd, site 35, and in Endrőd, site 6 and at the Endrőd 45 site (Square I) respectively. The sherd shown in *Plate 16,6* is evidently one of the earlier such specimens and still belongs to the Notenkopf stylistic period. It was clearly found in the ALP layer (this corresponds to a 60 to 85 cm depth in the eastern part of Square I) in association with ALP sherds which were completely similar to sherds recovered from the site of Endrőd 42. The other such piece (*Plate 16,5*) was found at the same spot, however, at a depth of between 30 to 60 cm. Finally, the third sherd (*Plate 16,4*) was found in the western part of Square III. It definitely displays Zseliz characteristics and it very likely belongs to the Szakálhát layer, in spite of the depth (60 to 80 cm) it was recovered from. The sherd could have moved down this deep, for example, when Grave 1 was dug. The Zseliz type decoration of this small sherd was not noticeable at the time of excavation since it was covered by a synter layer. A further piece, an early Zseliz type sherd (*Plate 16,3*) was discovered at the site of Endrőd 6 in Square II at a five spade depth. That part of the site was characterized by late ALP finds corresponding to the synchronous layer of Endrőd 42 (excavation and communication by Dénes Jankovich).

A quantity of very late ALP sherds with shiny black polish (especially on the inside) were found at the site of Endrőd 36. This location is only at about a 100 m distance from the site of Endrőd 35. The Endrőd 42 site is located to about 900 m from the area under discussion.[8] It is quite remarkable that the fine, shiny black polish and the broad band pattern with so called red 'crusted' painting are common in several assemblages from the Endrőd 36 site. At the same time, ALP and the early Szakálhát incised pattern are both missing, although a few imported (or in this case borrowed) pieces occur. This situation shows that the microchronology of the closely located ALP sites under discussion here may only be clarified by excavations carried out at a number of sites.

---

[7] J. Pavúk: Chronologie der Želiezovce-Gruppe. *SlA* 17:2 (1969), Abb. 12,7-8, as well as 10.

[8] *MRT 8* (see footnote 1), the map of Endrőd.

Some of the painted sherds from Endrőd 42 display an inventory of typical ALP patterns: incised lines are accompanied on both sides by dark (black or brown) painted bands (*Plate 14,6–7*). Others, which dominate at this site, are associated with the Esztár type painted ware. These pieces may be covered by red or even by purple, finely polished slip. They are often worn and no longer have their original dark painted bands (*Plate 14,13-23*). Sherds in another group are covered by a light brown slip with dark bands or other designs (*Plate 14,8-10*). Finally, sherds with well-polished purple surfaces are worth mentioning here (*Plate 14,11-12*). The sherds are usually fragments of small spherical cups, while only two pieces from a major vessel were encountered: they belong to the lower part of a bomb-shaped pot with a flat bottom. The side of this specimen is decorated by bands grouped in a "V" shape which start at the bottom and are broaden upwardly. The motif was divided into four segments on the vessel's side (*Plate 15,16*).

Special attention should be paid to two fragments which belong to the same vessel with an unknown shape but a roundish body. This specimen was finely made with both the inner and outer surface covered by a light yellowish brown slip and a medium polish. It was decorated with 2 to 3 cm broad bands composed of incised lines typical of the ALP style. These bands were arched and their inner side as well as the contact areas may have been decorated by arrowhead patterns or some variations of this. The areas inside the bands were covered by a red crusted paint following firing, while the other, intermediate areas between the bands were left unpainted. They retain the original color and polish of the surface. A broad red band painted similarly after firing may be seen on the inside of both pieces (*Plate 15,11-12*). There is no doubt that the immediate precursor of the most important Szakálhát stylistic element may be recognized on these two sherds: an alternation between crusted painting and unpainted bands on the incised linear pattern within the bands and outside the bands respectively. In more developed Szakálhát phases a finely polished version of the unpainted bands occurs as well. This technique may first be observed in the Furugy type which should be considered the latest developmental phase of the ALP. It also occurs in the ALP- Szakálhát transitional period as evidenced by the pedestalled bowl brought to light at the site of Békésszentandrás - Furugy itself.[9]

Among the other vessel forms, the core of a pedestal is worthy of mention. Its wall was decorated by four, symmetrically located vertical oblong holes which had been cut through the pedestal (*Plate 14,5*). An unusually stout, well divided pedestal ring with a concave bottom and a line design running around its neck is similarly interesting (*Plate 15,14*). The neck part of a bottle-shaped vessel had a single spout or a sieve-like spout with numerous holes while a handle broke off of the line decoration which followed the contour of the rim (*Plate 14,1*). Finally, the rim fragment of an upside down truncated cone shape strainer vessel deserves attention (*Plate 15,8*). A wagon-axle form bead (*Plate 15,15*) and a fragment of a small vessel with an asymmetrically arched belly were also found (*Plate 15,1*).

Stemmed handles resembling schematic representations of human (or animal?) heads (*Plate 15,5-7*) may have been positioned on the shoulder of thick-walled vessels or the rim of cups. Two such handles, which were more probably attached to a neck section rather than to a vessel's shoulder or belly, really show human faces. Both of them were located on the vessel's wall in a slightly oblique position (assuming that the upper plane of their stem was horizontal). One of the faces has a deltoid shape and is entirely smooth, decorated only by four narrow engraved lines indicating the eyes, mouth and the medial division chin (*Plate 15,3*). The other face, more exactly possible head, has a triangular shape with one point directed downwards and slightly

---

9 Ibid., Plate 5,3.

rounded sides. The obvious elements of the face (mouth, eyes, nostrils and beard) are accompanied by the well known symbolic frontal motif. This is a strange pattern starting at the left eye and ending beyond the right eye with a double "V" form. It is accompanied by two notches above the root of the nose as well as a perforation in the upper left section of the right forehead. The pattern is completed by another motif above the left eye (*Plate 15,2*).[10] As far as we know, this is the first such representation found south of the Körös river. This extremely specific frontal decoration occurs beginning with the earliest, so-called Szatmár II phase of the ALP[11], and its somewhat distorted varieties survived until the late phase of the ALP-Szakálhát developmental sequence, that is, the Tisza Period.[12] Consequently, it may be regarded as a definitive ethnic trait of the ALP culture block, although we do not have the faintest idea what this pattern symbolized or meant.

In summary, the Endrőd 42 site belongs to the kind of ALP settlements which were made up of a few scattered dwellings (pit houses, as far as is known today). It is not possible to tell, however, if these were contemporaneous or followed each other diachronically. It is even difficult to tell if such units followed each other in an uninterrupted sequence within a site (were they built by the same small community?). If interruptions occurred, the question is whether the original population or someone else re-occupied the same settlement. This problem could be solved only by the excavation of several pit dwelling and refuse pits at at least two or three sites located in each other's proximity. At the site under discussion here no house or pit dwelling was found as we expected on the basis of the ceramic occurrences on the surface. From a chronological point of view this assemblage is the possibly most "regular" since the ALP type ceramics are accompanied by Esztár type painted ware (these latter cannot be further sub-divided), the very early occurrence of the Furugy type and pieces which may either be regarded as Transdanubian LP import specimens or their copies. From this point of view, our chronological table published in 1984 (in which Zseliz ceramics were treated as parallel with the classical phase of the Szakálhát period) does not have to be revised. This classical phase represents a time period when ALP characteristics or sherds with ALP decoration are missing as well as the Furugy types).[13] One may not accept, however, an alternative dating which implies that the Notenkopf decorated ceramics was not only fully coeval with the Szakálhát phase, but its late (so-called Zseliz-Notenkopf) phase even survived this latter culture.[14]

---

[10] N. Kalicz and J. Makkay: Gefässe mit Gesichtsdarstellungen in der Linienbandkeramik aus Ungarn. In: *Idole, Prähistorische Keramiken aus Ungarn.* Wien, 1972, 9-15, with figure as well as N. Kalicz and J. Makkay, *op. cit.* (see footnote 4), 61-64, with map and figure.

[11] Tiszavalk - Négyesi határ: N. Kalicz and J. Makkay, *op. cit.* (see footnote 4), Site 448, 170 and most recently P. Raczky: Chronological framework of the Early and Middle Neolithic in the Tisza Region. In: *Neolithic of Southeastern Europe and its Near Eastern Connections* = VAH 2, 1989, Figure 5,1.

[12] J. Banner and J. Korek: Negyedik és ötödik ásatás a Hódmezővásárhelyi Kökénydombon (The fourth and fifth excavations at Hódmezővásárhely - Kökénydomb). *ArchÉrt* 76:1-2 (1949), Plate 15,2 = H. Quitta *op. cit.* (see footnote 6), Abb. 10, 9 and most recently *The Late Neolithic of the Tisza Region.* L. Tálas (general ed.), Szolnok-Budapest, 1987, Figure 9 on 19 showing Hódmezővásárhely – Kökénydomb.

[13] J. Makkay: Chronological links between Neolithic cultures of Thessaly and the Middle Danube Region. *ActaArchHung* 36 (1984), chronological table following 28.

[14] N. Kalicz and P. Raczky: *The Late Neolithic of the Tisza Region.* (see footnote 13), chronological table on p. 30. and also *Les agriculteurs de la Grande Plaine Hongroise.* Dijon, 1991, 32.

The catalogue of artifacts
(all from Endrőd 42 with exceptions indicated)

Plate 1: 1-3, 5-6: decorated pedestalled vessels; 4: bomb-shaped vessel with incised spiral decoration

Plate 2: 1: bottle; 2-3: decorated cups; 4, 6-7: undecorated cups, 5: flat bowl

Plate 3: 1-2: miniature vessels; 3-5: wide undecorated bowls of an upside-down truncated cone shape

Plate 4: decorated (except for 7) fragments of pedestalled vessels

Plate 5: fragments of decorated bottles (1-2, 5-6, 8, 12-13) and cups

Plate 6: fragments of cups, bottles, a pedestalled bowl (3) and of a large, undecorated vessel (4)

Plate 7: fragments of pedestalled bowls (4-22), bottle or jug (1) and cups (2-3)

Plate 8: fragments of decorated pedestalled bowls

Plate 9: fragments of large, usually undecorated vessels

Plate 10: fragments of large, usually undecorated vessels

Plate 11: fragments of large, usually undecorated vessels with oblique grooving

Plate 12: fragments of large vessels decorated with incised and engraved lines, pinching and surface coarsening

Plate 13: fragments of undecorated, cup-shaped vessels with occasional ribs (6) and linear pattern (10) and fired clay plastering (12)

Plate 14: 1-3: rim fragments of bottle-shaped vessels, one of them with the attachment point of a handle that broke off (1); 4: rim of a cup with handle; 5: core of a pedestalled vessel with angular holes cut into it at four places; 6-7: rim fragments of bottle- or cup-shaped vessels with remains of dark paint along the decorative linear pattern; 8-10: dark painting on light slip cover (or on a polished clay surface); 11-12: dark (originally black) bands painted on a purple base of the Esztár type; 13-23: fragments of Esztár type fine ware with a purple cover; 24: remains of painted band pattern on an eroded purple surface

Plate 15: 1: small bottle with an asymmetrically broadening belly; 2: human head showing the forehead pattern. Patterns formed of four zig-zag lines incised on the top and the neck indicate hair (?); 3: handle with the representation of a human face; 4: a string of limestone beads of the ALP found as a grave offering in Grave 1 at the site of Endrőd 158 in 1985; 5-7: undecorated handles of vessels; 8: the lower rim fragment of a strainer dish; 9: spherical vessel with an eroded black cover with an incised and smoothed linear pattern forming bands and sectors on the outside; 10: fragment of polished black fine ware showing high polish decorated with incised-smoothed parallel lines on the outside (bottom) as well as the inside; 11-12: fragments of the same vessel with crusted red paint on the outside among the bands of the ALP line pattern and dark red crusted bands painted inside; 13: fragment with incised pattern characteristic of Transdanubian Linear pottery; 14: fragment of a vessel with a very low pedestal ring; 15: tiny clay ball, bead; 16: bottom part of a large, bomb-shaped vessel with an Esztár type band pattern painted on it (the surface is eroded).

Plate 16: 1-2, 4-11: Endrőd Site 35: 1-2: vessel fragments with Esztár type painting from the ALP pit of Square III west at the depths of 30-60 cm (1) and 60-80 cm (2) respectively. Inventory numbers: BMMM 78.244.78 and BMMM 78.245.18. 4: Square III west, 60-80 cm, inventory number: BMMM 78.245.19; 5-6: Square I east, 60-85 cm (6) and 30-60 cm (5) respectively. Inventory numbers: BMMM 78.231.10 and BMMM 78.232.12; 7: Square III west, ALP pit 60-80 cm: fragment with incised pattern on the outside and dark, crusted painting on the inside; 9-11: Square III west, ALP pit 100-120 cm, inventory number: BMMM 78.3.26 (9); 90-110 cm, inventory number BMMM 78.3.27 (10); 100-120 cm, inventory number: BMMM 78.3.25 (11); 8: Square III west, grave good from Grave 1. Height: 8 cm, undecorated cup, Szakálhát group. Inventory number: BMMM 78.3.28; 3: Endrőd Site 6, 1975, Square II, five spade lengths depth accompanied by ALP artifacts. Excavation by Dénes Jankovich.

Plate 1.

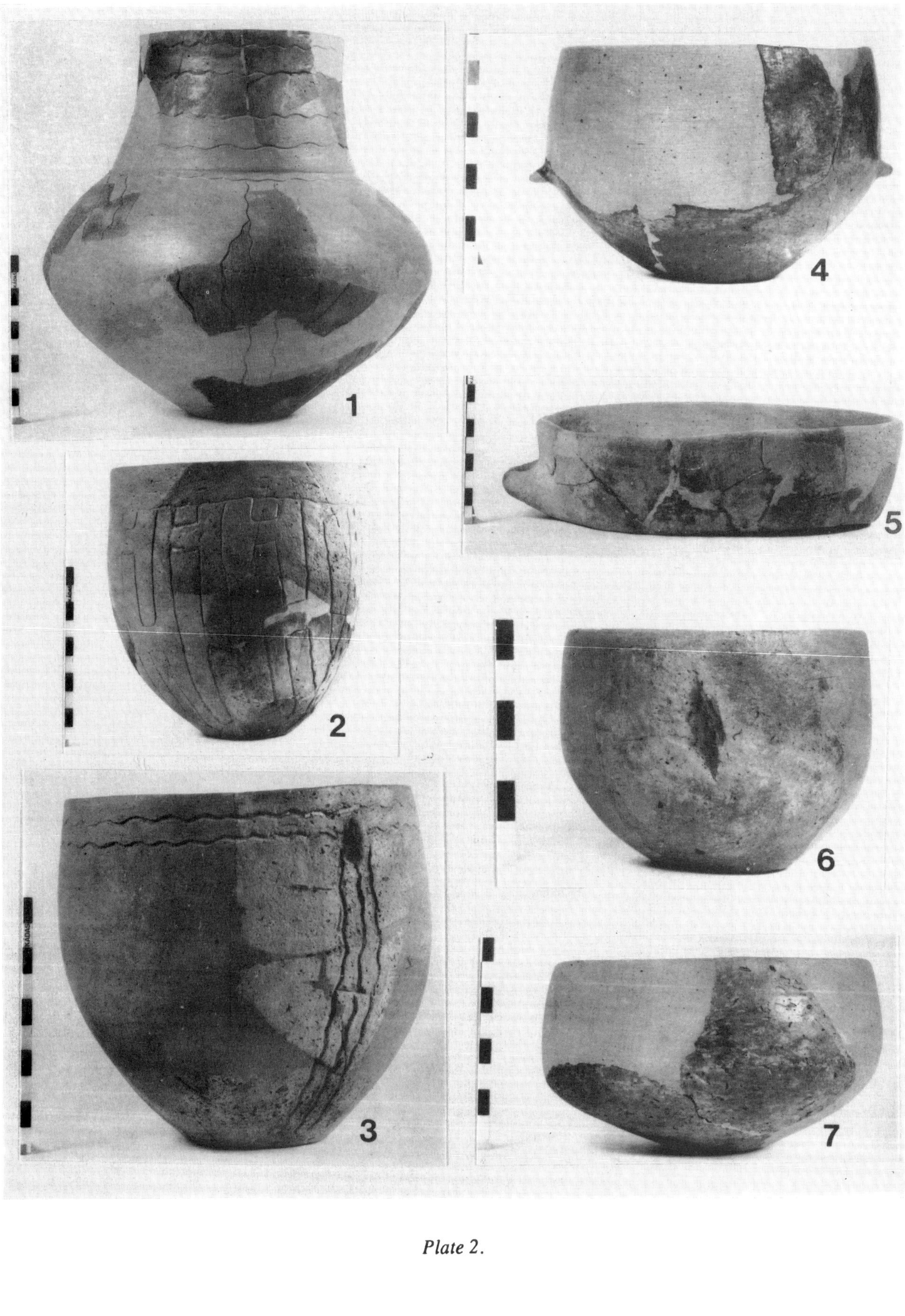

Plate 2.

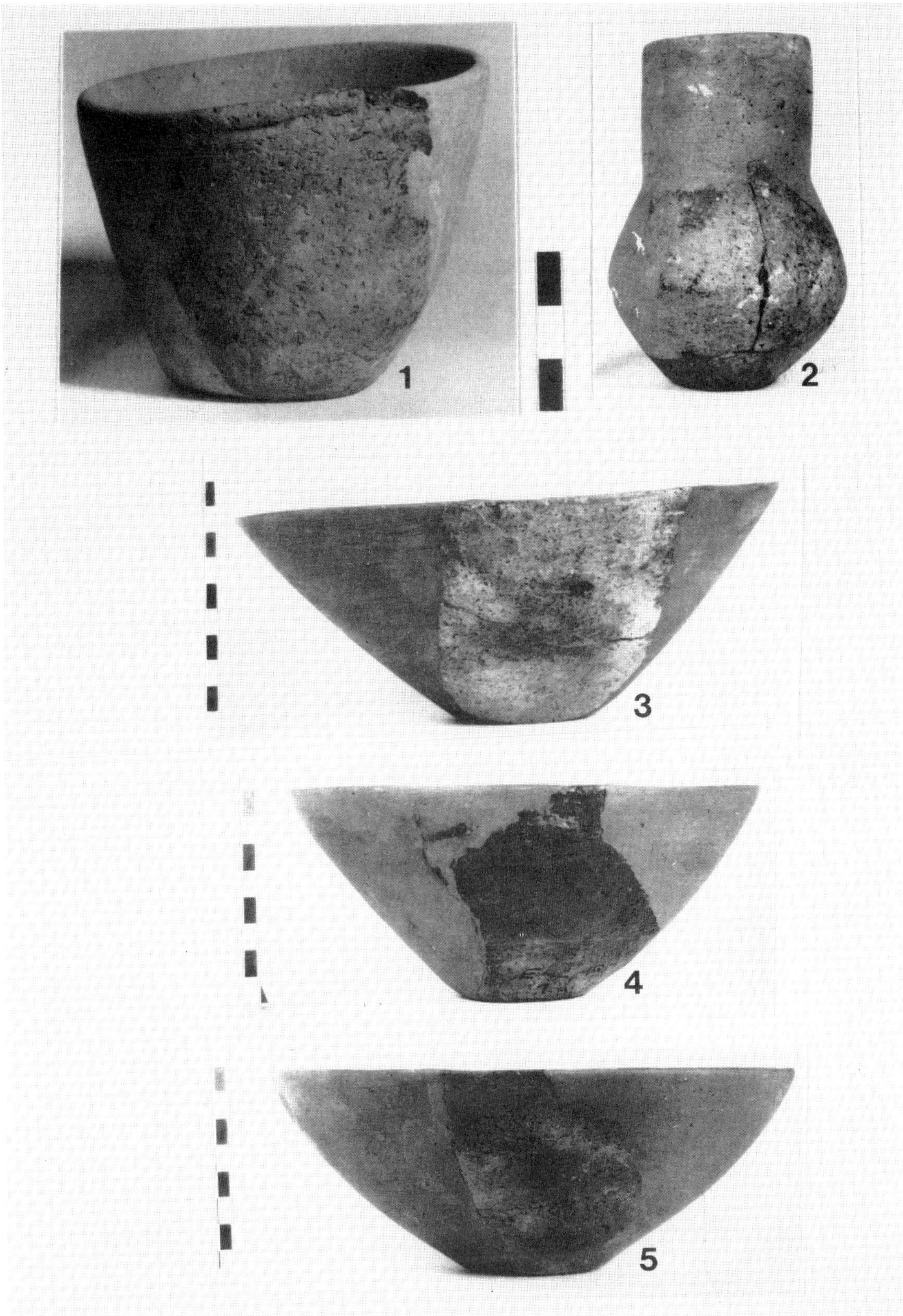

Plate 3.

Plate 4.

Plate 5.

Plate 6.

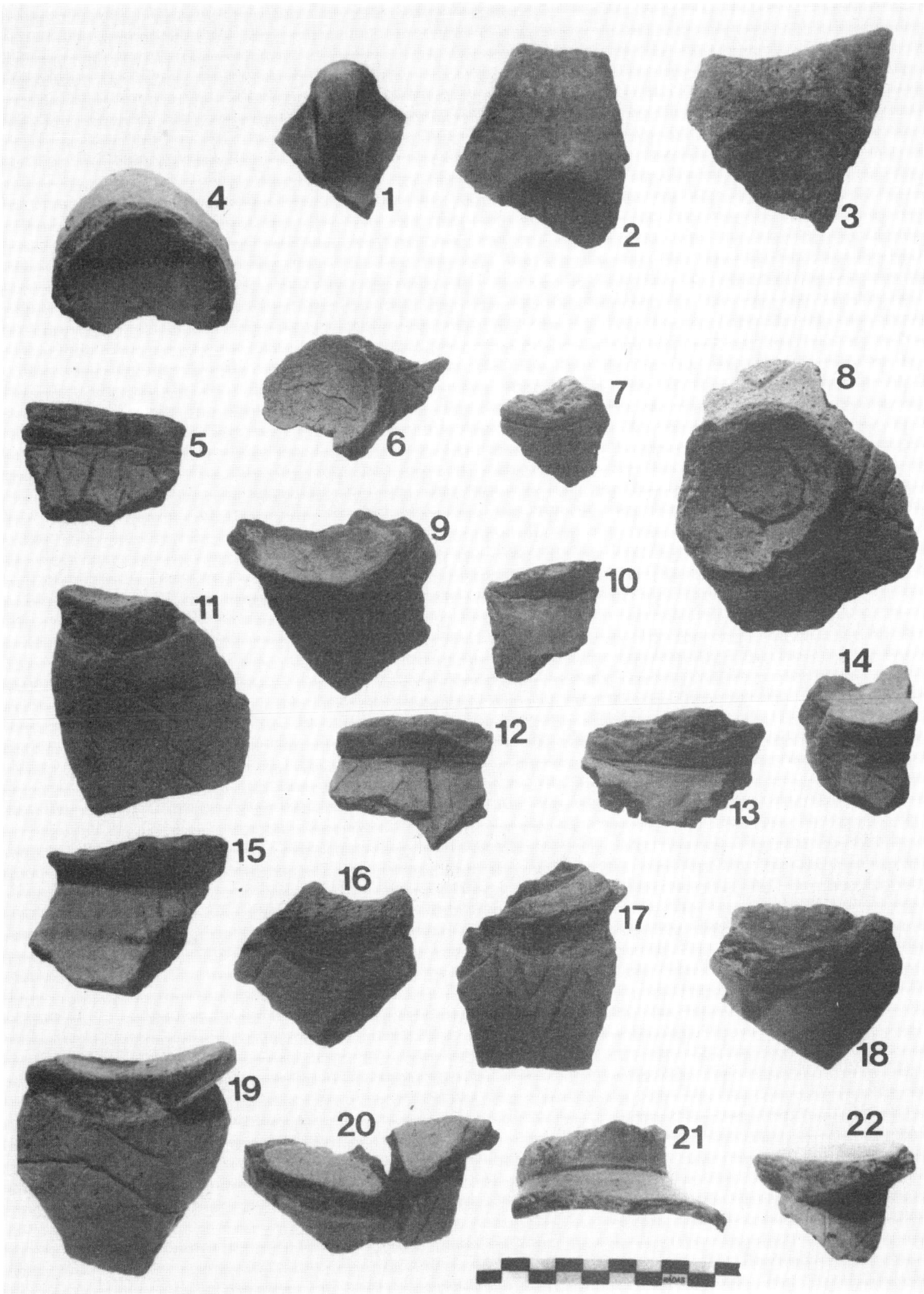

Plate 7.

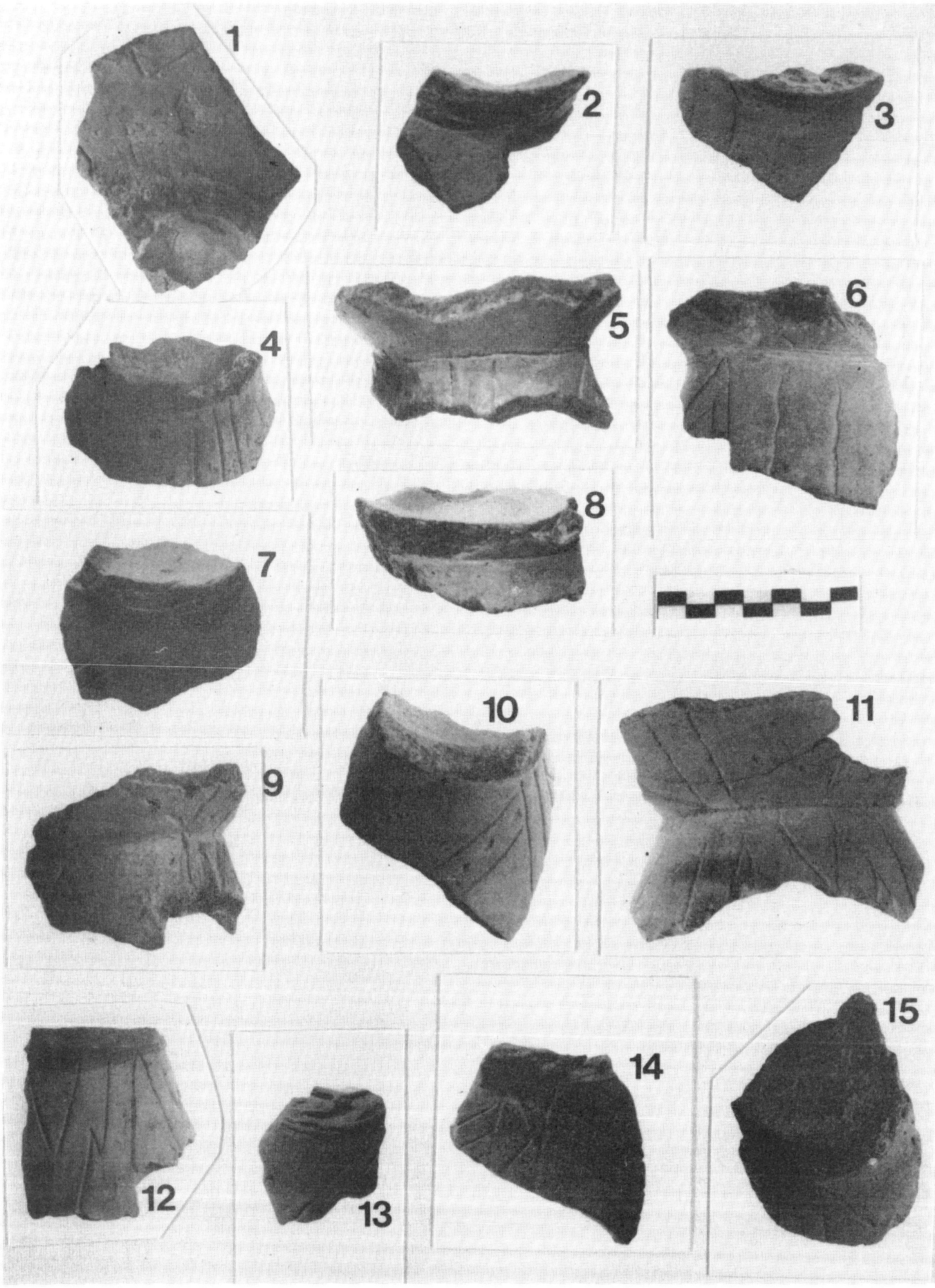

Plate 8.

Plate 9.

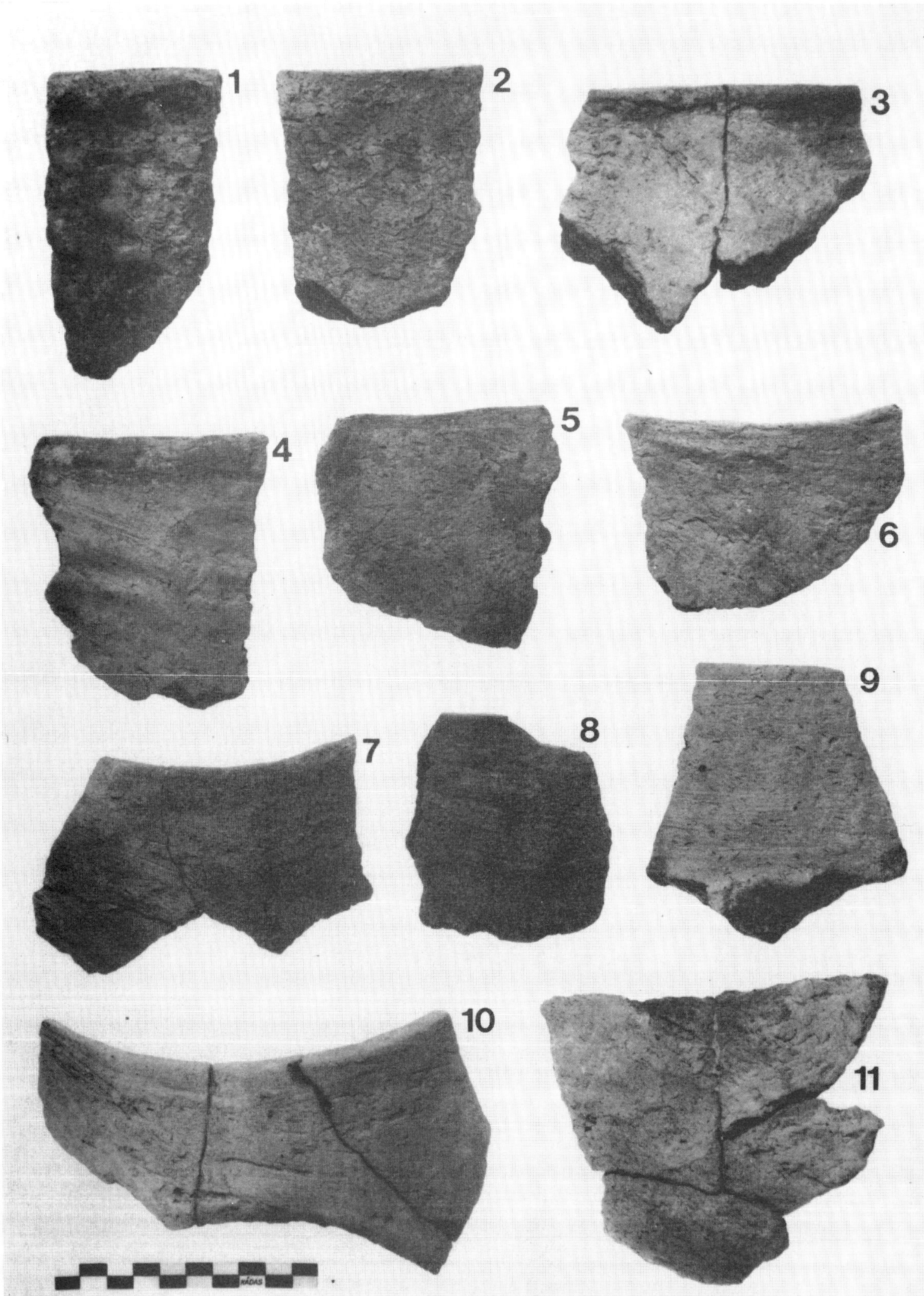

Plate 10.

Plate 11.

Plate 12.

Plate 13.

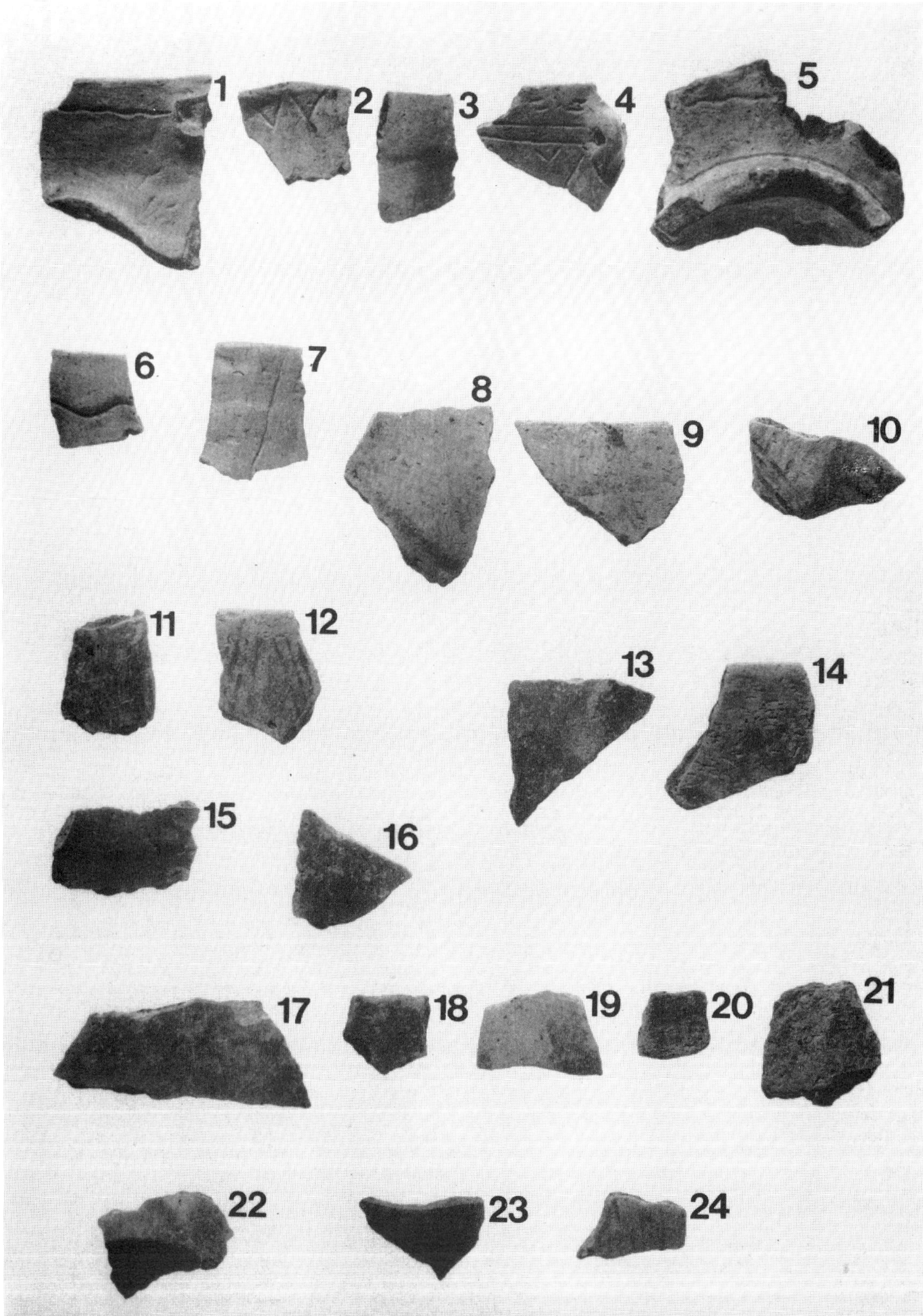

Plate 14.

Plate 15.

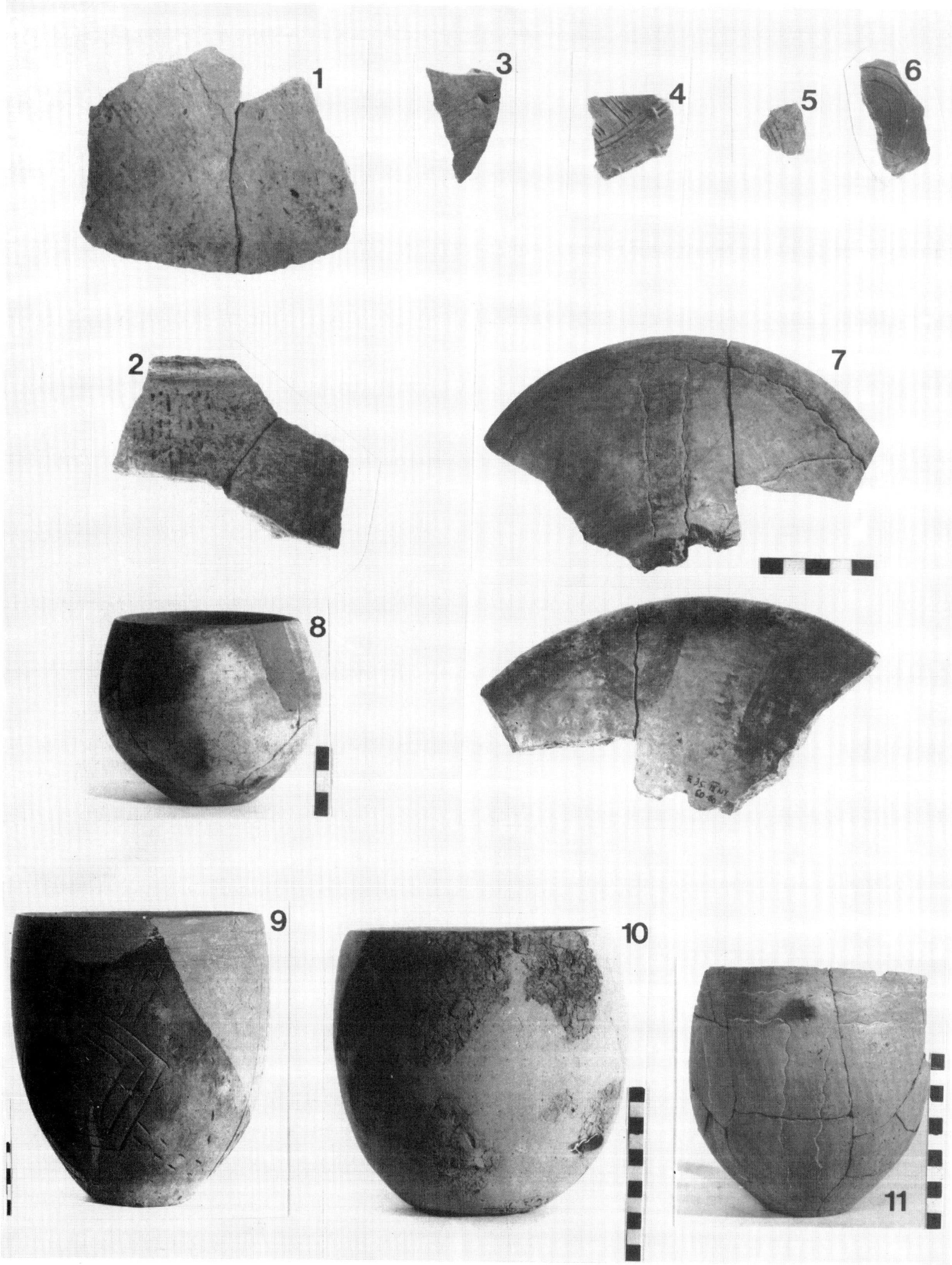

Plate 16.

The Endrőd 19 Project

The 1988 Season

BRUNO GENITO

Foreword

After two seasons of work at Gyoma 133, a Sarmatian settlement in South-Eastern Hungary (Genito 1988)[1], the Italian Archaeological Mission in Hungary initiated a new project in 1988. The operation at Gyoma 133 constituted a kind of introduction to the aims of our archaeological interest in Hungary (the study of the steppe peoples of Asiatic origin who arrived in the Carpathian Basin at various times bringing their own particular social patterns with them). The trial-trenches there did not represent a large-scale project, but were done rather to achieve, in the shortest possible time, the diagnostic capacity necessary for approaching Hungarian archaeological cultures which the Italian archaeologists had never dealt with before.[2]

The new project, based on a comprehensive topographic, geoarchaeological, geophysical and archaeological operation proposes to analyze settlement patterns over an approximately 600 m long and 400 m wide area. The area is located in the middle of the "Microregion", in South-Eastern Hungary, in Békés County, chosen for an integrated settlement-historical study in 1984 (Jankovich 1985), by the Archaeological Institute of the Hungarian Academy of Sciences. In the Microregion, hundreds of sites had been previously identified during an intensive investigation within the framework of the Hungarian Archaeological Topographic Survey.[3] According to the archaeological and historical interpretation of the topographic data gathered previously, the area chosen is believed to have been occupied, in different periods, by the Asiatic peoples to which the attention of the Italian Archaeological Mission is directed: the Prescythians/Scythians (8th-6th century B.C.) and the Sarmatians (1st-4th century A.D.). Different sites had been distinguished in the area among which the central ones are indicated in a numbered list as Endrőd 19, Endrőd 124, Endrőd 125, Endrőd 126 and Endrőd 57 (*Fig. 1.*). A huge Kurgan, called Papp-halom (the priest's Kurgan), and indicated as Endrőd 18, is located among them almost in the centre of the area. Among these sites, Endrőd 19 and Endrőd 124 were of special interest for us; on the basis of the topographic data, the first one was mainly related

---

1 The activity in those years led to the discovery of a settlement of the Sarmatian period. The archaeological remains consisted of pits, one house and more channels distributed in a very large area. The settlement had been previously partially destroyed by a quarry dug for a brick factory. The most significant finds were a small house in the easternmost part of the tested area where a silver coin was also found (Genito in press, b), two pits with dog skeletons (Genito 1986, Fig. 9) and a pit from the Late Bronze Age with many vessels of urn type. Two more articles are planned for publication regarding those campaigns, a general Second Interim Report (Genito in press, b), and another related to the reconstruction of the 7 large urns discovered in 1985 (Genito and Kemenczei 1990).

2 From an archaeological point of view the most important result of the campaigns at Gyoma 133 is that various systems of excavation have been experimented with in the field and, what is more, in a cultural area completely new for the Italian archaeologists. A vertical approach was tested in opening long sections, and a horizontal approach, opening long, narrow and intersecting trenches. The first one was facilitated by the still existing edges of the quarry, the second was suggested by the necessity of having the maximum result with the minimum effort (Genito, in press, b).

3 The "*Magyarország Régészeti Topográfiája*" (MRT) consisting of a series of volumes (up until now 8 volumes have been published) is planned to provide the Hungarian regions with the archaeological data collected during very accurate and intense topographic work. The series, which represents a very ambitious project, is divided by district in an alphabetical order, with numerous plates and figures ordered in chronological sequence; and is a very practical and useful instrument for research, being, besides a complete and detailed archive (Jankovich 1985).

to a period between Late Bronze and La Tène Age (11th-4th centuries B.C.) (Jankovich, Makkay and Szőke 1989, 136-137), and the second one to the Sarmatian Age (1st-4th century A.D.)(idem, cit., 167). The two sites were located near to the Kurgan Endrőd 18 – one to the east of it and one to the west – although they were not from the same period. The sites could therefore offer data from a considerable time span.

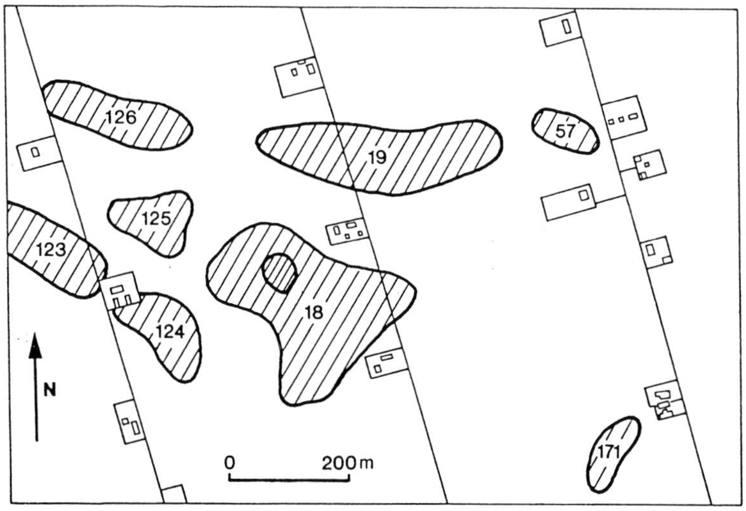

Fig. 1. The area of the sites, Military map (1:10000, issued in 1968).

1. Introduction

It is known that many waves of Asiatic peoples coming from Central Asia reached the Carpathian Basin at different times bringing their own specific ways of life (Erdélyi 1965 and Genito in press, a). These historical events underline the strategic importance of the Plain in relation to both East and West. This very characteristic aspect of the history of the Plain is explained by its geomorphology; the Plain represents, in fact, the westernmost extension of the Euro-Asiatic steppe-land (Pécsi et al. 1969). The Carpathian Basin, acting as a backwater, contained waves of peoples of Asiatic origin who, following the path of the steppes, brought their own cultures. The first known wave of steppe peoples to reach the Plain during the Copper Age (3200-2200 B.C.) is the so-called "Kurgan culture" (Ecsedy 1979)[4]. Between the Late Bronze Age and Early Iron Age (from here LBA/EIA, 11th-7th century B.C.) a new wave of people, coming from the steppe, reached the Plain (Gallus and Horváth 1939). The time-span between the LBA and Late Iron Age (from here LIA) in Hungary, to which Endrőd 19 seems to belong, is very interesting with respect to this 2nd wave of steppe-people. Generally, the wave

[4] The question of the ethnic identification of the peoples of "Kurgan culture" is very difficult. Traditionally, kurgan is considered as a burial system of oriental origin belonging to groups of nomadic peoples. Its geographical diffusion is, however, very wide, covering various and distant areas from Volga to Danube basins and to Western Europe (Gimbutas 1970). The close relationships between the Tripolye (Ukraina), Cucuteni (Romania) (Ellis 1984), Sredni Stog cultures, located in the area between the Dnieper and South Siberia (Telegin 1986), and the Kurgan culture have been more than once pointed out (Ecsedy 1979). The theories about the origin of the Kurgan culture intersect with those on the origins of the Indo-Aryan languages. Renfrew (1988, 188-210.) identifies the nomadic peoples of the steppes, bearing the Kurgan system of burial, as one of the vehicles for the spreading of the Indo-Aryan languages. Mallory (1989, 210-215.) does not completely agree with him. The question seems to be still open regarding the contributions by Häusler (1974 and 1976) and Anthony (1986) who refuse, in principle, the concept of "Kurgan culture", or Makkay (1987 and 1989).

is attributed to the Pre-Scythians/Cimmerians and to the Scythians. This period represents, in fact, one of the historical key-points for understanding the rate of evolution of the nomadic society previous to the Middle Ages, its inner social dynamics and the external relationships with the sedentary peoples. On the other hand, with the Sarmatian period (1st-4th century A.D.), to which most of the finds from Endrőd 124 seem to belong (a period, generally considered Roman in Hungary), the movement of Asiatic peoples starts again more intensively. This is known as the Migration Period which fully developed from 4th century to the 10th century (Huns, Avars and Magyars). Passing, at the moment, over the operation effected at Endrőd 124, which will be the object of a future work, let us go now to Endrőd 19 and to the historical phase, to which it belongs.

It is known that from the beginning of the 2nd Millennium B.C., a unified system of corporate kin-groups, socially organized in a structure of common patrilinear descent and economically based on complex livestock herding, started to establish full political control over the Eurasian steppes. At the end of the 2nd and the beginning of the 1st Millennium, this political control was completed with the emergence of pastoral nomadism, which had almost developed into equestrian nomadism. This stage of the evolution of the Eurasian pastoral nomadism was, in fact, characterized by the full exploitation of the horse as a mounted animal. For the first time, archaeological evidence for this period is being compared with ethnographic material and with the historical sources to examine political phenomenon of the emergent Steppe peoples (Artamonov 1949). So great was the impression they made on all the neighbouring states that the contemporary Assyrian, Chinese, Greek and Roman records provide a unique mosaic of ethnographic observation (Ecsedy Il. 1982 and 1984; Tosi and Seaman, in press). The evolution of the Eurasian nomadism, and its external relationship with sedentary peoples in Asia and in Hungary has been frequently discussed by numerous scholars.

From the point of view of the history of Central Asia the historical transition between LBA and EIA is related to the arrival of groups of nomads, mainly ethnically related to the Iranians.[5] Soviet scholars have collected an incredible amount of archaeological data on the topic (Grakov and Meljukova 1953 and 1954). These bibliographical references represent the basic starting point for research on both chronological and geographical grounds.[6] The inner social dynamics of the human society in the area between the rivers Volga and Danube has been, in the period we are dealing with, thoroughly analyzed from the standpoint of new historical documentation (Archeologija SSSR 1989).

The methodological approach of the archaeological research into this period is different from the previous one. The aim is to fit the historical data to the archaeological material. Numerous scholars have discussed the problems of bringing historical and archaeological

---

5 The coming of the Iranians to Iran and the relationships between those groups of peoples and the first nomads of identical ethnic character, is a very complex historical and archaeological question. It cannot be summarized in a few words. The enormous importance of the arrival of the Iranian people on the Iranian plateau stimulated the archaeological research to try to find out the paths and the various pauses during their movement through Central Asia. The main theories are summarized in Hinz (1973), Dyson (1973), and Crossland (1967). Deshayes (1969) hypothesized the Khorasan, in the south-eastern corner of the Caspian Sea, as the road to the Plateau. Ghirshman (1977), proposed Central Asia for the migration of the Indoeuropeans, and the Caucasus for the Iranians. Young (1967 and 1985), rejected the Caucasian migration in favour of a Central Asian path. Further attempts to make comparisons between the material documentation and linguistic evidence are seen in the contributions by Grantovski (1981) and Kuzmina (1986). At the same time, a general work to be considered as "summa" is that of Frye, 1984. As far as the two perspectives of interest followed in my work, i.e. the westernmost expansion of the Iranian tribes and their nomadic character are concerned, cf. the articles and works by Roska (1937), Benadik (1953), Crişan (1965), Meljukova (1955), Harmatta (1968).

6 Soviet scholars have been dealing with these problems for a long time. Specialized and basic works are those of Iessen (1953) and Meljukova (1955).

material together. The complex archaeological data when combined with an historical perspecitve which combines ethnic, geographical and political considerations can actually create more problems than it solves. It is very easy to misinterpret historical or archaeological data when trying to force them into a methodological straightjacket in order to prove that they can reciprocally informative. The question is not less complicated from the point of view of the history of the Great Hungarian Plain. Many scholars have approached the question from different points of view but, despite their numerous articles and works, many problems still remain unsolved.

The Cimmerian "problem", the Scythian or the Thracian "questions" constitute spatial-temporal historical and archaeological examples of the enormous difficulties met by scholars who try to combine historical sources with archaeological evidence.[7] The historical documentation bears witness to the existence and the importance of these peoples, while the archaeological records cannot certainly attribute the material culture to these groups. Such an approach, difficult in itself, develops into a quite impossible task when, in other contexts, information and data from the written sources on some peoples are compared to the cultures of others without written documents. The analyses and comparisons of the gold hoards, of Iron Age date, found in Romania, Hungary and Moldavia, which were discussed by Harmatta (1946/47), represent a very important contribution to the archaeological and historical debate about Cimmerians, just like those efforts completed more recently from an archaeological perspective (Sulimirski 1960) or in linguistics (Diakonov 1981). Various more recent efforts and numerous attempts by other scholars to make further attempts to compare the various material cultures attributed to these people, and found in areas where, according to the sources, they seem to have lived. This people is either dated to the middle of the 2nd Millennium B.C. (Terenožkin 1976) or to the middle of the 1st Millennium B.C. (Ghirshman 1983); still the material culture and the exact chronological slot of the Cimmerians cannot be defined. The complex character of the material culture attributed to the Thracians, is not helpful if one seeks to reconstruct an unified character for the people. The efforts to identify this people have increased the element of "puzzle" where ethnicity, culture, race and language constitute the various aspects of the same problem.[8]

---

7 These arguments have been, over a long period the subject of debate by the Hungarian archaeologists; the pioneering works by Fettich (1927 and 1928) were devoted to some specific gold finds of Scythian origin, meanwhile, his summary from 1931 tried to set up the theoretical basis for the study of the Scythians in the Carpathian Basin. Different and more ambitious is the volume by Gallus and Horváth (1939), who, together with Harmatta (1941 and 1946/47) tried to order the large amount of the material documentation of Asiatic character. With the booklet by Bottyán (1955), and with the contributions of Párducz (1952, 1954, 1955) as well as his general works from 1965a, 1965b, 1966, 1973 and the article by Harmatta (1966), the Prescythian and the Scythian period and its relationships with the Hallstatt and the Cimmerian cultures started to become more clear. Kemenczei (1963 and 1970) and Kovács (1966/67) started to examine the material of Late Bronze Age in North-Eastern Hungary and its numerous regional variants from respect of Asian cultures.

8 The Thracians have been dealt with many times by scholars, starting with Parvan (1926). The analyses and studies of cemeteries of Hallstatt period, cf. Vulpe (1967), or of Getian period cf. Budzugan (1968), contributed considerably to the enlargement of the scientific debate. The collective work of MIA 1969 was entirely devoted to the archaeological traces of the Thracians on the Black Sea coast, while the Thracians and the Hallstatt cultures were the topic of Lapušnjan 1972, the Thracians in Moldavia by Lapušnjan (1979). The relationships between the Scythians and the Thracians were discussed by Členova (1975), Meljukova (1979), Lang (1980) and Vulpe (1980).

The Scythian culture, identified in the Black Sea, Kuban and Caucasian areas[9], cannot be easily related to the contemporary cultures of the Carpathian Basin (Gazdapusztai 1963, 1965, 1967). The relationship between the LBA culture of Hungary and contemporary ones in Eastern Europe has been the topic of long debate (Kemenczei 1980, 1981, 1982) and has provided evidence of more than one common factor (Smirnova 1969, 1974 and 1975). The regional differentiation of the LBA in Hungary only partially fits the pre-Scythian influx (Gallus and Horváth 1939 and Párducz, 1973). The supposed Thraco-Cimmerian and Thraco-Scythian cultural unity, even after half a century of study[10], is also very far from being completely defined. One of the main difficulties met by the scholars analyzing these particular cultures, nomadic in character but also involved in the social and economic dynamics of the sedentary societies, is that the historically most significant meeting among those two socio-cultural entities occurred either between the LBA and EIA (the end of 2nd Millennium B.C. and 1st Millennium B.C.) or between the late-antique and the early-medieval age (4th-10th centuries A.D.) (two different Dark Ages). These periods represent, in fact, the most evident examples of multi-ethnic cultural contexts with, on one hand, the emergence of great states and of the first equestrian pastoral nomad communities and, on the other, the decline and fall of the great state formations and the emergence of the great nomadic empires.[11]

The approach of the archaeological research of these periods has never been easy since it is related to the nomadic world. Particular aspects of the nomadic community have been approached with the same methodology which has already been used for complex state societies where the characterization of social, economic, artistic and monumental aspects are more easily detected because they were more distinguishable. The nomadic societies, mainly because the direct information from the sources is missing, should, thus, be analyzed without attributing an ethnic identity to different aspects of the social production. Groups of different peoples, probably lived together in similar ways of life and the general historical outline derived from the sources often show more about the prejudices and ideological premises of the authors than about the groups of peoples they describe. They present, thus, an amount of information of partial use and in some cases, of negative value. The same phenomenon occurred in the case of other areas as well, during the same periods; for example in northwest Iran where the identification of the material culture and its ethnical attribution is also a very difficult task.

2. The Preliminary Campaign of 1987 at Endrőd 19

In May 1987, when the Italian Archaeological Mission was still working in Gyoma 133, some geophysical surveys were effected at Endrőd 19 in order to plan a future large-scale operation in an area already at the focus of interest (cf. above). The difficulties in chosing a site led us to use different methods of archaeological approach. The experience collected at Gyoma 133 (Genito in press, b) proved to be very useful. A big soil collection during the work on a new

---

[9] The bibliographic references are too numerous to be listed here. The basic volumes are MIA 1958, devoted to the Scythian antiquities of the Black Sea, MIA 1961, devoted to the general relationships between the Late Bronze and the Early Iron Age in the same area. MIA 1961, 1962a, 1962b and 1971 constitute the best and most complete collection of the Scythian/Sarmatian material culture. The volumes by Il'inskaja 1968 and that by Liberov 1962 give details of the Scythian finds in the Dnieper and Donau areas; the volume edited by Terenožkin (ed.) 1973 is more concerned with the Pre-Scythian cultures in the area of Dnieper.

[10] This aspect has often been discussed by scholars: on the Slovakian Thracian/Scythian unity cf. Dušek 1961; for that in Carpathian Basin cf. idem 1964 and 1975. For more western relationships see Foltiny 1963, and Harmatta 1968. A very useful and updated work is that by Meljukova 1989.

[11] The question is very difficult and historically very relevant. No specific study of this matter exists until now. Hazanov 1982 provided the theoretical basis for the approaches to the different regional analyses. The evolution from the Early Central Asian Empires to the Early Medieval ones was analyzed a long time ago by McGovern 1939. This is still an interesting and very useful study.

track for the Gyoma Railway Station uncovered, by chance, the major part of the site. This was located just outside of the previously identified topographic limits. No matter how accurate topographical field work is, it cannot be accepted as the only indicator for starting an excavation, although it may acquire more significance if followed by an archaeological test. The attribution of parts of a site to different chronologies, on the basis of the surface collection of pottery, did not provide sufficient preliminary knowledge with which to start an archaeological operation.[12] The small geophysical operation in 1987, carried out by Fondazione Lerici, was effected in order to achieve more data with which to start the archaeological operation.[13](cf. Appendix B)

3. The May 1988 operation

The area of the work begun in May[14] is visible in the south-western corner of the sheet indicated as Endrőd 609-231 in the map series (1:10 000 issued in 1968) whereas the outlines of the sites 124, 126, 125, 18 (Papp-halom), 19 and 57 are clearly evidenced. The geographic coordinates of the area are 20 45' and 46 53'. The area with its natural wavy profile, is very wide and was covered, at the time of the Italian operation, by a large field planted with maize (*Fig. 2*). Two dirt-roads delimit the area, one to the east lapping site 57 and the other to the west lapping the sites 124 and 125; the area was also crossed in the middle by another, N-S oriented road, which led to a couple of houses (tanyák) still visible in the map of 1968 (*Fig. 1*); this road has now disappeared.

---

[12] Prof. G. Johnson of Hunter College in New York closely examined the possibility of working with the topographic data of the Microregion, in order to set up a rank size ordering of settlements. The possibility of operating such an approach is based on the hierarchy of the sizes of the sites. The chronological attributions of the single parts of a site are still absent in Microregion.

[13] During the organization of the summer season at Gyoma 133, the Italian Archaeological Mission decided to start some geophysical sondages in Endrőd 19. The detection was made rather at random.

[14] The operation in Endrőd 19 occured at two different times, the first one in May, (from the 1st to the 12th) and the second one in July/August (from the 4th of July to 5th of August) (during the 1988 campaign, the excavation of the small size Kurgan Endrőd 122 was also undertaken by the Italian Archaeological Mission, cf. Genito et al. in press). The Italian Archaeological Mission, directed by the author, was composed of Dr. M. Cattani, archaeologist, Dr. S. Pracchia, archaeologist, Mr. M. Mascellani, topographer, the students, M. Bartolini, from University of Rome, I. Chiesi, from University of Bologna, F. La Manna and S. Laurenza, from Istituto Universitario Orientale, Naples. Dr. M. Cucarzi, geophysicist, scientific director of Fondazione Lerici and M. Cremaschi, geoarchaeologist also participated in May. In June, a team of Fondazione Lerici directed by Dr. Cucarzi accompanied by A. Rivolta, geologist, M. Boni, informatic and D. Gabrielli, technician completed a large geophysical operation on the sites Endrőd 19 and Endrőd 124, detected in April by the Italian Archaeological Mission. During the July operation Prof. M. Tosi, from Istituto Universitario Orientale, Naples, visited the Italian team providing useful and important suggestions; Mr. L. Spezzini, technician, Mrs. J. Merényi, topographer and Mrs. M. Paczona from the Institute of Archaeology of the Hungarian Academy of Sciences also participated to the work in July. On this occasion I would like to express my deepest thanks to the Italian Ministry of Education (60% MPI), Ministry of Foreign Affairs and National Council for Research on whose funds the activity was based. Special gratitude is also due to the whole Italian staff which helped me a great deal, sometimes working in very difficult conditions; without them I would have achieved much less. Thanks are also due to Prof. M.Tosi who was, even when at distance, always at my complete disposal, Prof. G. Gnoli, President of the IsMEO who provided me with a lot of support, Prof. A. Rossi, director of Department of Asian Studies of Istituto Universitario Orientale, Naples, who always followed my work with interest. The Director of the Italian Cultural Institute in Hungary Prof. Porsia and his vice Prof. Ventola must be mentioned particularly because they helped the whole Italian Archaeological Mission in many particular situations. I would like to express my deepest thanks also to S. Bökönyi, Director of the Archaeological Institute of the Hungarian Academy of Sciences, to D. Jankovich-B., scientific secretary of the Institute, Dr. J. Makkay for their precious help and encouragements, Mrs. M. Vicze, executive secretary for the Microregion project, who was very kind and generously helped me at any time I required it. Last but not least, I want to thank also the local Cooperative "Béke TSZ" in Gyomaendrőd and particularly Mr. G. Gubucz who helped the Italian Archaeological Mission giving a scraper and a tractor and Mr. J. Eliás, camera man, who worked with us in July.

A general topographical survey was necessary in order to provide a detailed map of the contour lines of the whole area, inhabited in different periods (*Fig. 3*). From the point of view of the settlement pattern, the area can be considered as a unified geoarchaeo- logical context where the different chronologies represent only different systems of settling. In a plain where the only archaeological traces are the materials spread over fields, this unified conception looks most convincing from a methodological point of view and most useful from a practical one.

Fig. 2. General view (IsMEO, Dep.Neg. L16701/29).

From this perspective the difference between the site areas and the 'siteless' areas (Dancey and Dunnel 1983) are less significant than expected. Of course, a site cannot be investigated in the same way as a 'siteless' area, a different approach and different methods are necessary. The analysis of the latter one, which does not represent the uninhabited area of a territory, provides the most complete amount of information about an anthropic soil. This approach is facilitated, in my opinion, by the Great Hungarian Plain. In this area, the perspectives and limits of a "field archaeology" have a right collocation because the once existing steppe-land is today, basically, an enormous lay-out of planted fields.

The steppe-land archaeology (theoretical premises, technical approaches and scientific aims)[16], as it developed in the whole Eurasian continent has a logical and consequent possibility of being carried out in Hungary, because the topographic work there provided enough data to start long-term projects. As it can clearly be seen on the map (*Fig. 3*), the contour lines individuated in our topographical work, clearly follow the elevations of the supposed archaeological deposit. It is probable that the elevations in most cases hide archaeological deposits, although the difference in elevations may not correspond exactly to their limits. From the data provided by the topographic work, one could see the central ancient fluviatile deposit, abandoned by the Körös river as it changed its course from South to North (cf. Cremaschi), which runs around the Kurgan 18 along in an east-west direction (*Fig. 3*).

---

[16] The steppe-land archaeology does not exist as an individual topic. The wide time-span and geographical distribution of the steppe cultures make them suitable for many different scientific approaches. Notwithstanding the large variation in the ethnic, historical, geographical and geomorphological aspects, a unified character can be found amongst the nomadic steppe cultures, in my opinion, only in the common method of production based on the pastoral nomadism.

The first work was to open long trenches just between the part of the area already identified as a site and the part considered empty of archaeological remains, located in the main fluviatile deposit. The preliminary aim was to verify archaeologically to what extent Endrőd 19, at that point, could be considered a site. The second aim was to find an archaeological deposit on which to start the geophysical detection and the archaeological work. This operation complements the work which the Hungarian scholars were simultaneously doing at other sites to prove the accuracy of the previous topographic work (Jankovich and Pattantyús 1990).

The trenches opened were numbered as sections No. 1, each one in Endrőd 19, Endrőd 124 and of Endrőd 125. The section No. 1 at Endrőd 19 is 50 m long, N-S oriented, with the points S and N at a distance of 155 m and 225 m from the top of Papp-halom (Endrőd 18). The trench was opened with the help of a mechanical excavator with a rear bucket with teeth (*Fig. 4*). The trench was deep enough (-1,15), to reach the yellow virgin soil, and sufficiently narrow to preserve as much as possible of the archaeological deposit, yet wide enough to allow its walls to be seen. The stratigraphic analysis was made previously to the geoarchaeological analyses in order to obtain the archaeological data as they were coming out directly from the fields, for

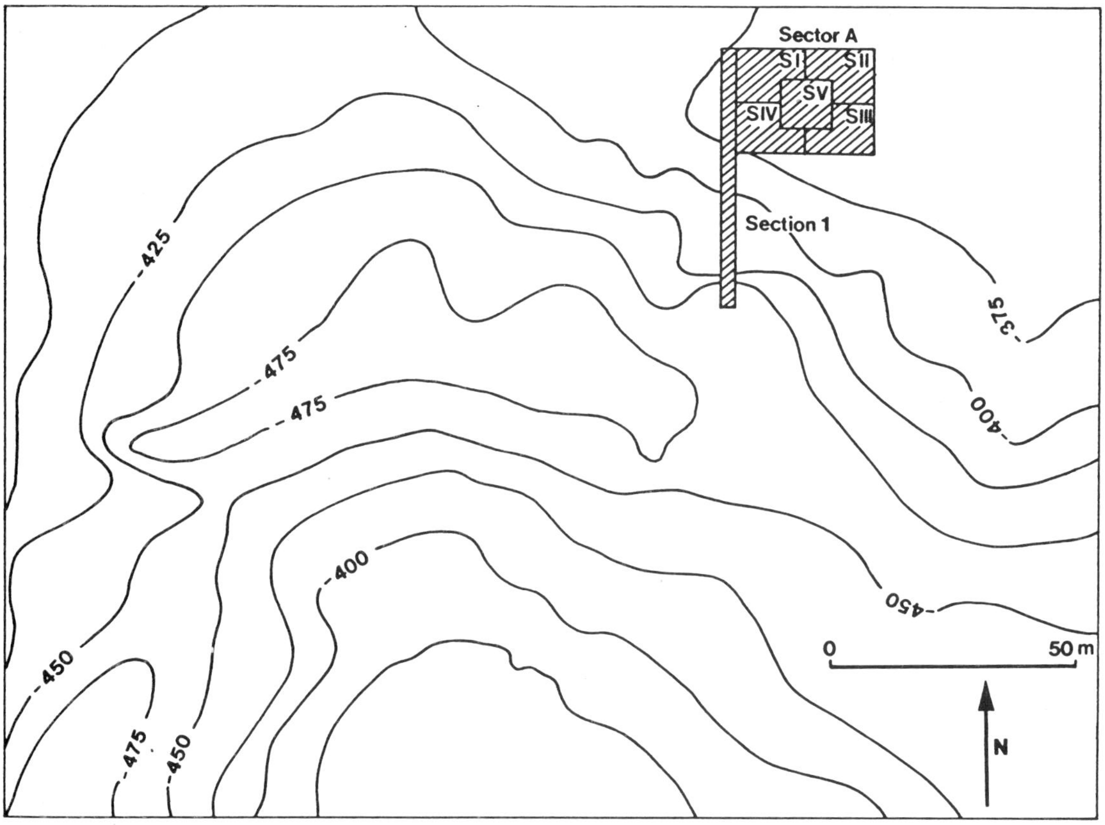

Fig. 3. General map with contour lines by M. Mascellani and I. Chiesi.

comparing them, later, with the soil analyses. There are three layers differentiated by the excavations. The first one corresponded to the agricultural soil, the second one to the ancient agricultural soil mixed with archaeological deposits and the third one to the yellow virgin clay soil, each one with further divisions (*Fig. 5*). After the first cut of the trench it turned out that its southern part constituted actually the fluviatile deposit; that point was reached at -1,60 m at least from the surface and the colour of the virgin soil was different from the rest of the surface.

Fig. 4. The Digger Loader used for the section (IsMEO, Dep.Neg. L16702/36).

The pottery fragments which came to light from this southern part had very abraded surfaces (*Fig. 6*). Both the colour and the surface degradation of the pottery fragments can be attributed to their presence in the stagnant water in the main bed of the river. The archaeological analysis of the trench has revealed, starting from 34 m ca from point S, a concentration of pottery fragments, which fits very well with the topographic indication of the limit between the site and the area empty of any archaeological deposit. This limit is located, in fact, in the area already identified by the topographical work in a depth varying from -3.75 to -4.00 m. This limit is not

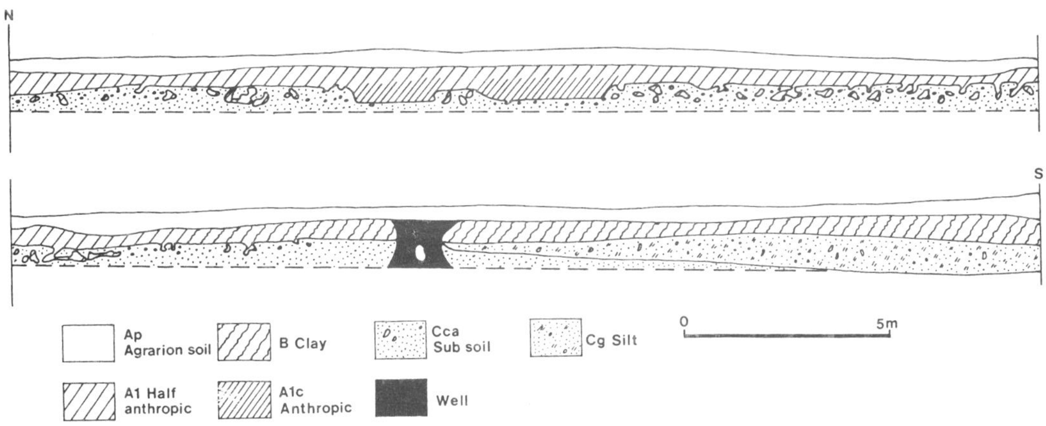

Fig. 5. Section No. 1 by M. Cattani, I. Chiesi and M. Cremaschi.

exactly identifiable from the archaeological analysis but it can be better identified by means of a future geoarchaeological analysis (Appendix A).

Northward from this point, some structural evidence came to light; the soil is different in colour from that of the rest of the section (*Fig. 7*). At the bottom of the trench two more structural elements came to light: a well (*Fig. 8*) at 15 m from point S, and an irregular rectangular-shaped structure, probably belonging to a grave, between 25 and 27.5 m (*Fig. 9*).

Fig. 6. The southernmost part of section No. 1 (IsMEO, Dep. Neg. L16698/4).

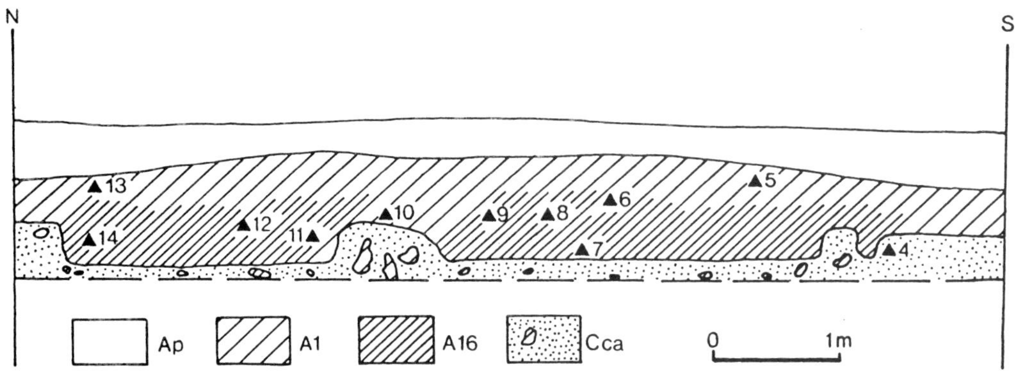

Fig. 7. Section No. 1 by M. Cattani, I. Chiesi and M. Cremaschi. (Legend: See Fig. 5.)

The archaeological material (pottery fragments and one fragment of obsidian) concentrated in the northern part of the section, confirmed the presence of the archaeological deposit. Looking at the section in detail, one can see the contour of the deposit which runs from South to North including at least two different structures. One seems to be located at the right side and the second, probably more complex one, at the left (*Fig. 10, 11, 12*). From the general profile it is not easy to have a clear idea of what these structures represent. While the first smaller one, could probably be a pit, the second, substantially larger, looks like a house or a

Fig. 8. The traces of a well in the area of section No. 1 (IsMEO, Dep. Neg. L16698/14).

Fig. 9. Grave No. 1 in the area of section no. 1 (IsMEO, Dep. L16885/16).

very large pit. At the extreme southern point a smaller feature, looking like a post-hole, could confirm the first hypothesis. In a geoarchaeological context like a Flood Plain, the possibility of finding certain, definite and clear architectural features appearing in the trench, is very slight; the consistency of the architecture was poor and the width of the trench was too narrow. The aim of this operation was, in fact, to have clearer evidence from the geoarchaeological survey of where to open a large-scale archaeological excavation. We assume that archaeological structures actually date to a chronological horizon close to that of the fill which they contain ; and that the latest material in the fill will provide a *terminus ante quem* for the structure. Starting from this point of view, the collection of pottery fragments was made not to identify a site, but to provide the archaeological evidence of a deposit, while the traditional survey either ignored or unsystematically recorded the artifacts (the surface artifacts) which do not appear to mark the location of a site of a potential excavation (Dancey 1974). The presence of certain fragments in stratigraphical position and not only on the surface could serve both aims: the identification of fragments and the identification of a deposit.

It is very important that 11 of the 14 pottery fragments were located in the northernmost part of the section (*Fig. 7*). The three fragments outside the deposit were located at 2.5 m from the point southern of the section in one case, while in the case of the second (which is, in fact, constituted by three fragments) and the third, between 18 and 19 m from the same point, very close to the well. Two of these 5 fragments belong to the LBA/Gáva horizon (black burnished hand-made pottery), one to Late Sarmatian period (gray burnished pottery) (both represented by spot No. 2 on the section); one is prehistoric (No. 3) and one seems to be of Scythian period (No. 1). The fragments nos. 4, 6, 7, 8, 12, 13 in the deposit can be dated between the LBA and the Iron Age; nos. 5 and 9 (*Fig. 13*) can be attributed to the LIA, while the nos. 10, 11, 14, are of Sarmatian date. As the collocation and also the distribution of the pottery fragments does not

Fig. 10. Archaeological deposit (IsMEO, Dep.Neg. L16698/10).

Fig. 11. Archaeological deposit (IsMEO, Dep.Neg. L16698 /12).

Fig. 12. Archaeological deposit (IsMEO, Dep.Neg. L16698/11).

help much in their interpretation, it should thus be verified through the archaeological test. With regard to the size of the extension of the trench to be opened, starting from the section, a geophysical campaign was held in June in order to identify the eastward extension of the deposit (cf. Appendix C).

4. The July operation

On the basis of the data collected during the May operation, the section was enlarged towards the east with a width of 20 m south from point B, in an area where the major part of the material was collected and where the structural elements came to light. The area chosen, called sector A, is 20 m wide (N-S) and 28 m long (W-E) (*Fig. 14*). A portion of sector A, the one which comprised section No. 1, was left as a boulk 1.5 metres wide in order to improve the correspondence between the data collected from the section and the ones from the excavated contiguous area. Its width was chosen on the basis of the archaeological analysis of the trench

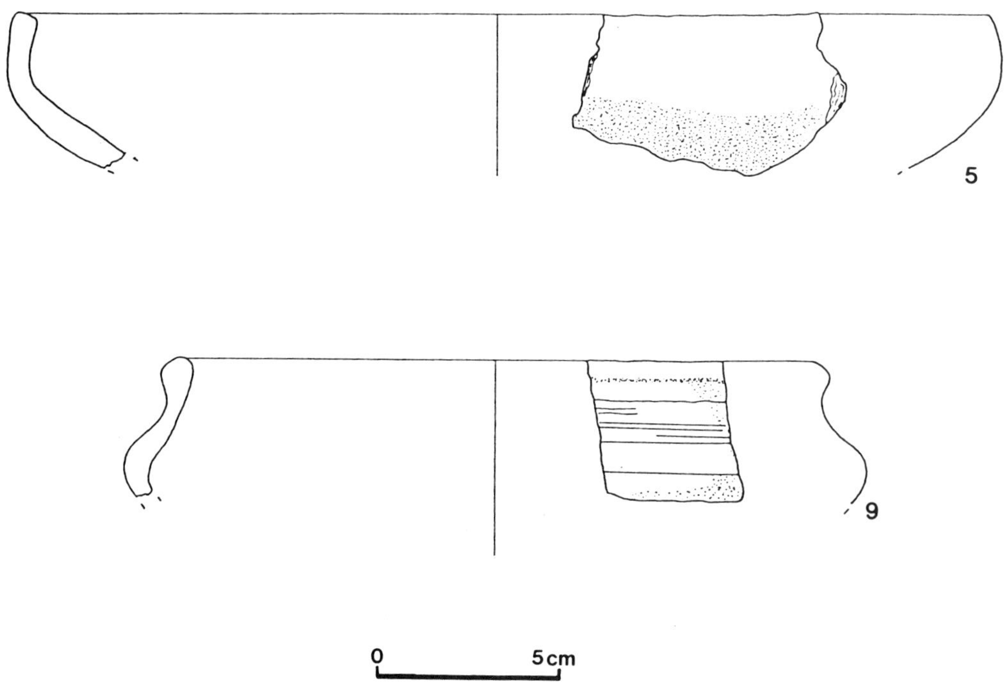

Fig. 13. Fragments No. 5 and No. 9 by I. Chiesi.

but its length was chosen on the basis of the geophysical analysis (cf. Appendix C). Before starting the work with a scraper the area was cleared of vegetation and field-walked. 80 pottery fragments came to light from the area divided into 5 different sectors (*Fig. 15*).

List of selected pottery fragments (Fig. 16)

No. 35, Surface : rim of a jar, globular-shaped with a low neck, coarse ware, heavily fired, blackish coloured. Slipped creamy surface outside and brownish inside. Dm. 14; width 0.7 (Late Bronze Age/Pre-Scythian)

No. 23, Surface V : flaring rim, coarse ware, medium fired buff-coloured, slipped creamy burnished surface Dm. 9 (Late Bronze Age/Pre Scythian)

Fig. 14. General view of sector A before the excavation (IsMEO, Dep. Neg. L16884/37).

	I.	II.	III.	IV.	V.
Modern	20		1	3	
Late iron age	10	2	2	1	10
Iron age	8	2	4	1	3
Late bronze age	2		4		4
	40	4	11	5	17

Fig. 15. Diagram of surface distribution of the pottery fragments by I. Chiesi.

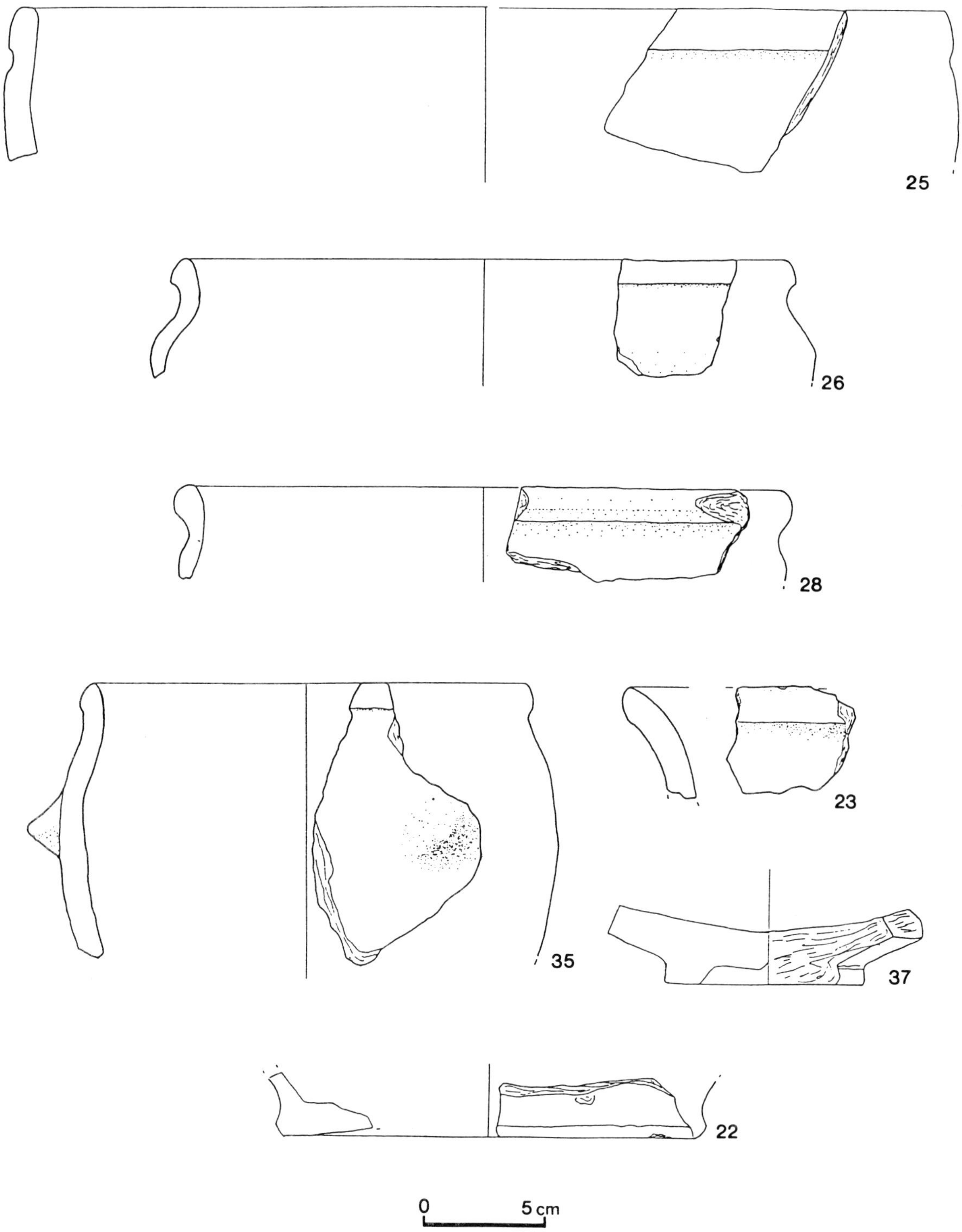

Fig. 16. Pottery fragments from the surface, by I. Chiesi.

No. 28, Surface V : everted and swollen rim of a bowl, medium ware, heavily fired dark buff coloured, slipped black burnished surface. Dm. 19.3; width 0.6 (Late Iron Age)

No. 26, Surface V : lightly everted rim of a jar, globular shaped, medium ware, heavily fired, blackish coloured, slipped black burnished surface. Dm. 17.5; width 0.5 (Late Iron Age)

No. 25, Surface V : swollen rim of a jar, globular-shaped medium ware, lightly fired red/orange coloured, slipped red burnished surface. Dm. 26.5; width 0.9 (Sarmatian Age)

No. 22, Surface I : Flat base, medium ware, heavily fired grayish coloured, slipped grayish burnished surface. Dm. 13.5 (Sarmatian Age)

No. 37, Surface V : Ring-shaped base; medium ware, red/orange coloured. Slipped red surface. Dm. 0.5 (Sarmatian Age)

A preliminary earth moving operation was made using a towed scraper. It consists of a box and a cutting edge mounted in a stout frame and towed by a crawler tractor. (*Fig. 17*). The various parts of the scraper are operated from the tractor by means of steel ropes and a complex system of winch and blocks. The towed scraper's wheels are not required for traction and are not, usually, deeply ribbed. The tractor's grouser tracks are likely to cause the greatest damage, particularly when the digging resistance is high. The earth moving operation is not new in the archaeological research and has successfully been experimented and theorized (Nichols 1954 and Petch 1968).[17] The system has been included in a plan which provided to open preliminarily sections and to effect better oriented geophysical detections in an area extended from the section. The system, planned in this way, has provided the possibility of reaching the archaeological levels in the shortest time by means of a less destructive system than that of a bulldozer, in that a bulldozer track follows the blade, whereas the towed scraper is only likely to leave light wheel marks. The depth to be reached by the scraper edge can be regulated according to the level of the deposit known from the stratigraphic analyses of the sections. This depth, in our case, corresponds, to the lowest level of the ancient agricultural soil. The thickness of the cut of the blade is regulated to 8 cm and for covering altogether ca. 40 cm foreseen that five cuttings should be done. Considering the large amount of soil to be taken out, the operational area was, of course, much more ample than the archaeological one in "strict sense". For this a huge area twice larger than Sector A was left parallel to it for collecting the removed soil.

The work with the scraper presented, naturally, some difficulties; the edges of sector A could not be equally and regularly because of big dimensions of the machine (*Fig. 18*) (they were left to be worked by hands). A further great difficulty was provided by the quality of the soil which is so hard and compact that it gave much resistance to the tractor. It was necessary to use the maximum power of the engine and its iron wheels went a little bit deeper than planned (*Fig. 19*). The first two cuttings did not give results, but from the third, a long and narrow curving shadowy feature came to light in the eastern part of the sector (*Fig. 20*). It probably belonged to a ditch. Numerous archaeological features came to light in the whole area, indicated by dense concentrations of pottery in some specific places. Apart from pits nos. 1, 13 (around them two more features were found: No. 15 and 23), 10, 11, 17, 20, 25, the features are concentrated in three different areas. The first one at the west which comprises spots No. 9, 28, 27, 19, 21, 4, a second one in the middle with the numbers 8 and 24 (two fire places), 12, 22, 29, 14, 2, 3, 7, 18 and the third one at east, almost isolated from the others by the ditch; numbers 16, 5, 6, 26 (*Fig. 21*). This third area is of special interest. A large dark brown shadowy feature probably of elliptical form, was surrounded by the ditch. The first area, at the west, is very particular because it presents two small post-holes which are indicated on the map as x and y (*Fig. 22*); these post holes suggest the presence of a living space. In the southern part a horse skull (spot No. 4) with

---

17 The earthmoving operation in the archaeological field has often been theorized and discussed. The volume of Pryor 1974 gives the basic instructions for operating in that way.

Fig. 17. Towed Scraper (IsMEO, Dep. Neg. L16885/23).

Fig. 18. The traces of the work of the Towed Scraper (IsMEO, Dep. Neg. L16885/17).

Fig. 19. The whole area of sector A after the third cut (IsMEO, Dep. Neg.L16885/37).

Fig. 20. Sector A, Eastern part, the channel (IsMEO, Dep.Neg. L16887/6a).

some vertebrae (c. infra) came to light and in addition a horse shoe (spot No. 9) (*Fig. 23*)[18]. Around spot No. 27 there is a very large brownish feature which cannot be very well delimited. The general impression is that the third area belongs to a central architectural feature surrounded by the ditch. Unfortunately the relationship between structure No. 26, which is clearly very late, the channel and the central "structure" which seem, on the basis of the material, contemporaneous, is not evident.

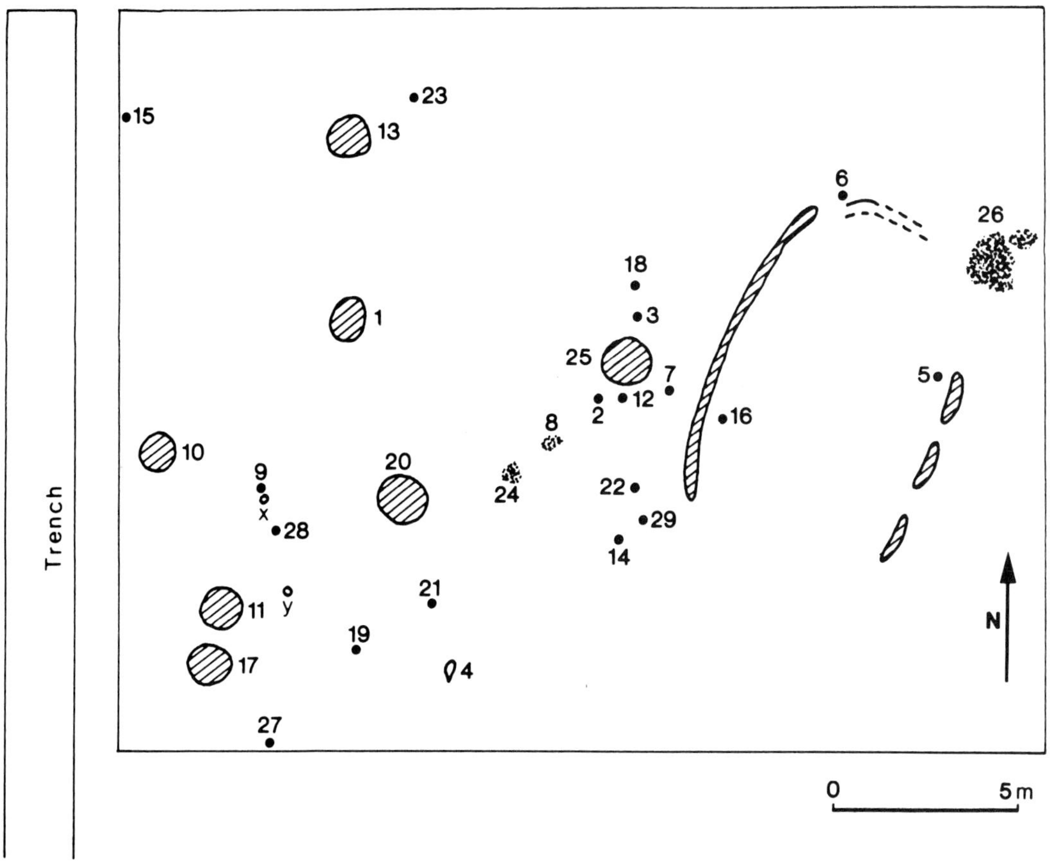

Fig. 21. General map of excavation by M. Cattani, I. Chiesi and S. Pracchia.

Spot fragments (Fig. 24)

No. 16, spot No. 2 : jar, flat base with a vertical ring-shaped handle; lightly everted rim. Coarse ware, medium fired, buff coloured. Light brown burnished surface. Dm. rim 18.5; Dm. base 9.0; width 0.9 (Late Bronze Age/Pre-Scythian)

No. 21, spot No. 5 : vertical rim of a carinated bowl, coarse ware heavily fired blackish coloured. Slipped brownish burnished surface. Dm. 13.0; width 0.7 (Late Bronze Age/Pre-Scythian)

---

18 This skeleton is incomplete and poorly preserved, it actually contains only the fragmented skull, 6 cervical vertebrae /incl. epistropheus/ and the hoof /phalanx distals/ of an adult mare. The skull is of average size and of fine structure. When compared to frontal indices /100 Greatest width: basal length/, the resulting 42.29 % value shows a relatively narrow head, closest to the mean value of Germanic horses /n = 19, X = 40.7%/ and Hungarian Conquest Period horses /n = 34, x = 43.7%/. (information by L. Bartosiewicz).

Fig. 22. Post-hole spot "y" (IsMEO, Dep. Neg. L16888/15).

Fig. 23. Horse skull, spot No. 4 (IsMEO, Dep. Neg.L16886/22).

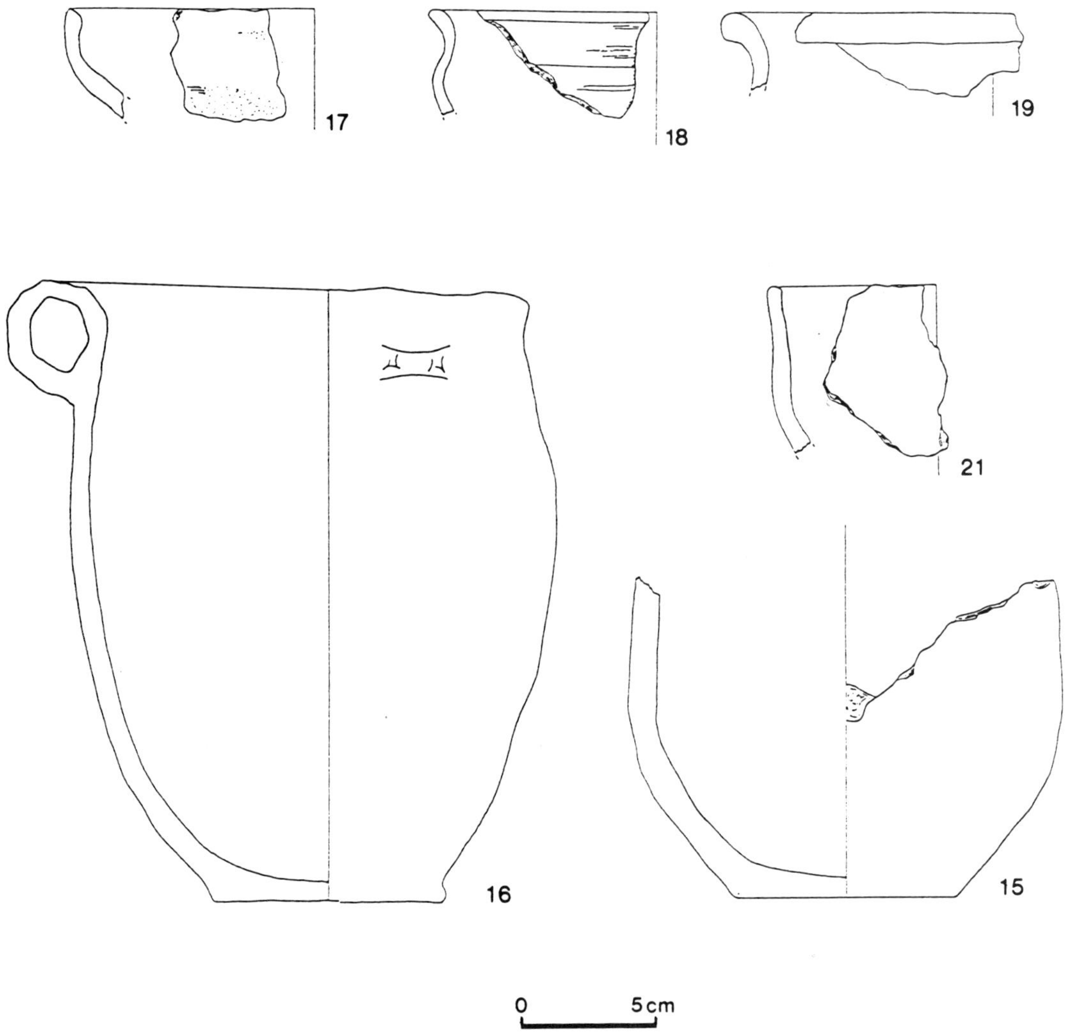

Fig. 24. Pottery fragments from sector A (Spots) by B. Dukay and I. Chiesi.

No. 15, spot No. 12 : flat base of a jar; coarse ware, heavily fired, blackish coloured. Slipped and burnished blackish surface. Dm. 8.5, width 1.0 (Late Bronze Age/Pre-Scythian)

No. 17, spot No. 2 : globular bowl; coarse ware, heavily fired, buff coloured. Slipped brownish burnished surface. Dm. 19.0, width 0.5 (Late Bronze Age/Pre-Scythian)

No. 18, Section No. 1 (sporadic): everted rim of a bowl, medium ware, heavily fired, blackish coloured, slipped blackish burnished surface. Dm. 18.0; width 0.6 (Late Iron Age)

No. 19, spot No. 21 : everted and swollen rim of a jar, medium ware buff coloured. Slipped dark brownish and burnished surface. Dm. 11.0; width 0.5 (Late Iron Age)

Appendix A

MAURO CREMASCHI

Geomorphological survey and the distribution of archaeological sites

The morphology of the research area resembles a gently undulating plain forming the hydrographic basin of the Körös river. The river transverses the center of the area in an east-west direction with an average altitude of 82 m above sea level.

The presence of numerous ancient riverbeds abandoned by the Körös river during its lateral migration from south to north is the main characteristic of this plain. Their strongly meandering courses suggest that this formation probably originated in the Pleistocene. Moreover, these shifts in the course of the river should be emphasized, i.e. the shifts from southeast to northwest are probably attributable to the Upper Pleistocene. The river turned in a northeast-southwest direction, as may be seen in the research area, where the present direction of its flow was finally set.

In regard to the lithology, the composition of sediments can be attributed to alluvial, lacustral and sometimes marshy deposits from the Upper Pleistocene and Holocene as well as to the important contribution of windblown dust which, however, was always resedimented due to the action of the water flow (infusion loess; Pécsi et al. 1982).

The characteristic morphological trait of the small sector examined here is the meandering track of the ancient river bed which is slightly depressed in comparison to the level of the surrounding countryside, from which a number of kurgans of various heights arise.

Pedostratigraphic conditions

The pedostratigraphy in the trench at Endrőd 19 is rather simple: there were two fundamental types of soil, chernozem and vertisols. These occur together in a topographic sequence which developed in accordance with the transition from the typical flat countryside into the depression of the ancient riverbed.

The flat countryside is not completely even but slightly rolling: the undulations may correspond to the ancient river lines of the alluvial system which gave rise to the plain.

The organic chernozem horizons are profoundly eroded in the area of elevations, while the CCa horizon (calcium carbonate accumulations) follows almost immediately under the plough zone.

Cernozem typically appears in the areas better protected from erosion. There it is composed of the following layers:

Ap -0.40 cm: clear silt, very dark brownish- grey 10 YR 3/2, with medium size and large, rounded polyedric particles slightly depressed, fine grained stones. Roots frequently occur here, the pores are ordinary. The humidity and resistance of this layer are low, its lower limit is clearly undulating.

A1 -40-90 cm: clearly silted, very dark brown 10 YR 2/2, with well defined lumpy aggregations. Stones are absent but roots commonly occur. The pores are ordinary. Humidity and resistance of this layer are low. Scanty yellowish brown 10 YR 5/4, spots occur in it. The lower limit is distinct.

A1C -90-110 cm: clear sandy silt loam, very dark greyish brown 10 YR 3/2, with somewhat distinct lumpy aggregations. Stones and roots are missing. Crotovinas causen burrowing rodents are common. The pores are ordinary. Humidity and resistance in this layer are low. The lower limit appears gradually.

CCa -110-130 cm: clearly sandy, yellowish- brown 10 R 5/4 to 5/6. Crotovinas are common. Stones and roots are absent. The pores are ordinary. Humidity and resistance in this layer are low. The calcium carbonate concretions are small and medium size occuring commonly to frequently. The lower limit of this layer was not reached.

In the middle of the ancient river bed, the pedological profile describes a vertisol with hydromorphological characteristics clearly recognizable within this territory:

Ap -43 cm: silty clay, very dark grey 10 YR 3/1, moderately distinct rounded polyhedric particles. In this layer, stones occur very rarely, boots roots are common. The pores are ordinary. The humidity and resistance in this layer are low and its lower limit undulates clearly.

(B) -43-120 cm: clayey, very dark grey 10 YR 3/1, with clearly distinct, medium size and large prismatic angular aggregations. Stones are absent the pores are common with clay skins covering their insides. Roots occur frequently. Pressure on these aggregates results in slickensides. The humidity in this layer is low, its resistance is great. Rarely a slight Fe-Mn patina may occur. The lower limit develops gradually.

Cg -120-150 cm: silty clay, olive browish 2.5 Y 4/4 changing to olive 5 Y 4/3, with medium clearly distinct angular polyhedric aggregations. Stones are absent while pores are common. Roots are scarce. Yellowish brown 10 YR 5/6 spots may also be seen in this layer. Minute Fe-Mn pellicles and nodules commonly occur. The lower limit was not reached.

These two types of soil display a lateral transition. The study of the ancient riverbed's pedological profile allows for a more precise examination of their formation. The clayey texture suggests that there are no waterborn deposits toward the middle of the ancient riverbed. This river bed was formed by the water cutting through the area. Stagnant water followed only in a later, secondary phase. Clay deposited subsequently by this water enriched the pedological profile.

Conclusions

As far as the wear of soil is concerned, research in this area indicated a great stability in pedological conditions.

Since the Neolithic, the area has looked predominantly like steppe or parkland steppe. There is no evidence for more heavily forested phases.

More profound pedological analysis in the laboratory and micromorphological examination in particular, are required to definitely clarify this point.

The ancient riverbed seems to have formed during the Neolithic, since a small Neolithic pit was found within it (Endrőd 125, Section 1). The accumulation of clay followed later, and therefore one may hypothesize a phase of stagnant waters following the Neolithic but preceding the Sarmatian Period.

It should be ascertained whether such a phase is indicative of an increasingly humid environment.

The beginning of intensive cultivation in the area and consequent erosion of the chernozem seem to have been very recent processes. They may have followed the Middle Ages.

Appendix B

MAURO CUCARZI

The geophysical operation at Endrőd 19, May 1987

A trial investigation, based on 3000 magnetic field intensity measurements, was made in an area of 2400 square metres in Endrőd 19. (*Fig. 25*). In this case a couple of proton G 856 AX magnetometers were used. The statistical distribution (*Fig. 26*) of the measurements ranges from -10 to 12nT, the normal value, wherein 66% of all values lay, ranges from -2 to 2nT. A remarkably limited interval, therefore, with low differential values. The histogram furthermore, shows that the values to be considered anomalous in this research are those between 6 and 14 nT. On this basis, the association between grey scales and intervals of differential magnetic field intensity values (see legend) was established. Due to the low scanning value, required to identify significant anomalies, it was necessary to operate with band-pass filters, to reduce the "salt and pepper" disturbances occurring in the first representation. The final thematic map (*Fig. 27*) shows two fairly big anomalies, indicated as A and B which, to judge by their wave characteristics, do not seem to be caused by buried anthropic structures, but rather by phenomena tied to the water situation.

Anomalies A and B (and X) have wave characteristics similar to those caused by archaeological features already found in previous tests. Anomaly X in particular has similar wave width and length to those of the Örménykút fire places (see Cucarzi in this volume).

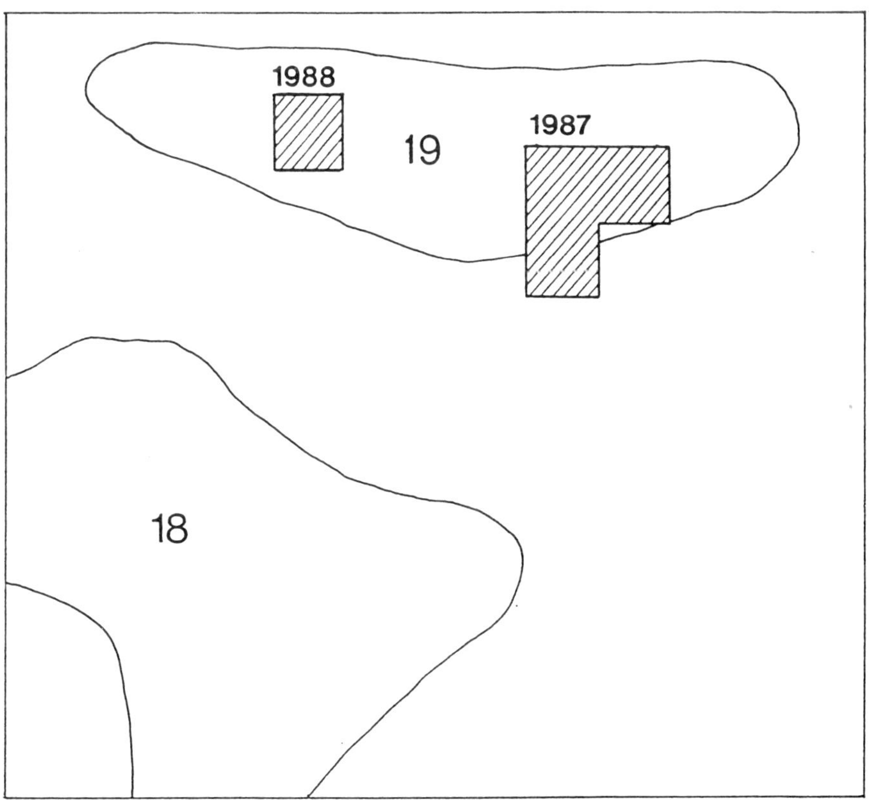

Fig. 25. Sites of Endrőd 18 and 19 with the areas of geomagnetic surveys (hatched).

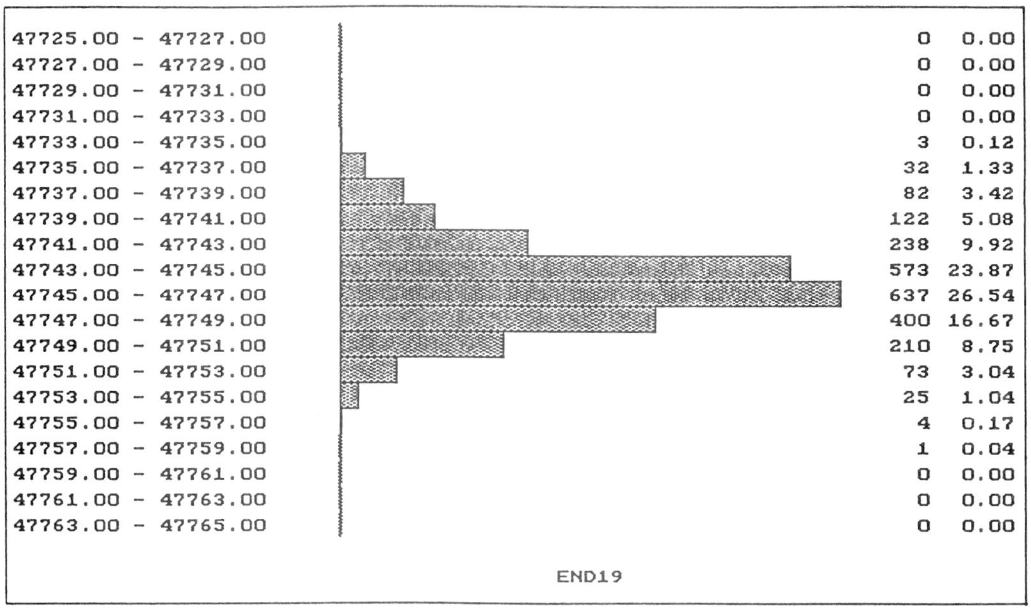

Fig. 26. Statistical distribution of magnetic values measured in 1987 at site Endrőd 19.

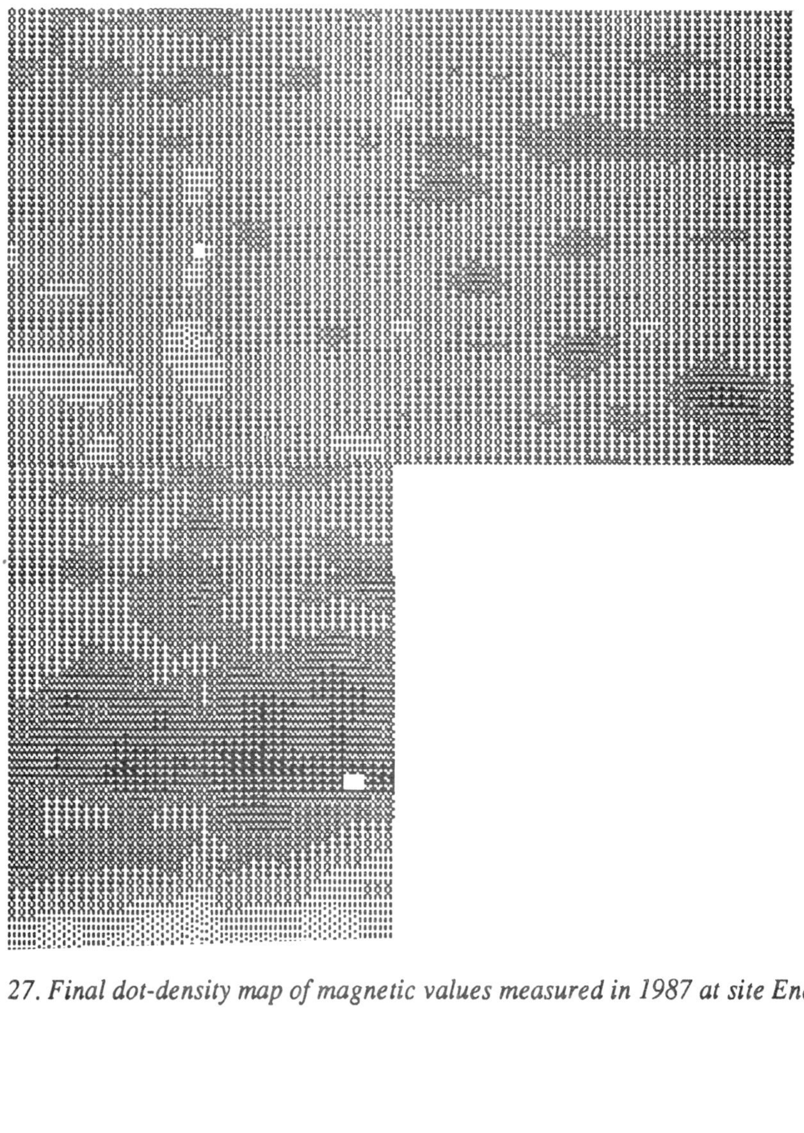

Fig. 27. Final dot-density map of magnetic values measured in 1987 at site Endrőd.

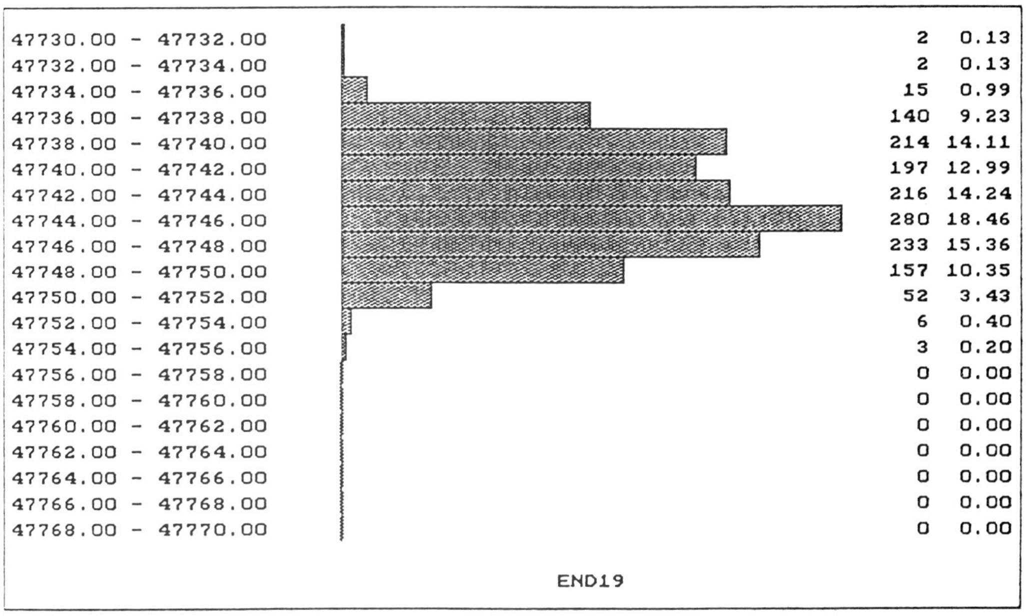

Fig. 28. Statistical distribution of magnetic values measured in 1988 at site Endrőd 19.

Fig. 29. Dot-density map of magnetic values measured in 1988 at site Endrőd 19.

Appendix C

MAURO CUCARZI

The geophysical operation at Endrőd 19, June 1988

In June 1988, a new geomagnetic campaign was carried out on a surface of 1600 square metres, in the western sector of the site contiguously the section No. 1 opened by the Italian Archaeological Mission (cf. supra) with the purpose of gathering further data in areas of more ancient settlements. The same procedure as used in Endrőd 124 (see Cucarzi in this volume), was applied for the measurements and the data treatment, and the histogram (*Fig. 28*) of ground magnetic field intensity shows a very limited distribution (98.12 % of the values lay in the interval between 47735 and 47742 nT). On the basis of this, the association between shades of grey and value intervals was programmed, in order to make up the thematic map (*Fig. 29*). The map shows the prospected area divided into two distinct sectors: one in the north, with relatively high values (47735-47742). This representation shows no "suspect" anomalies, except for a few high intensities in the southern sector. This is due basically to the fact that what may be anomalies produced by buried anthropic structures are small and do therefore not show within a vast area with a high field value. Hence, the data matrix was filtered to remove trend surface, with least-squares methods and map F, where a few anomalies worthwhile mentioning occur, was created with the residuals.

The first is the low intensity anomaly which crosses the whole area from north-east to south-west, and coincides with the separating line between the two distinct sectors mentioned previously.

Then there are some anomalies with medium-high values, generally ranging between 5 and 7 nT (the anomaly with the highest value is that of 10-12 nT, the semi-wave length 1/2 is averagely between 4 and 6 metres.

Considering the wave characteristics of the anomalies registered in various places in the Microregion, it may be estimated that those in Endrőd 19 are comparable with those in Gyoma 133, Örménykút and in the first part of Endrőd 19 (1988). However, to be able to formulate interpretational suppositions on their origin, it is considered necessary to carry out the stratigraphic drilling on a few of the anomalies.

The anomaly which obliquely crosses the prospected area has characteristics and a bearing very similar to the one in Endrőd 124, which may be recently made for agricultural purposes (a small canal, maybe filled with material of lower magnetic susceptibility).

The vast anomaly with relatively high field intensity values, shown in map F, later eliminated with the previously described procedure, seems to have characteristics similar to those registered in Endrőd 124, in the northern sector, and may have the same origin.

References

Arheologija SSSR 1981. *Stepi evropejskoj časti SSSR v skifo-sarmatskoe vremja*. Moskva

Anthony, D. W. 1986. The "Kurgan Culture". Indo-European origins and the domestication of the horse: a reconsideration. *Current Anthropology* 27, 291-314.

Artamonov, M. I., 1949. *Ėtnogeografija Skifii*. Učënye zapiski. Leningradskij Gosudarstvennyj Universitet 13, 129 ff.

Benadik, B., 1953. Skythsky problem ve světle novych archeologickych nálezu na Slovensku (Le problème scythique à la lumière de nouvelles trouvailles de la Slovaquie). *ARozh* 5, 672 ff.

Bottyán, Á., 1955. *Szkíták a magyar Alföldön* (Skythen in der großen Ungarischen Tiefebene). *RFüz* I:1.

Buzdugan, C., 1968. Necropola getică de la Slobozia. *Carpica* I., 77-94.

Členova, N. L., 1975. O svjazah severo-zapadnogo Pričernomor'ja i Nižnego Dunaja s vostokom v kimmerijskuju epohu. *Studia Thracica* S.1.

Crişan, I. H., 1965. Săpături şi sondaje în valea mijlocie a Mureşului (Probe- und Rettungsgrabungen im mittleren Mureschtal), *Acta Musei Napocensis* 2, 39-75.

Crossland, R. A., 1967. Immigrants from the North. *Cambridge Ancient History*. Vol. I, Chap. XXVII, 61.

Dancey, W. S., 1974. The archaeological survey: A reorientation. In: *Man in the North East* 8, 98-112.

Dancey, W. S. and Dunnel, R. C., 1983. The siteless Survey: A regional scale data. In: *Advances in Archaeological Method and Theory* 6, 267-287.

Deshayes, J., 1969. New Evidence for the Indo-Europeans from Tureng Tepe, Iran. *Archaeology* 22/1, 10-17.

Diakonov, I. M., 1981. The Cimmerians. *Acta Iranica*. Hommages et Opera Minora. Vol.7, Monumentum Georg Morgenstierne. I, 103-140.

Dušek, M., 1961. Die thrako-skythische Periode in der Slowakei. *SlA* 9, 155-157.

Dušek, M., 1964. Einfluß der Thraker auf die Entwicklung im Karpatenbecken. *Iswestija Sofija* 65 ff.

Dušek, M., 1975. Die Thraker im Karpatenbecken. *SlA* 22, 356.

Dyson, R., 1973. The archaeological evidence of the second millennium B.C. on the Persian plateau. *Cambridge Ancient History* 2, 16.

Ecsedy, I., 1979. *The People of the Pit-grave Kurgans in Eastern Hungary*. Fontes Archaeologici Hungariae. Budapest

Ecsedy, Il., 1972. Tribe and tribal society in the 6th century Turk Empire. *Acta Orient* 25, 245-262.

Ecsedy, Il., 1984. Ancient Turk (T'u-chieh) burial customs. *Acta Orientalia* 38, 263-287.

Ellis, E. 1984. *The Cucuteni Tripolye Culture*. BAR, IntSer 15, Oxford

Erdélyi, I., 1965. Résultats des recherches archeologiques relatives aux trouvailles des cavaliers nomades orientaux en Hongrie (1953-1964). *ActaAnt* 18, 373-378.

Fettich, N., 1927. Der Goldhirsch von Tápiószentmárton. *ArchÉrt* 41, 138-45, 312-18.

Fettich, N., 1928. *La trouvaille de Zöldhalompuszta*. ArchHung 3, 46.

Fettich, N., 1931. Bestand der skythischen Altertümer Ungarns. In: M. Rostovcev (ed.) *Skythien und der Bosporus*. 494-529, Berlin

Foltiny, S., 1963. Zur Frage des "skythischen" Einflüsses in Österreich und in Slowenien. *AAustr* 33, 23-63.

Frye, R., 1984. *The History of Ancient Iran*. München

Gallus, S. and Horváth, T., 1939. *Un peuple cavalier préscythique en Hongrie*. DissPann II:9 Budapest

Gazdapusztai, Gy., 1963. Beziehungen zwischen den präskythischen Kulturen des Karpatenbeckens und des Nordkaukasus. *ActaSzeged* 5, 5-39.
Gazdapusztai, Gy., 1965. Zusammenhängen des Kulturen des Ungarischen Tiefebene und des Nordkaukasus in der Skythenzeit. *ActaSzeged* 8, 107-111.
Gazdapusztai, Gy., 1967. Caucasian relations of the Danubian Basin in the Early Iron Age. *ActaArchHung* 19, 307-334.
Genito, B., 1986. The Italian Archaeological Activity in Hungary. *East and West* 36, 367-375.
Genito, B., 1988. The Archaeological Cultures of the Sarmatians with a Preliminary Note on the Trial-Trenches at Gyoma 133: A Sarmatian Settlement in South Hungary (Campaign 1985). *Annali dell'Istituto Universitario Orientale di Napoli* 81-126.
Genito, B., (in press a) The nomadic Peoples in the Carpathian Basin: A Western Backwater of the Eurasian Steppes between the First Millennium B.C. and the First Millennium A.D., In: Seaman, G.(ed.): *Ecology and Empire*. Nomads in the Cultural Evolution of the Old World. Los Angeles
Genito, B., (in press b) Trial-Trenches at Gyoma 133. A second interim report
Genito, B. et al. (in press) The excavation of Kurgan Endrőd 122.
Genito, B. and Kemenczei, T., 1990. The Late Bronze Age vessels from Gyoma 133. *ComArchHung*, 113-125.
Ghirshman, R., 1977. *L'Iran et la migration des indo-aryens et des iraniens*. Leiden
Ghirshman, R., 1983. Le Manuscript de Roman Ghirshman. *Le Cimmeriens et leurs Amazones*. Paris
Gimbutas, M., 1970. Proto-Indo-European Culture: The Kurgan Culture during the 5th to the 3rd millennia B.C., in G. Cardona, H. M. Koenigswald and A. Senn (eds.), *Indoeuropeans and Indoeuropean*, 155-98, Philadelphia
Grakov, B. N. and Meljukova, A., 1953. Dve arheologičeskie kul'tury v Skifii Gerodota. *SA* 18, 111-112.
Grakov, B. N. and Meljukova, A., 1954. *Ob ėtničeskih i kul'turnyh svjazah v lesostepnyh oblastjah evropejskoj časti SSSR v skifskoe vremja*. 39-93, Moskva
Grantovski, 1981. "Seraja Keramika" "raspisnjaja keramika" i indoirancy. In: Asimov et al. (ed.): *Ėtničeskie problemy istorii Central'noj Azii v drevnosti* (II tysjačeletie do n.ė.). *245-273*.
Harmatta, J., 1941. *Quellenstudien zu den Skythika des Herodot*. Budapest
Harmatta, J., 1946/47. Le problème cimmérien. *ArchÉrt* 7-9, 79-132.
Harmatta, J., 1966. Kimmerek és szkíták (Kimmerier und Skythen). *AT* 13, 107-116.
Harmatta, J., 1968. Die früheisenzeitlichen Beziehungen zwischen dem Karpatenbecken, Oberitalien und Griechenland. *ActaArchHung* 20, 153-157.
Hazanov, A. M., 1982. *Nomads and the outside world*. Cambridge
Häusler, A. 1974. *Die Gräber der älteren Ockergrabkultur zwischen Ural und Dniepr*. Berlin
Häusler, A. 1976. *Die Gräber der älteren Ockergrabkultur zwischen Dnepr und Karpaten*. Berlin
Hinz, W. 1973. Persia c. 1800-1550 B.C. *Cambridge Ancient History* 2, 256-288
Iessen, A. A., 1953. K voprosu o pamjatnikah VIII-VII vv. do n. ė. na juge evropejskoj časti SSSR, *SA* 18, 49-110.
Il'inskaja, V. A., 1968. Skify dneprovskogo lesostepnogo Levoberežja (Kurgany Posulia), Kiev
Jankovich-B., D., 1985. Archaeological Topography. Theoretical and practical lessons. *MittArchInst* 14, 283-292.
Jankovich-B., D., Makkay, J. and Szőke, B., 1989. *MRT* 8, Budapest
Jankovich-B., D. and Pattantyús-Á., M. 1990. Archaeological and geophysical survey of a multi-aged site Endrőd 170, South East Hungary *Prospezioni Arch*. 1/1990, 123-132.

Kemenczei, T., 1963. Angaben zur Geschichte der Spätbronzezeit in Nordungarn. *ArchÉrt* 90, 169-188.
Kemenczei, T., 1970. A kyjatice kultúra Észak-Magyarországon, *A Herman Ottó Múzeum Évkönyve* 9, 17-68.
Kemenczei, T., 1980. Szkíta kori lelet Balassagyarmatról. (Ein skythenzeitlicher Fund aus der Umgebung von Balassagyarmat) *FA* 31, 65-75.
Kemenczei, T., 1981. Ostungarn in der Zeit der Frühhallstattkultur. *Symposium Steyr* 1980, Linz 1981, 79-92.
Kemenczei, T., 1982. Die Gava Kultur. Poludniowa strefa kultury luzickiej i powiazania tej kultury z poludniem 282, Kraków/Prznyse
Kovács, T., 1966/67. Eastern connections of north-eastern Hungary in the Late Bronze Age. *FA* 18, 27-58.
Lang, A., 1980. Zur Chronologie frühskythischer Funde im Karpatenbecken und an der unteren Donau. *Actes du II Congrès International de Thracologie* I. Bucureşti, 247.
Lapušnjan, V. L., 1972. Raskopki pamjatnikov kultury frakijskogo galstata v Moldavii. *Arheologičeskie issledovanija v Moldavii*, 1968-69, 92.
Lapušnjan, V. L., 1979. Rannie frakijcy X - načala IV v. do n. ė.. v lesostepnoj Moldavii. Kišinev
Liberov, P. D., 1962. Pamjatniki skifskogo vremeni bassejna Severnogo Donca, *Kul'tury skifskogo vremeni*, 35, Moskva
McGovern, W. M., 1939. *The Early Empires of Central Asia.* Chapel Hill, North Carolina
Mallory, J. P., 1989. *In Search of the Indoeuropeans.* London
Makkay, J., 1987. The Linear Pottery and the Early Indo-European. In: Skomal, S. N. and Polome, E., (eds.) *Proto-Indo-Europeans: The archaeology of a linguistic problem,* Washington D.C. 176-177.
Makkay, J. 1989. Cultural groups of SE-Europe in the Neolithic: The PIE homeland problem and the origin of the Protogreeks. *Atti del Convegno su "Linee e Momenti di Preistoria Linguistica dell'Eurasia"*, Napoli, 16-17 Maggio 1988, 117-137.
MIA 1958. *Pamjatniki skifo-sarmatskogo vremeni v Severnom Pričernomore.* Moskva
MIA 1961. *Pamjatniki epohi bronzy i rannego železa v Severnom Pričernomore.* Moskva
MIA 1962a. *Pamjatniki skifo-sarmatskoj kul'tury.* Moskva
MIA 1962b. *Lesostepnye kul'tury skifskogo vremeni.* Moskva
MIA 1969. *Drevnye frakijcy v Severnom Pričernomore.* Moskva
MIA 1971. *Problemy skifskoj arheologii.* Moskva
Meljukova, A. I., 1955. K voprosu o pamjatnikah skifskoj kul'tury na territorii Srednej Evropy, *SA* 22, 239-253.
Meljukova, A. I., 1979. *Skifija i frakijskij mir.* Moskva
Meljukova, A. I., 1989. Predskifskij (kimmerijskij) period v stepi i lesostepi Vostočnoj Evropy. In: Kimmerijcy, skify i ih sosedi v stepi i lesostepi Vostočnoj Evropy. Meljukova, A.I., (ed.), *Stepi evropejskoj časti SSSR v skifo-sarmatskoe vremja*, 10-35.
Nichols, H. L. jr., 1954. *Moving the Earth, the workbook of excavation.* Princeton
Párducz, M., 1952. Le cimetière hallstattien de Szentes-Vekerzug. *ActaArchHung* 2, 143-169.
Párducz, M., 1954. Le cimetière hallstattien de Szentes-Vekerzug II. *ActaArchHung* 4, 25-89.
Párducz, M., 1955. Le cimetière hallstattien de Szentes-Vekerzug III. *ActaArchHung* 6, 1-18.
Párducz, M., 1965a. Graves from the Scythian Age at Ártánd. *ActaArchHung* 17, 137-231.
Párducz, M., 1965b. Western relations of the Scythian Age Culture of the Great Hungarian Plain, *ActaAnt* 13, 273-301.
Párducz, M., 1966. The Scythian Age cemetery at Tápiószele. *ActaArchHung* 18, 35-91.
Párducz, M., 1973. Probleme der Skythenzeit im Karpatenbecken, *ActaArchHung* 25, 27-63.
Pârvan, V., 1926. *Getica. O protoistorie a Daciei.* Bucureşti

Pécsi, S. et al. 1969. *A tiszai Alföld.* (The Tisza Plain). Budapest
Petch, D. F., 1968. Earthmoving machines and their employment on archaeological excavations. *Journal of the Chester Archaeological Society* 55, 15-28.
Pryor, F., 1974. *Earthmoving. On open archaeological sites.* Peterborough
Renfrew, C., 1988. *Archaeology and Language.* New York
Roska, M., 1937. Der Bestand der skythischen Altertümer Siebenbürgens. *Eurasia Septentrionalis Antiqua* 9, 167-202.
Smirnova, G. I., 1969. Poselenie Magala - pamjatnik drevnefrakijskoj kul'tury v Prikarpat'e. Drevnie frakijcy v Severnom Pričernomore. MIA, 150, 26-30.
Smirnova, G. I., 1974. Complexele de tip Gava-Holihrady – o comunitate cultural-istorică. *SCIVA* 25, 359.
Smirnova, G. I., 1975. Gavsko-holigradskoj krug pamjatnikov v vostočnokarpatskom bassejne. *Arheologičeskij Sbornik Gosudarstvennogo Ermitaž* 17, 20-21.
Sulimirski, T., 1960. The Cimmerian Problem. *Bulletin of the Institute of Archaeology, University of London* 2, 45-64.
Telegin, D., 1986. *A Settlement and Cemetery of Copper Age Horse Keepers in the Middle Dnieper.* BAR, IntSer 287, Oxford
Terenožkin, A. I., 1973. *Skifskie drevnosti.* Kiev
Terenožkin, A. I., 1976. *Kimmerijcy.* Kiev
Tosi, M. and Seaman, G. (ed.) (in press), *Theoretical considerations on the origin of the Pastoral Nomadism, Ecology and Empire.* Nomads in the cultural evolution of the Old World. Los Angeles
Vulpe, A., 1967. *Necropola hallstattiană de la Ferigile.* Biblioteca de arheologie 9, Bucureşti
Vulpe, A., 1980. Problèmes actuels de la protohistoire thrace. *Actes du II Congrès International Trachologie.* Bucureşti
Young, T. C. jr., 1967. The Iranian Migration into Zagros. *Iran* 5, 11-34.
Young, T.C. jr., 1985. Early Iron Age revisited: Preliminary suggestions for the re-analysis of old constructs. *De l'Indus aux Balkans, Recueil à la memoire de J. Deshayes.* 361-378.

The excavation of a settlement and a cemetery in the outskirts of Gyomaendrőd

(Preliminary report)

ISTVÁN ERDÉLYI

These investigations were carried out within the framework of the Microregion Research Project between 1985 and 1989. The primary aim was to discover traces of a late Avar Period settlement which would warrant further excavation. Thus, following results of an 1984 field survey, a test excavation was started at site 60 on the outskirts of Endrőd. The site is located in the courtyard of farmstead No. 254/1 owned by L. Bacsa (*Fig. 1*).

Fig. 1. The Bacsa farmstead and its environs with the excavated squares.

A few sherds from wheel-thrown vessels with wavy decoration and a clay pot fragment were brought to light in the neighbouring kitchen garden. Investigations covered the area of farmsteads in the immediate neighbourhood as well as the ploughed fields of Sándor Bula, but no other Avar Period remains were discovered. In 1985, one of the test trenches revealed remains of a fireplace within a depression. Sherds of a clay vessel found by this fireplace could be reconstructed into a broad bowl with a high rim (diameter of the mouth: 29 cm, wall height: 7,5 cm). The fireplace was associated with an oval pit dwelling indicated only by its faint outline.

In 1986, research was concentrated on the northern side of the dirt road which passed by the farmstead. Here a 25 by 2 metre trench was opened at the top of a slight elevation. It lay parallel with the long axis of the road. A child's skeleton (Inf. I) was discovered at an 11 metre distance from the northern end of this trench. It was found at a depth of 40 cm without accompanying artifacts. The plastered bottom of an 'oven' was identified in the immediate proximity of this burial. The infant skeleton lay by a fragment of wall from this feature. This part is rather obscure but may represent the oven's flue. It is possible that the dome of this oven was destroyed by deep ploughing, but no burnt clay fragments were found in the area. One should not rule out the possibility that these remains belong to a primitive baking pit of a type used in the territory of Southern Russia during the 8-9th centuries. No sherds were brought to light in the proximity of the infant skeleton. Likewise, no pottery fragments were found either inside or around the oven.

Since (with the exception of a few vessel fragments from the Sarmatian Period) neither this nor subsequently opened squares yielded any early medieval finds worthy of more detailed examination, further investigations were conducted in the periphery of Endrőd in the direction of Szarvas. It must be mentioned in relation to this site, that a very narrow slightly curving (possibly circular), ditch segment was discovered in S. Bula's ploughed field, which may potentially belong to a Sarmatian burial mound. This investigation could not, however, be continued due to planting. Another excavation was started at site 157. At this spot, a cemetery was discovered during the opening of a clay extraction pit related to the reconstruction of the Timár farmstead (now called the Szigetvári farmstead) in 1938. László Bacsa directed our attention to this site. According to his account, two skeletons were found at that time.

First a 10 by 1.5 metre test trench was opened parallel to the farmstead's building. This was located towards the road on the western side of a row of mulberry trees.

Grave 1 was found at the northwestern corner of this trench. The depth of this burial was 0.65 metres. It was oriented northwest. The grave contained a man's skeleton lying on its back with the arms tightly arranged along the body. The length of the skeleton was 140 cm in the grave. The feet lay slightly higher and the skull turned to the right. No grave goods were included.

Grave 2 is probably the remainder of the infant burial disturbed by the clay extraction pit in 1938. Only a few femur and tibia fragments were preserved at a depth of 0.85 metres around the middle of the square. The orientation of this grave could not be identified.

Fragments of a dark grey, handmade pot found at the southern end of this square may also belong to the graves disturbed by the previously mentioned construction pit. This set of sherds was labelled *Grave 3*. It was found at a depth of 1.40 metres.

Parallel with the first test trench another, similar, square was opened at a 1.5 metres distance to the west. Here, *Grave 4* was recovered at a depth of 1.06 metres. It was oriented to 39 degrees northwest. The grave contained a small girl's skeleton lying on her back with arms tightly arranged along the body. The skull rested on its right side and was corroded. No grave goods were found.

Grave 5 was found in the same square in the proximity of the previously described burial at a depth of 1.11 metres. It was oriented to the northwest. The grave contained a woman's skeleton lying on her back with arms arranged so tightly along the body that the bones of the forearm were forced slightly under the pelvis. The skull faced left.

Grave goods included a pig's leg bone by the lateral side of the woman's leg and a heavily corroded iron buckle in the region of the waist.

The excavation was continued in 1987. First a 5 by 5 metre square was opened north of the farmstead building. The eastern edge of the square fell in line with the eastern wall of this building. The other square was located along the western side of the same building. It measured 16 by 4 metres and lay 7.20 metres from the wall (*Fig. 2*).

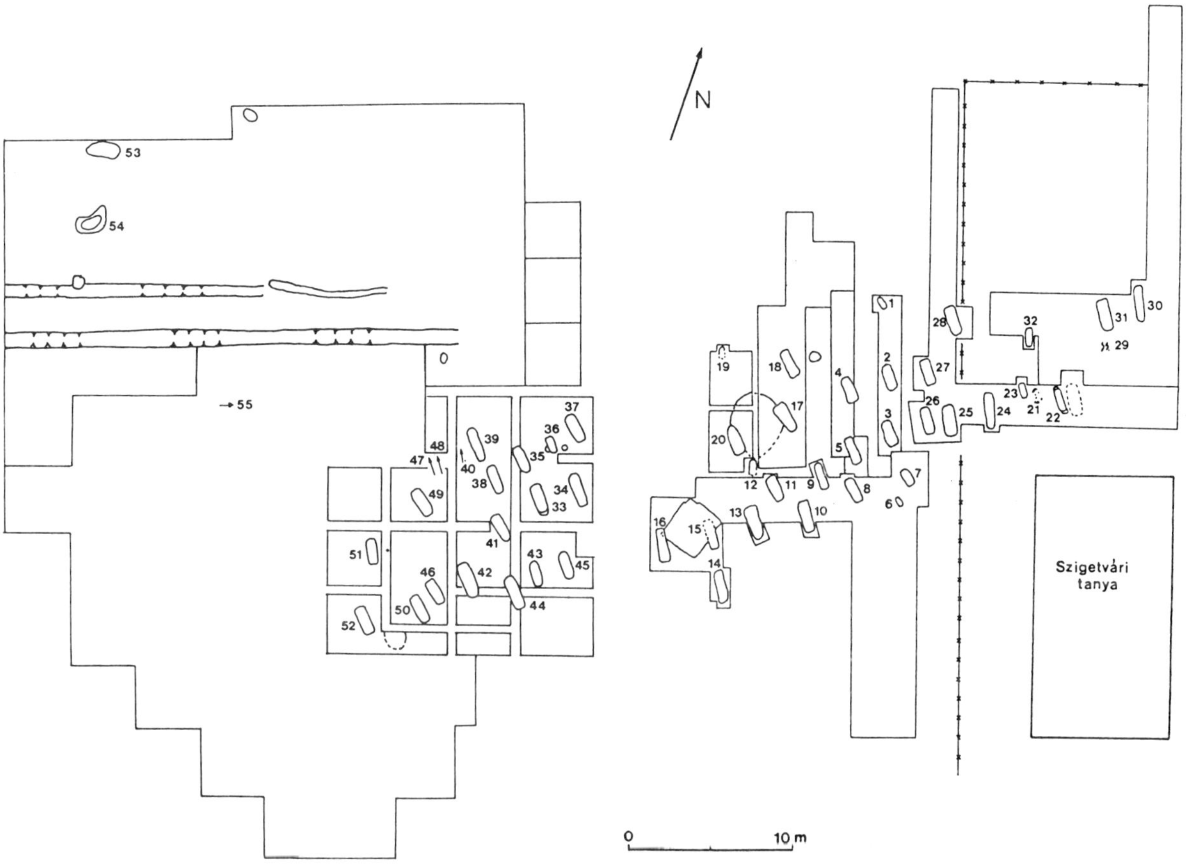

Fig. 2. Plan of the cemetery excavated near the Szigetvári farmstead.

The 5 by 5 metre square contained no burials. On the other hand, *Grave 6* was found in the larger square. It was 0.73 metres deep and oriented to the northwest. The grave contained an infant's skeleton (a child, approximately 18 months old) lying on its back with arms arranged tightly along the body. The skull is fragmented, and slightly displaced between the claviculae. The length of the grave pit is 1.90 metres, the width is 0.50 to 0.60 metres. No grave goods were found.

Grave 7 was found at the northern corner of this square. It contains an infant's skeleton lying on its back at a depth of 0.97 metres. The skull is turned to the left. Orientation: 40 degrees northwest. The length of the grave pit is 1.16 metres, the width is 0.50 metres. The length of the skeleton is 0.77 metres. Grave goods include a broken iron knife by the lateral side of the femur and an iron sickle placed between the right and left feet.

Grave 8 was 1.04 metres deep and oriented 40 degrees to the northwest. It is a well preserved skeleton of an adult lying on its back. Skeleton length: 1.56 metres. The skull was fragemented. The arms were arranged tightly along the body while the bones of the leg were found in a slightly contracted position. The only grave offering was a fragmented iron buckle found under the right side of the pelvis.

Grave 9 was 1.54 metres deep. It contained the well preserved skeleton of an adult lying on its back with the skull slightly turned to the left. Orientation: 40 degrees to the northwest. The arms were extended. No grave goods were found.

Grave 10 was 1.01 metres deep. It contained the very well preseved skeleton of an adult in an extended position. The skull was turned gently on to the right shoulder. The arms were arranged quite tightly along the body. The skeleton was 1.66 metres long. The length of the grave pit was 2.15 metres, the width was 0.58 to 0.61 metres.

Grave goods included an iron knife near the medial side of the right wrist and a fragmented iron buckle on the sacral bone.

Grave 11 was 1.50 metres deep. It contained the well preserved skeleton of an adult lying on its back. The skull turned slightly to the left, the arms were arranged tightly along the body. The skeleton was 1.65 metres long. The grave was oriented 40 degrees to the northwest.

Grave goods include a round iron buckle by the sacral bone and an another iron buckle near the bottom of the vertebral column similarly on the left side. An iron knife was found around the middle of the left arm on its lateral side.

Remains of an 11th century hut sunk half-way into the earth and accompanied by a plastered oven were found among the following set of graves. Consequently, the square had to be extended. This house cut through some of the graves.

Grave 12 contained a poorly preserved skeleton of an infant at a depth of 0.96 metres. The skull was slightly turned to the left. The skeleton was 0.96 metres long and oriented 10 degrees to the northwest. The length of the grave pit was 1.10 metres, the width was 0.50 to 0.45 metres.

The only accompanying grave offering was a fragmented iron buckle located above the left side of the pelvis.

Grave 13 was 1.40 metres deep. It contained a man's skeleton that lay on its back. The jaw was slightly dislocated. The skeleton's legs lay et some distance from one another. The grave was oriented 3 degrees to the north. The skeleton length was 1.73 metres. The length of the grave pit was 2.10 metres long, the width was 0.66 to 0.76 metres.

Grave goods include a bronze button on the chest and an iron knife by the medial side of the left forearm.

The next square was opened in a position adjacent to the north side of the two narrow test trenches investigated during the previous two seasons.

Grave 14, however, was still found in the previously discussed square. It revealed a well preserved woman's skeleton at a depth of 1.23 metres. She lay on her back oriented 30 degrees to the northwest. The skeleton was 1.64 metres long.

Grave goods included a row of eight glass beads located on the right side of the vertebral column. Five of these had the shape of melon seeds while the remaining three were cylindrical. A fragment of a pig jaw was found on the lateral side of the tibia.

Grave 15 was cut through during the construction of the early medieval house. The skeleton was missing above the knee. It belonged to an adult.

Grave 16 contained the skeleton of an adult man lying on its back. Its arms were arranged tightly along the body. Grave depth: 1.07 metres. Orientation: 22 degrees to the northwest. Skeleton length: 1.71 metres. Length of the grave pit: 2.00 metres, width of the grave pit: 0.62 metres. The only grave offering was a broken iron knife that lay between the femora in a parallel position.

Grave 17 contained the skeleton of an adult man lying on its back. His arms were arranged tightly along the body and the skull turned to the right. The left leg (together with the feet) was found on the right side. Grave depth: 1.34 metres. Orientation: 42 degrees to the northwest. Length of the grave pit: 1.96 metres, width of the grave pit: 0.70 metres.

The only accompanying find was an animal bone by the left foot.

Grave 18 contained the skeleton of a woman lying on its back. The arms were arranged tightly along the body. Orientation: 30 degrees to the northwest. Length of the skeleton: 1.50 metres. Length of the grave pit: 1.77 metres, width of the grave pit: 0.58 metres.

No grave goods were found.

Grave 19 contained an infant's skeleton with a flattened skull. Grave depth: 0.50 metres. Orientation: 13 degrees to the north-northwest. Length of the grave pit: 0.80 metres, width of the grave pit: 0.74 metres.

No grave goods were included.

Grave 20 contained the skeleton of a man lying on its back. The arms were arranged tightly along the body, with the skull tipped over. Orientation: 30 degrees to the northwest. Length of the skeleton: 1.74 metres. Length of the grave pit: 2.25 metres, width of the grave pit: 0.75 metres.

Grave goods included a belt set with bronze mountings: the buckle was found upside down near the proximal end of the femur. The main strap end was located pointing upward near the left wrist. In addition, four smaller strap ends, a so-called propeller-shaped mount and ten other belt mounts (three of them with pendants) were found. A hole reinforcement found here belonged to the same belt. The loop belonging to the buckle was replaced by an iron plate. An additional iron buckle was found by the caudal edge of the left pelvis. In the same area an obsidian blade and remains of leather were identified. An iron knife was found under the right hand. Some of the small mounts that decorated minor straps occurred sporadically. Eight such pieces could be recovered. All these bronze mounts are decorated in a style characterized by flat tendril patterns.

Excavations in 1988 resulted in the discovery of additional graves in the 8.70 by 4.90 metre trench opened at a three metre distance from the farm building's northern wall (and adjacent extensions).

Grave 21 was found at the western edge of the first square. It was disturbed by the planting of a coniferous tree at the beginning of the 1980's (which subsequently died). This grave was only 0.45 metres deep and oriented 17 degrees to the north-northwest. It contained the contracted skeleton of a 16-17 year old lying on its left side. The arms were bent in such a way that the hands could be placed in front of the face. The grave pit was impossible to detect.

No grave goods were found.

Grave 22 was located east of Grave 21 in a parallel position. It contained the extended skeleton of a young woman (?) in a very narrow pit with her left arm bent onto the pelvis. The skull turned to the left. Depth of the grave: 1.00 metres. Skeleton length: 1.28 metres. Orientation: 31 degrees to the northwest. Length of the grave pit: 1.59 metres, width of the grave pit: 0.43 metres.

An iron knife was placed by the right hand.

Grave 23 contained an extended infant skeleton with a fragmented skull. Grave depth: 0.77 metres. Orientation: 32 degrees to the northwest. Length of the grave pit: 0.95 metres, width of the grave pit: 0.30 metres. Skeleton length: 0.71 metres.

Grave goods included a melon-seed shaped bead broken in two pieces under the skull and a round white calcedony bead around the 2nd and 3rd vertebrae. A small, handmade clay vessel was placed at the feet. (*Fig. 3*).

Fig. 3. Clay vessel from Grave 23 of the Avar cemetery.

Grave 24 contained the very poorly preserved skeleton of a man with a flattened skull lying on its left, facial side. The left knee was slightly bent. Grave depth: 0.93 metres. Orientation: 29 degrees to the northwest. Length of the grave pit: 2.25 metres, width of the grave pit: 0.93 metres.

No grave goods were included.

Grave 25 contained the skeleton of an adult lying on its back with the arms arranged tightly along the body. The skull turned to the right. Grave depth: 0.95 metres. Orientation: 30 degrees to the northwest. Skeleton length: 1.61 metres. Length of the grave pit: 2.07 metres, width of the grave pit: 0.72 metres.

Grave goods included a round and an oblong iron buckle found to the left of the last lumbar vertebra. A white bead occurred at the edge of the right ilium. Under the right ilium lay an iron knife.

Grave 26 contained a woman's skeleton lying on its back with slightly contracted arms. Her hands were placed on the lower part of the pelvic region. The skull turned to the right while the jaw was found upside down near the waist. The sacrum pointed in the opposite direction from its caudal end. Grave depth: 1.20 metres. Orientation: 35 degrees to the northwest. Length of the grave pit: 1.78 metres, width of the grave pit: 0.62 metres. Skeleton length: 1.45 metres.

Grave goods included bronze earrings found under the nasal region and at the right shoulder as well as on the chest. A small iron ring (?) occurred by the left hand.

Grave 27 contained the skeleton of an adult individual lying on its back. The right arm was under the pelvis while the left arm lay above it. The skull turned to the right. Grave depth: 1.58 metres. Orientation: 25 degrees to the northwest. Skeleton length: 1.46 metres. Length of the grave pit: 1.78 metres, width of the grave pit: 0.65 metres.

No grave goods were found.

Grave 28 contained the extended skeleton of an adult individual with arms arranged along the body. Grave depth: 1.80 metres. Skeleton length: 1.44 metres. Orientation: 30 degrees to the northwest. Length of the grave pit: 1.68 metres, width of the grave pit: 0.63 metres.

No grave goods were included.

The remains of *Grave 29* were discovered in a secondary position under a small mound of yellow clay. This layer was created subsequently and is independent of the burial. The grave was located some 10 cm below the original surface. No grave pit could be identified.

Grave goods were not recovered.

Grave 30 contained the skeleton of an adult lying on its back with extended arms. Grave depth: 1.08 metres. Orientation: 30 degrees to the northwest. Length of the skeleton: 1.74 metres. Length of the grave pit: 2.20 metres, width of the grave pit: 0.62 metres.

The only grave offering was an oblong iron buckle discovered left of the left femur's proximal end.

Grave 31 contained the extended skeleton of an adult, with the arms arranged along the body. Orientation: 45 degrees to the northwest. Skeleton length: 1.59 metres. Length of the grave pit: 2.10 metres, width of the grave pit: 0.65 metres.

Grave goods included a bronze buckle around the middle of the chest, an iron buckle in the region of the waist and a fragmented iron knife by the right hand.

Grave 32 contained the skeleton of an infant lying on its back. The jaw was found in the abdominal region. Skeleton length: 0.93 metres. Grave depth: 0.72 metres. Orientation: to the northwest. Length of the grave pit: 1.30 metres, width of the grave pit: 0.50 metres.

No grave goods were included.

The excavation continued in 1989.

Grave 33 contained two skeletons. One of these belonged to an adult woman, the other to a child. This latter lay on the woman's left side in the region of the breast. Grave depth: 1.22 metres. Orientation: 32 degrees to the northwest. Grave goods included a white bead found in the right side section of the grave pit and a clay spindle whorl that occurred under the left clavicula of the woman's skeleton.

Grave 34 contained the skeleton of an adult individual. It lay on its right side with the skull turned to the right. Grave depth: 1.04 metres. Orientation: 35 degrees to the northwest. Grave goods included an oblong iron knife and an iron buckle.

Grave 35 contained the skeleton of a woman (?) lying on its back. The skull was turned slightly to the left and the left hand was placed on the edge of the left pelvis. Grave depth: 1.03 metres. Orientation: 38 degrees to the northwest. No grave goods were found.

The dimensions of *Grave 36* were indicative of an infant's burial, however, no bones were found. It contained a green, melon seed-shaped glass bead. Grave depth: 0.96 metres. Orientation: to the northwest.

Grave 37 contained the skeleton of a man with the left hand under the pelvis and the right hand on it. The skull tilted on to its left side. Grave depth: 0.81 metres. Orientation: 38 degrees to the northwest.

No grave goods were included.

Grave 38 contained the skeleton of an adult man lying on its back with arms extended along the body. Grave depth: 1.24 metres. Orientation: 32 degrees to the northwest. The right radius displaced and two of his ribs were from their original position.

No grave goods were found.

Grave 39 contained the skeleton of a woman lying on its back with the arms arranged along the body. Grave depth: 1.05 metres. Orientation: to the northwest. Grave goods included a bronze earring found to the left of the skull accompanied by two thin bronze plates. Their form suggests that these latter two pieces may have been mounted onto the edge of a wooden cup. An iron buckle was found in the region of the pubic bone.

Grave 40 contained the skeleton of a man lying on the back. This burial was oriented to the northwest.

Grave 41 contained the skeleton of a woman lying on its back with the arms arranged tightly along the body. Grave depth: 1.08 metres. Orientation: to the northwest. Grave goods included an iron knife found on the medial side of the left femur directly above the knee. A narrow, slightly bent iron band was found under the skeleton's lumbar region.

Grave 42 contained the skeleton of a woman lying on the back, oriented to the northwest.

Grave 43 contained the skeleton of a children lying on its back. Wooden rests probably belonging to a coffin were found under the skeleton. No grave goods were included. The skeleton oriented to the northwest. Grave depth: 1.10 metres.

Grave 44 contained the skeleton of a man lying on its back and oriented to the northwest.

Grave goods included a belt set with cast bronze mounts decorated by tendril patterns. The narrow mounts were accompanied by small pendants. It may be observed that originally these were fastened in groups of four side-by-side. *(Fig. 4)*.

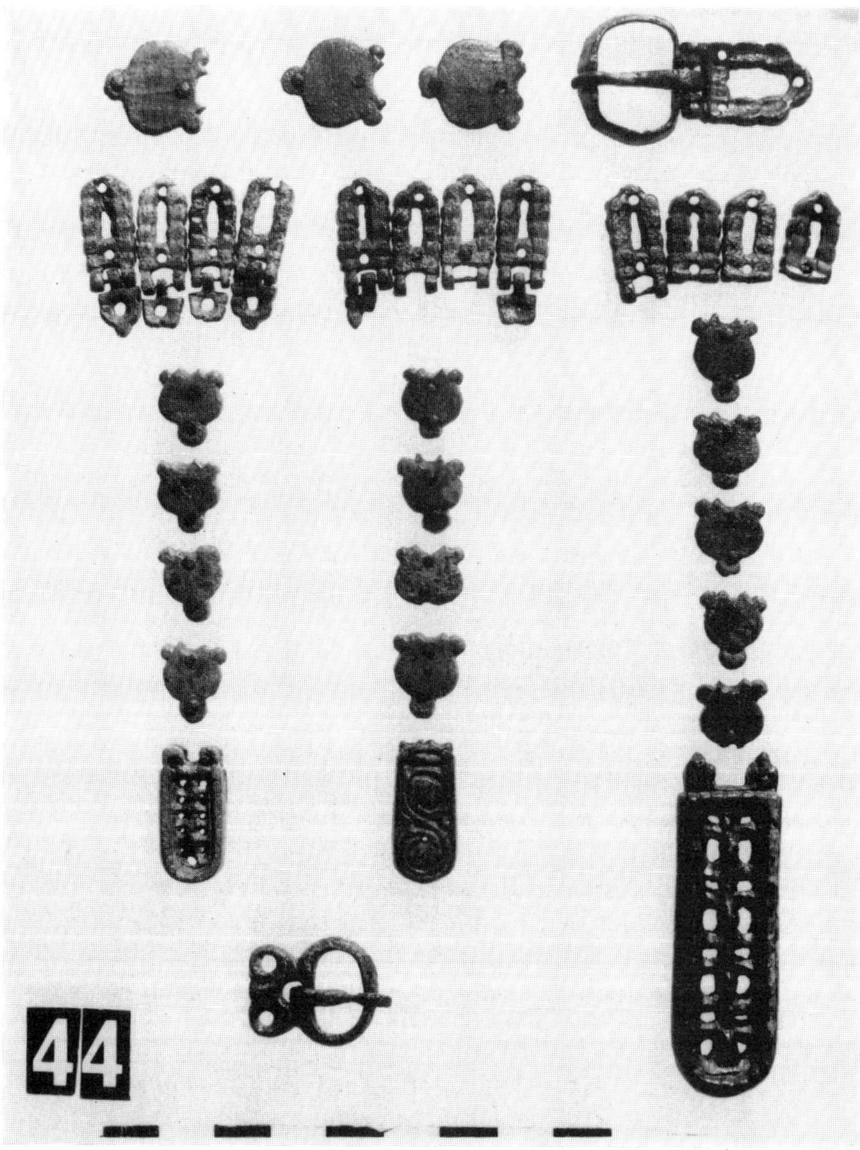

Fig. 4. Avar period mounted belt set from Grave 44.

Grave 45 contained the skeleton of a man lying on its back and oriented to the northwest. No grave goods were included. Grave depth: 1.02 metres.

Grave 46 contained the skeleton of a man lying on the back and oriented to the northwest. No grave goods were included. Grave depth: 0.47 metres.

Graves 47 and *48* each contained an infant's skeleton. These burials were extremely close to each other and quite shallow. Orientations: to the northwest. The outline of these graves could not be identified.

No grave goods were included.

Grave 49 contained the skeleton of a man lying on its back and oriented to the northwest. This grave cut into a Copper Age pit which contained a large, deep bowl with a big carbonized seed in it. No grave goods were included. Grave depth 0.98 metres.

Grave 50 contained a skeleton lying on its back. The skeleton oriented to the northwest. Grave goods included an iron knife and an iron buckle. Grave depth: 0.73 metres.

Grave 51 contained the skeleton of an infant lying on its back and oriented to the northwest.

Grave goods included 29 glass beads (one of which was of a melon seed-shape) scattered around the claviculae and the upper region of the chest. The fragment of a thin iron ring was found above the laterally pointing end of the left clavicula.

Grave 52 contained the skeleton of a man lying on its back. A thin layer of decaying wood could be detected along the lateral side of the left tibia. The burial was oriented to the northwest. Grave depth: 1.27 metres.

Grave goods included two thin, undecorated mounts of bronze plating (buckle loop-like) and an iron buckle in the region of the waist.

During four excavation seasons it was possible to identify three sides of this cemetery. It seems clear that the graves were placed in four or five, slightly irregular and distant rows that follow a westward trend. The major part (probably more than half) of this cemetery still remains to be excavated. (*Fig. 5*)

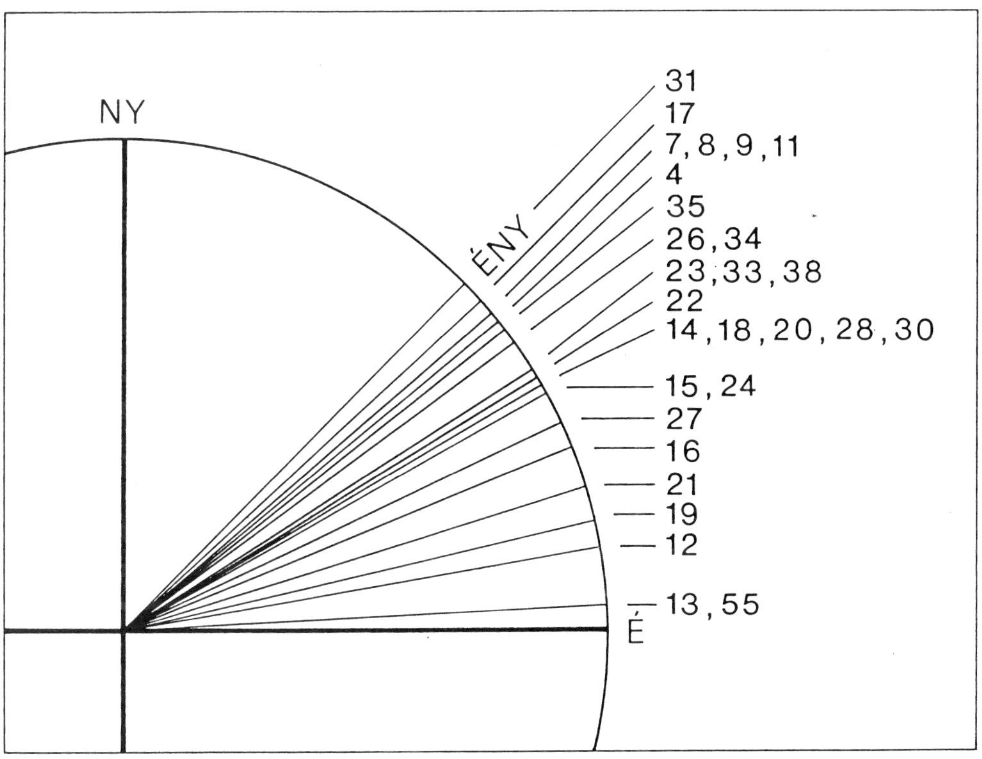

Fig. 5. The distribution of grave orientations in the Avar cemetery.

In addition to the graves, the remains of a place of sacrifice were also found. This was actually a small fireplace which contained fragments of a small, heavily burnt and undecorated pot from the 8th-9th centuries.

The beads and incomplete belt sets date this cemetery to the 8th century. The poverty of the community that used this cemetery is noteworthy, especially in light of the rich cemetery (even containing mounted horse burials) which has been excavated by I. Juhász on the right side of the Szarvas road. That site is located only about seven kms from the cemetery under discussion here and belongs to the same time period. She also excavated a cemetery near the railway station of Szarvas which is also located seven km from the Endrőd cemetery. This late Avar cemetery contained more than 800 graves and was date to the 8th century. This shows that groups in very different economic positions occupied nearby habitation areas simultaneously. The cemetery under discussion is located south of the main stream of the Körös river, left of the Endrőd-Szarvas road and the old dirt road. Prior to river regulation a narrow oxbow occurred here which flooded periodically. The graves are located along this oxbow on a slight elongated elevation where the water did not reach them. The water table also limited the depth of these burials. In addition to the shallowness of these graves it is remarkable how narrow they are. In numerous cases, the skeletons are almost squeezed into the graves. It is possible that these bodies were wrapped tightly in some sort of a textile. Frozen ground may alternatively explain why some of the grave pits are so narrow. A summarized evaluation of grave orientations would also be instructive.

No remains of Avar Period occupation (that is a settlement potentially associated with the cemetery) were found. Sporadic traces of a village-like settlement from the period of the Árpád Dynasty were identified on the previously mentioned elongated elevation and south of another elevation. These areas were occupied following the departure of the Avars.

A remarkably small number of clay vessels were found in this cemetery: only three were recovered from a total of 52 burials. This seems to point to the nomadic character of the economy. At the same time, this observation is contradicted by the fact that none of the graves contained sheep, horse or even cattle bones but rather pig remains, that is, the extremities of pork among the grave goods may be observed. Only two of the burials contained mounted belt sets characteristic of free nomadic (semi-nomadic) warriors and family heads.

Further questions may be answered only following the physical anthropological evaluation of the material. It seems already clear at this point, however, that there is no marked difference between the proportion of men and women and that infant mortality was not particularly high. As far as racial characteristics are concerned, skeletons in the cemetery under discussion show a great deal of heterogeneity. In some graves very tall individuals were buried, especially in a certain part of the cemetery.

It must be mentioned that a short distance to the south, beyond a parallel range of elevations, the so-called Varjú-channel may be found (this drainage facility is named after the Varjú family). Three or four skeletons were encountered some twenty metres from its beginning in 1967 and 1968 during the opening of this trench. This find spot was located at the channel's end close to the archaeological site under discussion here. One may hypothesize that these latter burials belonged to the cemetery of the early Hungarian village.

A few refuse pits from the Sarmatian Period are also worthy of mention. They were excavated between some of the graves. However, no house remains from that time could be found. Other Sarmatian sherds were sporadically discovered as well.

A new Avar Period runic inscription from Szarvas

IRÉN JUHÁSZ

The site of Szarvas 75 (Rózsási-halom, of the Archaeological Topography of Hungary), is located in the northeastern part of the town, north of the Endrőd road. It forms a plateau oriented east-west, slightly raised above the surrounding flat area. A kurgan is located in the northern sector of this plateau, the top of the burial mound has been eroded flat. Until the beginning of winter in 1990, the whole site was cut through by a canal running east to west to the south of the Kurgan. This canal was dug in 1962, and in all probability, destroyed some graves (subsequently identified as Avar burials) and material from a Prehistoric settlement.[1] The canal was widened during the spring of 1973, which led to additional disturbances in both the prehistoric settlement and the Avar cemetery. Contemporary rescue excavations uncovered 29 Avar burials. The majority of casted bronze fittings from one of the graves (Grave 3) were retrieved from the backfill shovelled onto the canal bank.

Unfortunately, excavations at this cemetery could not be continued until 1987, and were conducted partly within the framework of the Microregion Research Project. This work was carried out on a low budget, at short annual intervals. Excavations within a 60 m long and 5 m wide section opened parallel to the canal on its southern bank, have revealed 79 graves by 1989. Another rescue excavation had to be organized between October 3 and 13, 1990, since the canal was to be filled in. (Due to the lack of financial resources, no planned excavations could be carried out that year). The canal's sloping sides were examined, since the excavation of a burial (presently labelled Grave 80) had been abandoned there in 1987. Evan after mechanical removal of earth from the sloping sides of the canal both this grave and most of the burial sites excavated under rescue conditions in 1973 could be identified. In this way, the two sections of the cemetery could be connected. In addition to these burials, new graves were found on the canal bottom. Earlier recovery of these had been prevented either by the water level or else because they were overlain by silt from the canal banks. At present, 97 graves from this cemetery have been identified.

The known area of the cemetery is of elongated shape with the long axis running in a northwest-southeast direction. Eleven rows of burials could be distinguished from each other. The graves are oriented northwest to southeast with the exception of six horse burials, and three Neolithic graves that contained skeletons in a contracted position, which were oriented southeast to northwest. The depth of graves ranges between 75 and 240 cm. Differences in depth are not particularly associated with either sex. It should be mentioned, however, that graves of a depth between 75 and 105 cm predominantly contain either children or horse skeletons in two examples, and include a few burials with no grave goods. The rest of the horses were also buried only to a depth of 130 cm, while burials with metal belt fittings reached a depth of between 140 and 210 cm.

With the exception of the three Neolithic burials, the excavated part of this cemetery contained Late Avar graves, three of which were double burials. Preliminary examinations suggest a proportion of 40 males to 20 females, while 28 children were found. The sex of five individuals remained unidentifiable. Thirty-seven of the burials contained no grave goods, and

[1] This site is presented in detail in *MRT* 8 (Békés county IV/2) Budapest, 1989, 431-432.

metal belt fittings were found in 11 cases. All the horses were buried in harnesses decorated with mounts. One of them was even accompanied by an iron saddle bow in addition to the stirrups and cheek pieces. One of the belt fittings was decorated by the griffon and meander motif. The rest had meandering ornaments and clusters of three pendant fittings each. Among the belt fittings, the set found in *Grave 88* should be mentioned separately because of its large strap end bit.

Grave 88 was located in the northern section of the cemetery's excavated area. It was found in a row which, at present, has yielded four more graves with metal belt fittings. The belt fittings from neighbouring *Grave 80* included a large strap end bit of identical type. The distance between these two burials was no more than 120-130 cm and they had the same orientation. They differed only in terms of their depths (*Fig. 1*).

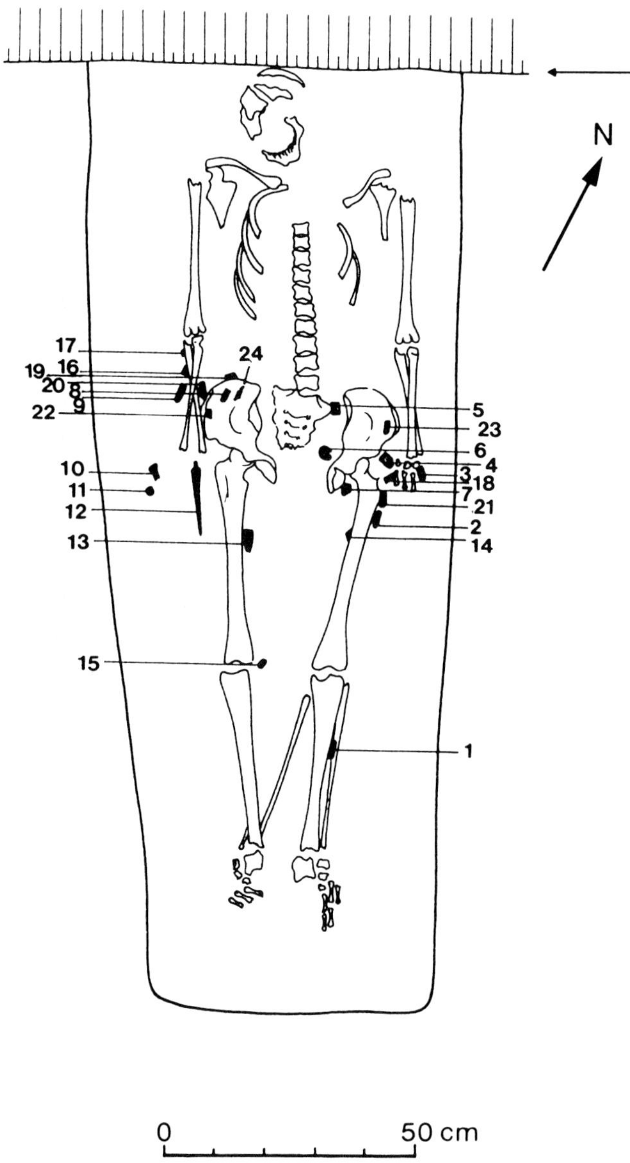

Sloping wall of the canal

M = 1:10, October 9, 1990
Depth: 160 cm
Orientation: Northwest-Southeast 36.5'
(25.23 degrees = 25 degrees)

1 small strap end bit
2 pendant fitting
3 fitting without pendant
4 large strap end bit with a small hole reinforcement
5 small iron plate
6 large hole reinforcement fitting
7 small hole reinforcement fitting
8 fitting without pendant
9 small strap end bit
10 pendant fitting (pendant missing)
11 small hole reinforcement fitting
12 iron knife
13 iron plate
14 iron buckle loop
15 small hole reinforcement fitting + little clamp iron
16 bronze ring
17 little clamp iron?
18 small hole reinforcement fitting
19 bronze buckle
20 pendant fitting (pendant pointing toward the skull)
21 small strap end bit (pointing toward the feet)
22 fitting without pendant (2) pointing toward the skull, plus small hole reinforcement fitting under the right pelvis)
23 pendant fitting (under the left ilium with pendant pointing toward the skull)
24 pendant fitting (pendant missing)

Fig. 1. Szarvas 75: Grave 88.

Grave 88: This burial contained a medium tall but robustly built man with a height of about 162 cm. The age of this individual was difficult to determine since most of the axial skeleton was poorly preserved – only the limb bones survived – and in addition a few teeth and a fragment of mandible. The age of death, estimated from the dental evidence, was about 30 years. The burial was oriented 36.5' northwest-southeast. It measured 55-70 cm wide by 200 cm in length, with a depth of 160 cm. The skeleton lay on its back in an extended position. The bones of the chest carity and the central pelvic area were disturbed. The cast bronze fittings occurred mostly in the pelvic region, around the ilia, hands and forearms. The fact that none of them was found within the pelves may be explained by the disturbance in this area. The fittings were arranged in three piece clusters. The buckle is smooth, while the pendant fittings appear in the form of small strap end bits accompanied by spade-shaped pendants. The three small strap end bits are decorated by a perforated lily pattern similar to that of the large strap end bit. The location of the fittings does not permit an exact reconstruction of the belt. The large strap end bit was found by the lateral edge of the left ilium in line with bones of the carpal region. On November 15, 1990, during the conservation of metal, an amazing observation was made: one side of the large strap end bit had incised runic signs on it. This discovery meant that in the Szarvas area, two finds with runic script have been found in Avar cemeteries located some 8 km, as the crow flies, from each other.[2] Graphic documentation of this inscription was carried out immediately. It could be identified by naked eye, however, a magnifying glass and a light microscope were used in addition. Using these tools, 18 runic signs could be recognized. As restoration of the large strap end bit advanced, weekly drawings were made and photographs taken.

The large strap end bit found (*Fig. 2.1* and *2.2*) in Grave 88 is 6.3 cm long (including the attachment piece). It is 2.2 cm wide, while the width of the sides is 0.9 cm. It is made out of tinned bronze cast. It is decorated by a perforated lily pattern and was attached to the belt by two, flattened, triple spherical attachment pieces. Both sides of this bit have a bead-like decoration. The side located near the left side of the front surface is covered by runic signs incised in a 0.9 cm wide band. These are clearest in the region near the strap end bit's upper end where the tinning is better preserved. At the lower, rounded end the strap end bit has been heavily damaged by corrosion. In spite of this, signs incised to a greater depth are still visible although not all of them are clearly legible. The signs occupy an approximately 4 cm long and 0.5 cm wide band in the middle of the slightly convex strap end bit. The side of this strap end bit is notched along both of its edges while this pattern is followed by the bead-like framing ornament. Some of the signs commence with a sharply defined dot in the upper notch, while the lower part of the signs often display places where the tool slipped. Sometimes thc lower, thinning line of the signs even runs into the bead-like framing ornament. Because of this phenomenon, in the case of several signs a variety of interpretations is possible. The script runs right to left from the upper to the lower end of the end bit.

---

2 The first find with runic script was brought to light in Grave 67 of the Avar cemetery at Szarvas 68 on April 27, 1983. Runic incriptions were made on all four sides of a bone needle case with a rectangular cross-section. To date, several attempts have been made to decode this text both in Hungarian and Turkish. That find is described in: Juhász, I.: A szarvasi avar kori rovásírásos tűtartó (Avar period needle case with runic inscription from Szarvas). *MTud* 30:2 (1985) 92-95; Ein awarenzeitlicher Nadelbehälter mit Kerbschrift aus Szarvas. Budapest, *ActaArchHung* 35:3–4 (1983) 373-377. Decoding attempts: Róna-Tas, A: A szarvasi tűtartó rovásírásos feliratáról (On the runic iscription of the Szarvas needle case). *MTud* 30:2 (1985) 95-98; A szarvasi tűtartó felirata (Inscription on the Szarvas needle case). *NyK* 87 (1985) 225-248; Vékony, G: Késő népvándorláskori rovásfeliratok II (Late Migration Period inscriptions II). *Életünk* 22:2 (1985) 146–69; A szarvasi feirat és ami körülötte van (The Szarvas inscription and what surrounds it). *Életünk* 22:2 (1985) 1133-1145. Harmatta, J.: A magyarság őstörténete (The ancient history of Hungarians). *MTud* 35:3 (1990) 243-261.

1

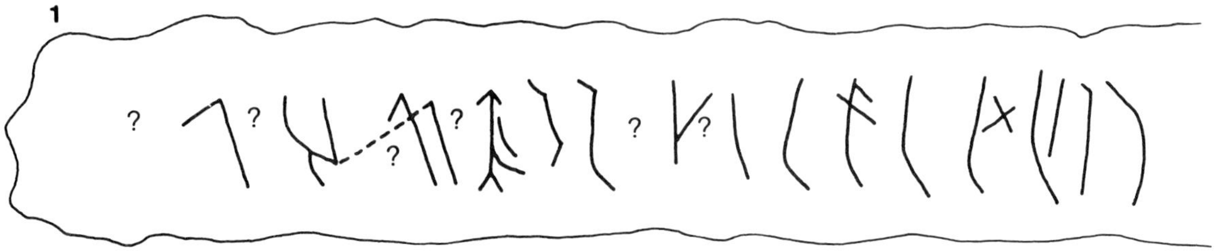

2

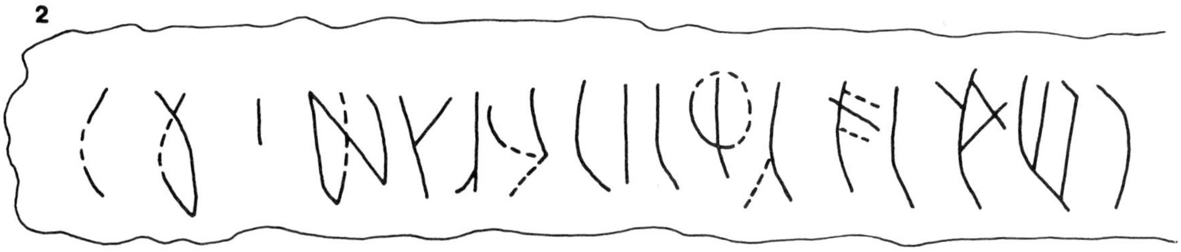

Fig. 2.1. Drawing prepared before restoration by details from photographs.
Fig. 2.2. Drawing prepared after restoration enhanced by details from photographs.

Documentation of these signs by the author was carried out using light microscope aided by free hand drawings and enlarged photographs as well as additional microscopic studies which show that these signs contain no roundish elements. The script on this Szarvas strap end bit bears most resemblance to the signs found on the bow handle plates discovered at Környe. The first sign, without doubt, represents the vocal equivalent of "N". The others may be assigned a variety of vocal values depending on the runic alphabets compiled to date.[3] It is suggested here, that the first 12 to 14 signs can be exactly identified, thus allowing the interpretation of the second runic inscription from Szarvas in a (possibly Turkic?) language.

[3] Alphabets published during the decoding attempts of the text on the needle case were taken into consideration here during the identification of vocal values for the newly discovered find. This attempt only aimed at the more precise identification of the signs themselves: linguists should be entrusted with the linguistic interpretation of these signs.

EDITORIAL

Many persons were engaged in the production of this volume and the Editors wish to express their thanks to every one of them.

Special acknowledgement has to be made for the work of the following persons:

In first place our thanks go to Erzsébet Jerem who acted as Assistant Editor and who was responsible for all kinds of editorial work, e.g. for the lay-out of the text and the illustrations, for the cover design and for seeing the volume through all stages of the production.

Thanks are due also to László Bartosiewicz and Alice Choyke for the English translation of articles originally written in Hungarian, to Katalin Simán and Lyn Sellwood who have checked the final version of the English manuscript.

The Drawing Department of the Archaeological Institute of the HAS headed by Endre Egyed greatly contributed to the volume by preparing illustrations for the camera-ready copy.

We are similarly thankful to András Kardos for word processing, to Ildikó Perjés for desktop editing and to György Fritz for producing special characters of PostScript writing.

György Csáki was involved in taking the aerial photographs – among others the cover picture – in the Gyomaendrőd area. He also digitalized the contour map included as general map in colour print. His continuous effort is sincerely appreciated.

Finally, we would like to thank Judit Solti for her help in proof reading and for compiling the list of Abbreviations.

The word and picture processing and the desktop editing was done on DTK Computers generously provided by KIN-PEX Electronic Ltd. Their invaluable assistance is gratefully acknowledged.

ARCHAEOLINGUA

A Publication Series of the Archaeological Institute
of the Hungarian Academy of Sciences

ARCHAEOLINGUA is a new publication series jointly edited by the Archaeological Institute of the Hungarian Academy of Sciences and the Linguistic Institute of the University of Innsbruck. It shall be dedicated to interdisciplinary research in Archaeology, Linguistics and related fields and will consist of two parallel series, a major series dedicated to works of more substantial character (monographs as well as collective volumes) and a subsidiary series for the publication of lectures and small monographs.

The following titles have already been published or are in course of publication:

Main Series

1: **Cultural and Landscape Changes in South-East Hungary. I: Reports on the Gyomaendrőd Project.** Edited by Sándor Bökönyi. 1992. 384 pp, with numerous maps, diagrams and illustrations. ca. DM 140.- ISBN 963-7391-60-6.

2: Stefan Schumacher: **Die rätischen Inschriften. Geschichte und heutiger Stand der Forschung.** 1992. 291 pp, including maps and illustrations. DM 100.- ISBN 963-7391-61-4.

3: **Onomasticon Provinciarum Europae Latinarum.** Edited by Barnabás Lőrincz and Ferenc Redő. Vol. I. 1992. ca. 320 pp, with maps and illustrations. ca. DM 96.-

Lectures and Small Monographs

1: Wolfgang Meid: **Gaulish Inscriptions. Their interpretation in the light of archaeological evidence and their value as a source of linguistic and sociological information.** 1992. ca. 60 pp, with many illustrations. ca. DM 32.-

2: Winfred P. Lehman: **Die gegenwärtige Richtung der indogermanistischen Forschung.** 1992. ca. 70 pp, ca. DM 32.-

3: Sándor Bökönyi: **Pferdedomestikation, Haustierhaltung und Ernährung. Archäozoologische Beiträge zu historisch-ethnologischen Problemen.** 1992. ca. 40 pp, ca. DM 24.-

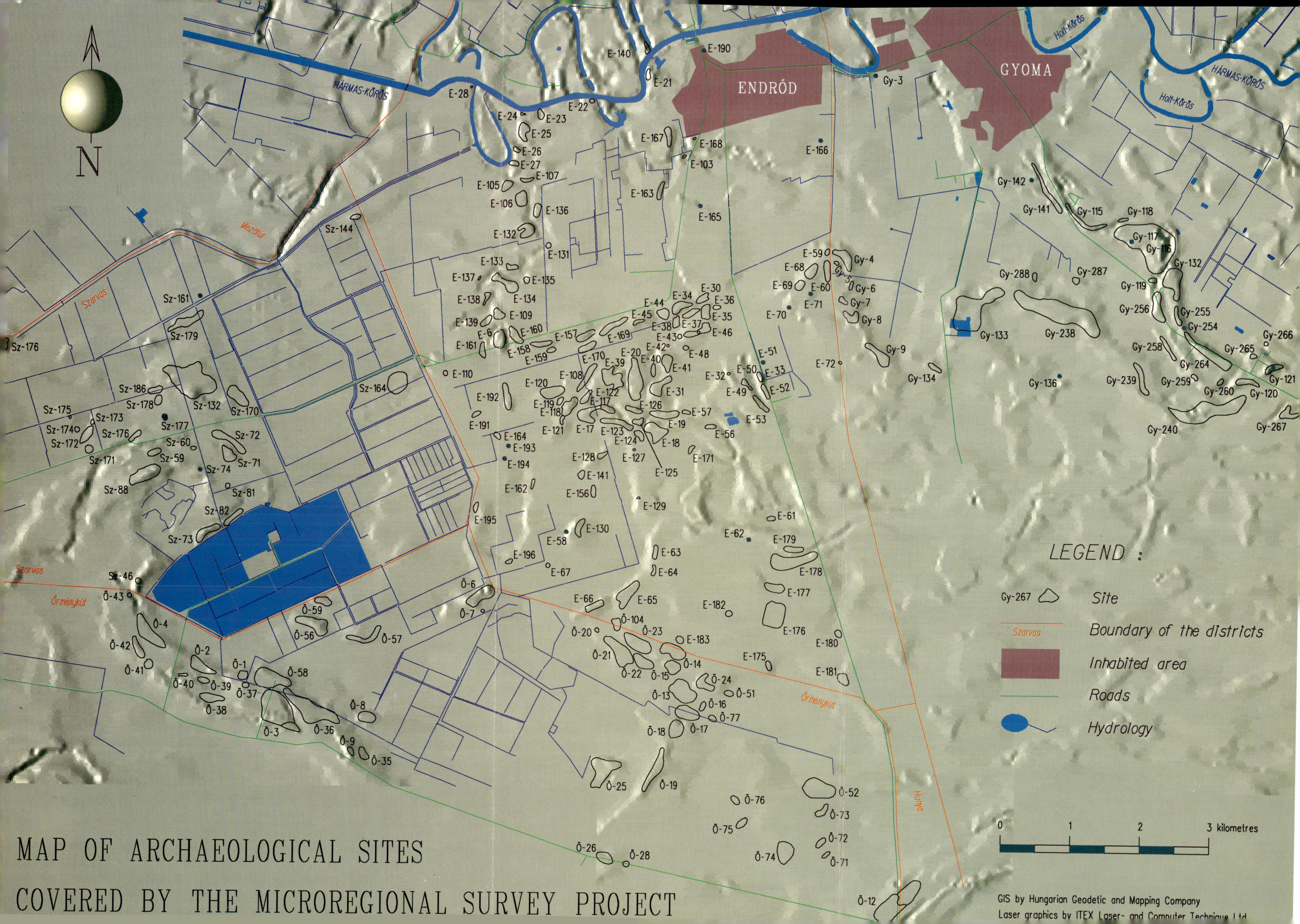

MAP OF ARCHAEOLOGICAL SITES
COVERED BY THE MICROREGIONAL SURVEY PROJECT